Masaaki Yoshida

Hypergeometric Functions, My Love

Edited by Klas Diederich

*A Publication of the Max-Planck-Institut für Mathematik, Bonn

Masaaki Yoshida

Hypergeometric Functions, My Love

Modular Interpretations of Configuration Spaces

Professor Masaaki Yoshida
Graduate School of Mathematics
Kyushu University 33
Fukuoka 812
Japan

Mathematics Subject Classification:
11 F, 14 D, 14 J, 20 F, 32 G 34, 33 C, 51 F 15, 53 B

Cover design: Wolfgang Nieger, Wiesbaden
Printed on acid-free paper

ISBN 978-3-322-90168-2   ISBN 978-3-322-90166-8 (eBook)
DOI 10.1007/978-3-322-90166-8

To my beloved cats late Shima (– 18 Oct. 1993), late Oshin (– 2 Jan.
1995), late Bat (– 3 Nov. 1995), late Tabi (– 30 Dec. 1995), Aka, Chisha,
Heiter, Nezu, Warabi, Sabi, Chloé, Zoro, and my dog Fuku, who came
to my house from nowhere to live with me

Preface

There are various interesting stories about the hypergeometric differential equation

$$E(a, b, c) : x(1 - x)u'' + \{c - (a + b + 1)x\}u' - abu = 0$$

and its solution, the hypergeometric function. Among others, I love the following story: when $(a, b, c) = (1/2, 1/2, 1)$, the ratio $u_1(x) : u_2(x)$ of linearly independent solutions defines a multi-valued map

$$\phi : X = \mathbb{P}^1 - \{0, 1, \infty\} \longrightarrow \mathbb{H} = \{\tau \in \mathbb{C} \mid \Im\tau > 0\},$$

its monodromy group is $\Gamma(2) = \{g \in SL(2, \mathbb{Z}) \mid g \equiv I_2 \mod 2\}$, the map ϕ induces the isomorphism $\varphi : X \xrightarrow{\sim} \mathbb{H}/\Gamma(2)$, and the inverse φ^{-1} can be expressed in terms of the theta functions (zero-values) defined on $\mathbb{H}$. Since the space X can be thought of as the configuration space $X(2, 4)$ of 4 distinct points on the projective line $\mathbb{P}^1$, I call this story a *modular interpretation of the configuration space $X(2, 4)$*.

You might ask why this story attracts me so much. Before answering this, may I pose a question to you? Can you give a logical answer to the question of why your friend (wife, husband or some such person) attracts you so much? Your answer may be "I just like her/him." My answer is similar, but if you insist that I explain further, I (a man) would add "she has many nice friends, who make my life more enjoyable."

I fell in love with the story of the modular interpretation of the configuration space $X(2, 4)$. This story has many friends, i.e. it is related to various kinds of mathematics such as differential equations, differential geometry, configuration spaces, invariant theory, elliptic curves, K3 surfaces and their moduli, uniformization, geometry of bounded symmetric domains, arithmetic subgroups, modular forms, and combinatorics.

This story was originated by Gauss and Jacobi. Other modular interpretations of $X(2, 4)$ were given by H.A. Schwarz. Terada and Deligne-Mostow later made several modular interpretations of the configuration spaces $X(2, n)$ ($5 \leq n \leq 8$) of n points on $\mathbb{P}^1$. These interpretations have been studied by a number of authors. I do not like too much (although I do not hate, and sometimes I enjoy) to share my girl friend with so many boys.

In this book, I tell my love story with a *modular interpretation of the configuration space $X(3, 6)$* of 6 points on the projective plane $\mathbb{P}^2$. You might wonder why don't I make a story for the general configuration space $X(k,n)$ of n points on $\mathbb{P}^{k-1}$. Natural question. If there were similar stories for general $X(k,n)$, and $X(3,6)$ were just another configuration space, then I would not have gotten the idea of publishing my story of $X(3,6)$. However, I know ([SYY]) that there are no such stories other than these for $X(2,n)$ $(n \leq 8)$ and $X(3,6)$.

When I describe my friend to somebody, I do not give her date or place of birth, height, nobility of her family, nor the estate she owns (please visit the tax office), but rather her sense of humor, her friends, her cooking technique, her likes and dislikes for various things including food.

So in my story, I have tried to avoid abstract notion, elaborate formulations and often rigorous proofs in order to make the story naked and let the reader really see the happenings. I made the story as plain and casual as possible. Furthermore, to help the reader understand the statements, I have included many pictures; please enjoy them. Those who think my description old-fashioned, non-polished and not sufficiently rigorous are surely polluted by "modern" mathematics. By the way, to understand a fact, a proof is neither sufficient nor necessary. A proof consisting of a chain of logic was originated in Greece and grew up in Europe, but I do not think it is the unique way to describe mathematics, at least in Japan it was not so: the most important thing is to "guess". I would like to tell you that mathematics is not a thing to suffer but a thing to enjoy.

The pre-requisite for reading and understanding this book is almost nothing, so that any undergraduate student should be able to enjoy the presentation.

This book is divided in three parts. Part 1 (Chapters I – IV) is devoted to the classical theory of modular interpretations of $X(2,4)$. I start from a very elementary level and introduce the configuration space $X(2,4)$ in Chapter I. A classical theory of elliptic curves is presented in Chapter II. Modular interpretations of $X(2,4)$ are made through use of the hypergeometric differential equation in Chapter III. The hypergeometric integrals and the associated loaded cycles are introduced in Chapter IV.

Part 2 (Chapters V and VI) is a survey of modular interpretations of $X(2,n)$. In order to give the reader an idea of the spaces $X(2,n)$, a combinatorial topological study of $X(2,5)$ is made in Chapter V.

Part 3 (Chapters VII – IX) is the main thrust of this book: a modular interpretation of $X(3,6)$. A geometric study of the (complex 4-dimensional) space $X = X(3,6)$ is made, and in particular a compactification $\overline{X}$ of X is given in Chapter VII. In Chapter VIII, the hypergeometric functions (the integrals and the system $E(3,6;\alpha)$ of differential equations) of type $(3,6)$ defined on X and associate loaded cycles are introduced, and a set of generators of the monodromy group and the invariant form are obtained. In Chapter IX, I show that the multi-valued map

$$\phi : X \ni x \mapsto u_1(x) : \cdots : u_6(x) \in \mathbb{P}^5$$

defined by six linearly independent solutions of the system $E(3,6;1/2)$ induces the isomorphism

$$\varphi : \overline{X} \to \bar{\mathbb{D}}/\Gamma_A(2)$$

between $\overline{X}$ and the Satake compactification of an arithmetic quotient of a symmetric domain $\mathbb{D}$ in the quadratic hypersurface of $\mathbb{P}^5$ defined by an integral symmetric 6×6-matrix A, which is essentially the invariant form obtained in Chapter VIII. The domain $\mathbb{D}$ is isomorphic to

$$\mathbb{H}_2 = \{\tau : 2 \times 2 \text{ complex matrices} \mid (\tau - {}^t\bar{\tau})/2i > 0\}.$$

In the last section, the inverse map of φ is expressed in terms of the theta functions defined on $\mathbb{H}_2$.

The involution $*$ on X defined in Chapter VII plays the clown throughout Part 3.

Part 1 stems from lectures [Yos2] given at the University of the Philippines during the winter semester of 1991. The present book can be thought of as a revised and enlarged version of [Yos1], which includes a good portion of Part 1 and Part 2 but lacks Part 3. The facts presented in Part 3 are found in [MSY1] and [Mat2], where a good number of results are proved by brute-force. Efforts were made to make some of these transparent and smooth; a method of getting a set of generators of the monodromy group is obtained in [MSTY1], and the theory of intersections of loaded cycles is established in [KY]. But a few proofs remain un-smoothened and therefore these clumsy proofs are not given in this book.

I am deeply grateful to Professor Takayuki Oda, who kindly informed me a key idea of this work, and to the co-authors of my papers: Professors F. Apéry, M. Kita, K. Matsumoto, T. Sasaki, J. Sekiguchi, J. Stienstra and N. Takayama, who tolerantly permitted me to quote many results of our joint papers. I would like to thank Dr. G. Paquette for polishing the English expressions.

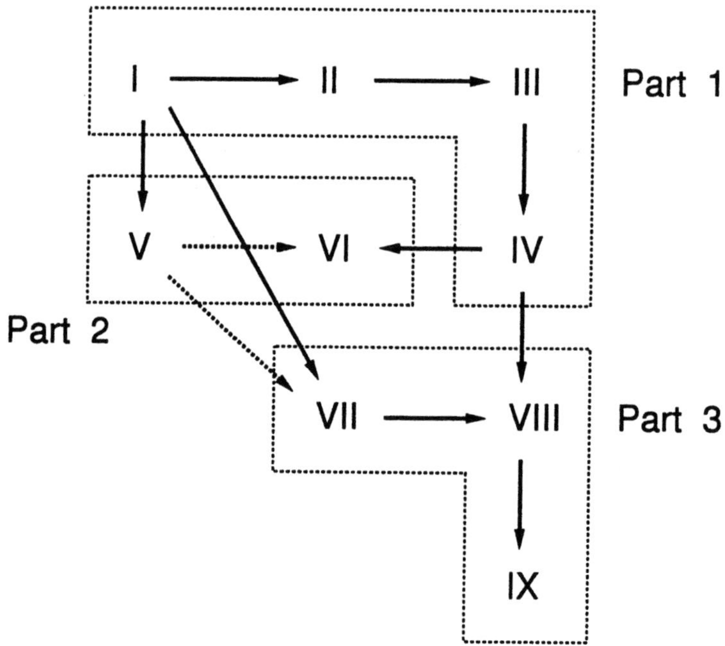

FIGURE 0.1. Relations of Chapters

Contents

Chapter II. Elliptic Curves   29

1.   Lattices in $\mathbb{C}$ 30

2.   Elliptic Curves as Quotients of $\mathbb{C}$ by Lattices   30

3.   Isomorphism Classes of Elliptic Curves   31

4.   Realization in Terms of the Weierstrass $\wp$ Function   34

5.   A Realization in Terms of the Theta Functions   43

Chapter III. Modular Interpretations of $X(2,4)$ 59

1.   The Hypergeometric Series   59

2.   The Hypergeometric Differential Equation   60

3.   Another Solution around the Origin   62

4.   Symmetries of the Hypergeometric Equation   64

5.   Time to Pay   65

6.   The Schwarz Map and Schwarz Triangles   67

7.   Schwarz's Reflection Principle   69

Part 2
The Story of the Configuration Space $X(2, n)$ of n Points on the Projective Line

Part 3
The Story of the Configuration Space $X(3,6)$ of Six Lines on the Projective Plane

Part 1

The Story of the Configuration Space $X(2,4)$ of Four Points on the Projective Line

CHAPTER I

Configuration Spaces – The Simplest Case

1. Classifications and Equivalence Relations

To begin with, I would like to introduce two important concepts in mathematics: equivalence relation and quotient space. It is crucial for mathematicians to be familiar with these concepts.

Let X be a set. Let us make a *classification* of X, i.e., divide X into several pieces. Each piece is called a class, and any element in a class is called a *representative* of the class to which it belongs. Let us fix a classification and call two members x and y of X equivalent, written $x \sim y$, if the two belong to the same class.

EXAMPLE 1.1 (SEX).

$$\begin{aligned}
X &:= \text{all students in a university.} \\
X_1 &:= \text{all female students in the university.} \\
X_2 &:= \text{all male students in the university.}
\end{aligned}$$

Thus here, two students are equivalent if they have the same sex.

The relation $\sim$ obviously satisfies the following three conditions:

(1) $x \sim x$

(2) $x \sim y \quad \Rightarrow \quad y \sim x$

(3) $x \sim y, \quad y \sim z \quad \Rightarrow \quad x \sim z$

Any relation which satisfies these conditions is called an *equivalence relation*. Conversely, the classes of a set can be recovered from the equivalence relation as follows: Pick an element x_0 of X and form the subset of X

$$[x_0] = \{ x \in X \mid x \sim x_0 \}.$$

Pick an element $x_1 \in X - [x_0]$, if any, and form the subset

$$[x_1] = \{x \in X \mid x \sim x_1\}.$$

Pick an element $x_2 \in X - [x_0] \cup [x_1]$, if any, and form

$$[x_2] = \{x \in X \mid x \sim x_2\},$$

and so on. Notice that if $i \neq j$ then $[x_i] \cap [x_j] = \emptyset$ from condition (3) above. Therefore, giving a classification of X and giving an equivalence relation on X are effectively identical.

EXAMPLE 1.2 (FRIENDS). On the set

$$X := \{\text{all human beings}\},$$

define $\approx$ by

$$x \approx y \iff x \text{ is a friend of } y.$$

Everyone loves oneself, so condition (1) is satisfied. If you have a friend, he or she must think that you are his or her friend, or let us believe so. Thus assume (2). However a friend of your friend can be your enemy, so (3) does not hold. Hence $\approx$ is not an equivalence relation. I do not give up. Let us define a new relation "acquaintance" $\sim$ as follows:

$$x \sim y \iff \text{there are } x_1, \ldots, x_r \in X \text{ such that}$$
$$x \approx x_1 \approx \cdots \approx x_r \approx y.$$

You can easily check that $\sim$ is an equivalence relation. The relation $\sim$ is called the equivalence relation *generated by* $\approx$. (See Figure 1.1.)

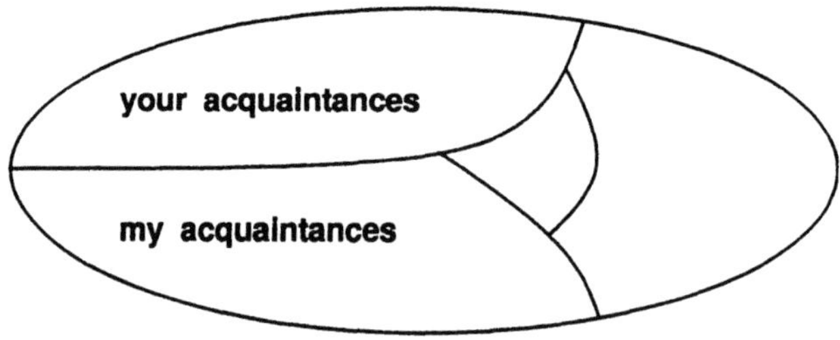

FIGURE 1.1. You and I have no common friend's friend's ...

2. Quotient Spaces

The set of the classes of the set X is called the *quotient space* of X with respect to the equivalence relation $\sim$ and will be denoted by $X/\sim$. There is a natural map

$$\mathrm{Pr} : X \longrightarrow X/\sim$$

called the *projection* which sends x to the class $[x]$.

For Example 1.1, the quotient space consists of two elements:

$$\{\text{male}, \quad \text{female}\}.$$

EXAMPLE 2.1 (RATIONAL NUMBERS). On the set $X = \mathbb{Z} \times (\mathbb{Z} - \{0\})$, define $\sim$ by

$$(x, y) \sim (x', y') \quad \Longleftrightarrow \quad xy' = yx'.$$

The class to which (x, y) belongs is denoted by x/y. The quotient space $X/\sim$ is the set of rational numbers. How does one define addition for rational numbers? If we propose the following

$$\frac{x}{y} + \frac{x'}{y'} = \frac{x + x'}{y + y'},$$

the operation is not well-defined because the result depends on the choice of representatives within a class, which is inconsistent with the definition of $\sim$:

$$\frac{2}{3} + \frac{1}{2} = \frac{3}{5}, \quad \frac{2}{3} + \frac{2}{4} = \frac{4}{7} \neq \frac{3}{5}.$$

In music, we use times such as $1/2, 2/4, 3/6, \ldots$. However, these expressions do not represent rational numbers because in $6/8$ you march, while in $3/4$ you dance.

EXAMPLE 2.2 (PROJECTIVE SPACES). You probably learned about ratios in elementary school. These are written as $a : b$, and for example, the following relations hold:

$$1 : 2 = 5 : 10, \quad 0 : 5 = 0 : 1, \quad \text{etc.}$$

However, as you will recall, $0 : 0$ is prohibited. Why? Well, this is a good question, because this is difficult. Please forgive me, and allow me to answer simply, "I have a right not to consider $0 : 0$". In mathematics, we call the set of ratios the *projective line* or the *1-dimensional projective space*. In general, for a field $K = \mathbb{Q}$, $\mathbb{R}$, $\mathbb{C}, \ldots$, the k-dimensional *projective space* $\mathbb{P}^k_K$ over K is defined as the quotient space of $K^{k+1} - \{0\}$

under the equivalence relation: $(x_0, \ldots, x_k) \sim (y_0, \ldots, y_k)$ if and only if there is a non-zero $a \in K^\times := K - \{0\}$ such that

$$x_j = ay_j \quad \text{for all} \quad j.$$

The equivalence class to which $(x_0, \ldots, x_k)$ belongs shall be denoted by $x_0 : \cdots : x_k$.

3. Realizations

The definition of quotient spaces I gave above is simple, but quotient spaces are abstract objects, and perhaps difficult to understand. Considering Example 1.1 (sex), there can be various love stories between actual boys $x_i \in X$ and actual girls $y_j \in X$, while in $X/\sim$ there is only an abstract philosophical love between male and female (cf. *The Symposium* by Plato). Quotient spaces are difficult not only for students but also for professional mathematicians, or it may be best said that these are by nature difficult for human beings. In order to understand these abstract objects, we often try to make *realizations*. Roughly speaking, a realization of a quotient space $X/\sim$ is a way to think of the space as a subset of a well-understood space S, say, $\mathbb{R}^n, \mathbb{C}^n, \mathbb{P}^n, \ldots$. More precisely it is a map $f : X \to S$ which

 (1) takes the same value on each class, and
 (2) takes different values on different classes.

That is,

 (1) $f(x) = f(y)$, $x \sim y$,   and
 (2) $f(x) \neq f(y)$, $x \not\sim y$.

The map f induces the following injective map $\bar{f}$ of $X/\sim$ to S.

$$
\begin{array}{ccc}
X & \xrightarrow{\;f\;} & S \\
{\scriptstyle \text{Pr}} \searrow & & \nearrow {\scriptstyle \bar{f}} \\
& X/\sim &
\end{array}
$$

EXAMPLE 3.1. Let X be the complex number field $\mathbb{C}$; in particular, it is an additive group. Let us consider, for a positive number π, a subgroup $G := 2\pi\sqrt{-1}\mathbb{Z}$ of $\mathbb{C}$ and define an equivalence relation on X by

$$x \sim y \iff x - y \in G.$$

The quotient space looks like a cylinder.

EXAMPLE 3.2 (QUOTIENT BY A GROUP). Let G be a group acting on a set X; that is, we are given a map

$$G \times X \ni (g, x) \mapsto gx \in X$$

such that $id_G x = x$ (the identity element of G is denoted by id_G) and

$$g(hx) = (gh)x, \quad g, h \in G, \quad x \in X.$$

Define an equivalence relation as $x \sim y$ if and only if there is an element $g \in G$ such that $x = gy$. By the definition of groups, one can easily check that this is an equivalence relation. The classes with respect to this equivalence relation consists of the G-*orbits*

$$Gx = \{gx \mid g \in G\}, \quad x \in X.$$

The quotient space of X with respect to this relation is denoted by $G\backslash X$ or X/G and often called the *orbit space* of X under G. A realization of X/G is a map $f : X \to S$ which has the properties

$$f(gx) = f(x), \quad g \in G, \quad x \in X$$

(f is invariant under the action of G) and

$$f(x) \neq f(y) \quad \text{if} \quad Gx \neq Gy$$

(f separates the G-orbits).

4. The Exponential Function

The *exponential function* is by definition given by the following power series converging absolutely everywhere on $\mathbb{C}$:

$$\exp x = \sum_{n=0}^{\infty} \frac{x^n}{n!}.$$

REMARK 4.1. For notational simplicity, I sometimes write e^x in place of $\exp x$. However this is only notation. Please do not entertain the idea that we are multiplying x times a number $e = 2.7\ldots$ This has no meaning.

PROPOSITION 4.1. *The function* $\exp x$ *is the unique solution of the initial value problem*

$$dz/dx = z, \quad z(0) = 1.$$

It is easy to see that the function $\exp x$ satisfies the equation. Uniqueness is due to Cauchy's fundamental theorem.

PROPOSITION 4.2.

$$\exp(x + y) = \exp x \cdot \exp y.$$

PROOF. If you admit the binomial theorem

$$(x + y)^n = \sum_{k=0}^{n} \binom{n}{k} x^k y^{n-k}, \qquad \binom{n}{k} = \frac{n!}{k!(n - k)!},$$

proof is straightforward:

$$\sum_{k=0}^{\infty} \frac{x^k}{k!} \sum_{l=0}^{\infty} \frac{y^l}{l!} = \sum_{n=0}^{\infty} \sum_{k+l=n} \frac{x^k y^l}{k!l!}$$

$$= \sum_{n=0}^{\infty} \sum_{k=0}^{n} \frac{x^k y^{n-k}}{k!(n - k)!}$$

$$= \sum_{n=0}^{\infty} \frac{1}{n!} \sum_{k=0}^{n} \binom{n}{k} x^k y^{n-k} = \sum_{n=0}^{\infty} \frac{(x + y)^n}{n!}.$$

$\square$

Another proof: Set $y = \alpha$, (this change of variable has no meaning mathematically, but has a psychological effect) and $a = \exp \alpha$. Then check that both sides of the equality in question solve the initial value problem:

$$dz/dx = z, \quad z(0) = a.$$

COROLLARY 4.3. *For any complex number x, $\exp x \neq 0$.*

PROPOSITION 4.4.

$$|\exp \sqrt{-1}\, t| = 1, \quad t \in \mathbb{R}.$$

PROOF. Let i represent $\sqrt{-1}$. We have $\overline{\exp x} = \exp \bar{x}$, since the coefficients of the series are real, and complex conjugation is a continuous operation. Thus we have

$$\begin{aligned}
|\exp it|^2 &= \exp it \cdot \overline{\exp it} = \exp it \cdot \exp \overline{it} \\
&= \exp it \cdot \exp(-it) = \exp(it - it) \\
&= \exp 0 = 1.
\end{aligned}$$

$\square$

Let us define the sine and cosine functions:

$$\sin x := \frac{e^{ix} - e^{-ix}}{2i}, \quad \cos x := \frac{e^{ix} + e^{-ix}}{2}, \quad x \in \mathbb{C}.$$

We have

$$\frac{d}{dx} \sin x = \cos x, \quad \frac{d}{dx} \cos x = -\sin x,$$

and

$$\cos x + i \sin x = e^{ix},$$
$$(\cos x)^2 + (\sin x)^2 = 1.$$

Note that if t is real, since $e^{-it} = \overline{e^{it}}$, we have

$$\sin t = \Im \exp it, \quad \cos t = \Re \exp it, \quad t \in \mathbb{R},$$

where $\Im$ and $\Re$ represent imaginary part and real part.

LEMMA 4.5. *The solution of the initial value problem*

$$u'' = -u, \quad u(0) = 0, \quad u'(0) = 1$$

has a zero in the interval $(0, +\infty)$.

PROOF. Assume on the contrary that $u(t) > 0$ for all $t > 0$. Since $(u')' = u'' = -u < 0$, the function $u'(t)$ is monotonically decreasing. If

$$\lim_{t \to +\infty} u'(t) = \omega < 0,$$

then the equality

$$u(t) = \int_0^t u'(s)ds$$

implies that $u(t)$ must be negative for sufficiently large t. Thus $\omega \geq 0$, and so $u(t)$ is monotonically increasing. On the other hand, since

$$u'(t) = 1 + \int_0^t u''(s)dt$$
$$= 1 - \int_0^t u(s)ds,$$

we have

$$\int_0^\infty u(s)ds = 1 - \omega,$$

which is impossible. $\square$

Since the function $\sin t$ solves the initial value problem above, there is a positive number t such that $\sin t = 0$, and so $\cos t = \pm 1$. If t_0 is the smallest such number, we have $\cos t_0 = -1$, because $\sin t$ is decreasing at t_0, and $(\sin t)' = \cos t$. Now we can define the number π.

DEFINITION 4.1. The smallest positive real number such that $\exp ix = -1$ is called π.

Note that 2π is the smallest positive number such that $\exp ix = 1$.

REMARK 4.2. Evaluation of the value of π is another problem; here we do not care about it.

From this definition, we see that the exponential function has period $2\pi i$, i.e.

$$\exp(x + 2\pi in) = \exp x, \quad n \in \mathbb{Z}.$$

The map

$$\mathbb{C} \ni x \mapsto z = \exp x \in \mathbb{C} - \{0\}$$

gives a homomorphism from the additive group $\mathbb{C}$ onto the multiplicative group $\mathbb{C}^\times := \mathbb{C} - \{0\}$, thanks to Proposition 4.2. This map can be thought of as a realization of the quotient space $\mathbb{C}/2\pi i\mathbb{Z}$. You should note that a realization of this simple-looking quotient space requires a transcendental function, because $2\pi i\mathbb{Z}$ is an infinite group. The map can be visualized as in Figure 4.1.

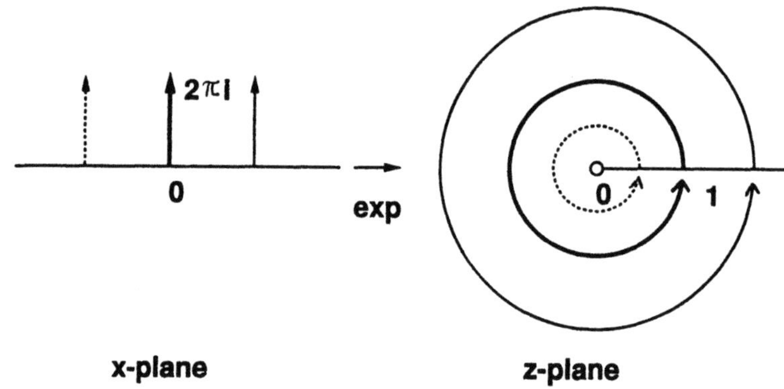

FIGURE 4.1. exp

EXERCISE 1. Let $\zeta = \exp 2\pi i/N$ be an N-th root of unity and $G = \{1, \zeta, \ldots, \zeta^{N-1}\} \subset \mathbb{C}^\times$ the finite multiplicative group acting on $\mathbb{C}$ as

$$\zeta^j : x \mapsto \zeta^j x, \quad x \in \mathbb{C}.$$

Find a realization of the quotient space $\mathbb{C}/G$.

5. The Logarithmic Function

Let C_z be a path in $\mathbb{C} - \{0\}$ starting at 1 and ending at z, i.e., a smooth map $[0,1] \longrightarrow \mathbb{C} - \{0\}$ such that $C_z(0) = 1, C_z(1) = z$. I define the *logarithmic function* log by the integral

$$\log C_z := \int_{C_z} \frac{d\zeta}{\zeta} \quad \left(= \int_0^1 \frac{C_z'(t)}{C_z(t)} dt \right).$$

If you deform the path C_z continuously in $\mathbb{C} - \{0\}$, fixing the initial and terminal points, then by Cauchy's theorem the value of the integral does not change. Although $\log C_z$ is usually written as $\log z$, (since its value depends not only on z but also on the path to z,) we shall avoid this notation for a while.

If the path C_z remains in a simply connected part D of $\mathbb{C} - \{0\}$, say,

$$D = \mathbb{C} - (-\infty, 0],$$

then $x = \log C_z$ is a single-valued function of $z \in D$, and this function solves the initial value problem

$$\frac{dx}{dz} = \frac{1}{z}, \quad x(1) = 0.$$

This result, together with Proposition 4.1, shows that $\log C_z$ is the inverse function of $z = \exp x$ near $x = 0, z = 1$. In particular, when r is real and positive, $\log C_r$ is real, and coincides with the logarithmic function $\log r$ that you learned in high school. If $|z| < 1$, we have the following power series expression

$$-\log C_{1-z} = \sum_{n=1}^{\infty} \frac{z^n}{n},$$

which can easily be checked by using the differential equation $dx/dz = 1/z$ satisfied by $\log C_z$.

PROPOSITION 5.1. *Let us define the product $C_z C_w$ of two curves C_z and C_w by $t \mapsto C_z(t)C_w(t)$. Then we have*

$$\log C_z C_w = \log C_z + \log C_w.$$

PROOF.

$$\log C_z C_w = \int_{C_z C_w} \frac{d\zeta}{\zeta} = \int_0^1 \frac{(C_z(t)C_w(t))'}{C_z(t)C_w(t)} dt$$
$$= \int_0^1 \frac{C_z'(t)}{C_z(t)} dt + \int_0^1 \frac{C_w'(t)}{C_w(t)} dt = \log C_z + \log C_w.$$

$\square$

For a given path C_z, define the two paths $|C_z|$ and $C_{z/|z|}$ by

$$|C_z| : t \mapsto |C_z(t)|, \quad C_{z/|z|} : t \mapsto C_z(t)/|C_z(t)|,$$

so that

$$C_z(t) = |C_z(t)|C_{z/|z|}(t).$$

(See Figure 5.1.) By applying the above proposition to the product
$C_z = |C_z|C_{z/|z|}$, we have

$$\log C_z = \log |C_z| + \log C_{z/|z|}$$
$$= \log |z| + \log C_{z/|z|}.$$

The first term of the right-hand side is real and is a single-valued function
of z, while the second one is imaginary and is not a single-valued function
of z. I define the (real valued) *argument function* arg by

$$\arg C_z := \frac{1}{i} \log C_{z/|z|}.$$

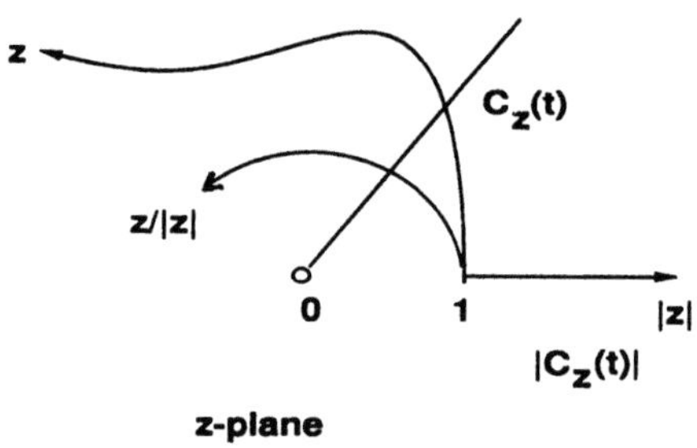

FIGURE 5.1. log and arg

Now consider what happens if z travels out of D. In particular, when C_1 is the positively oriented unit circle, by the definition of the number π together with the fact that $\log C_z$ is the inverse of $\exp x$, we have

$$\int_{C_1} \frac{d\zeta}{\zeta} = \log C_1 = 2\pi i.$$

(See Figure 4.1.) Since any given path C_z from 1 to z not passing through 0 can be deformed into the composition of an n-fold unit circle (for some $n \in \mathbb{Z}$) and a path in D, we have

$$\log C_z = \log C_z' + 2\pi i n, \quad \arg C_z = \arg C_z' + 2\pi n,$$

where $C_z' \subset D$. This implies that describing a path C_z up to deformations (fixing the initial and terminal points) is equivalent to giving z and the number n. Therefore instead of writing $\log C_z$,

I shall write $\log z$ *together with the value of the argument.*

6. Power Functions

The *power function* z^a is defined as

$$\exp(a \log C_z),$$

where C_z is a path defined in the previous section. This is a fairly complicated multi-valued function in z near 0, while it is holomorphic in a everywhere. When z goes around the origin once in the positive sense, its value is multiplied by $\exp 2\pi i a$. Clearly, without restricting domains of definition, the expression

$$(zw)^a = z^a w^a$$

is false or has no meaning. Of course the following is true:

$$\exp(a \log C_z C_w) = \exp(a \log C_z) \exp(a \log C_w).$$

7. Projective Spaces

In Example 2.2 (projective spaces), I gave a definition of projective spaces in terms of undergraduate-level language. Now I will give a definition of these in terms of graduate-level language. Do not be afraid; the objects are the same. They are just described in a little bit sophisticated language.

Let V be a $(k+1)$-dimensional vector space over a field K. The multiplicative group $K^\times$ acts on $V - \{0\}$ in the obvious manner. The

k-dimensional *projective space* $\mathbb{P}^k = \mathbb{P}^k_K = \mathbb{P}(V)$ is defined as the quotient space $V - \{0\}/K^\times$. Let us fix a basis $e_0, \ldots, e_k$ of V. Then every element $v \in V$ can be written as

$$v = x_0 e_0 + \cdots + x_k e_k.$$

For an element $[v]$ of $\mathbb{P}^k$, the ratio $x_0 : \cdots : x_k$ is called (a system of) *homogeneous coordinates* of $[v]$ with respect to the basis e. If x_0 is not zero, then $1 : x_1/x_0 : \cdots : x_k/x_0$ is also a representative of $[v]$. $(x_1/x_0, \ldots, x_k/x_0)$ is called the 0-th inhomogeneous coordinate. The j-th *inhomogeneous coordinate* is defined in the obvious manner. Let $f_0, \ldots, f_k$ be another basis of V and write

$$v = y_0 f_0 + \cdots + y_k f_k.$$

Then there is a non-singular $(k+1) \times (k+1)$ matrix P such that

$${}^t(y_0, \ldots, y_k) = P^t(x_0, \ldots, x_k),$$

where t represents the transpose operator. I introduce here the notation of a matrix group over a ring R:

$$GL(r, R) := \{r \times r\text{-invertible matrices whose elements are in } R\}.$$

This is called the *general linear group* of rank r over R. Remember that an $r \times r$ matrix is invertible if and only if its determinant is an invertible element of R. If R is a field, this implies that the determinant is non-zero, and if $R = \mathbb{Z}$, this implies that it is 1 or -1.

8. Projective Transformations

For a fixed basis, the above operation ${}^t(y_0, \ldots, y_k) = P^t(x_0, \ldots, x_k)$ can be considered as the action

$$x_0 e_0 + \cdots + x_k e_k \mapsto y_0 e_0 + \cdots + y_k e_k$$

of the group $GL(k+1) = GL(k+1, K)$ on $\mathbb{P}^k$. Since a non-zero constant multiple of a matrix P induces the same effect on $\mathbb{P}^k$ as P itself, the quotient group

$$PGL(k+1) = PGL(k+1, K) := GL(k+1)/K^\times$$

embodies the set of transformations on $\mathbb{P}^k$. This group is called the *projective linear group*. In order to familiarize ourselves with the projective space, let us fix a basis and have a look at 1-dimensional and

2-dimensional projective spaces, often called the projective line and the projective plane, respectively.

EXAMPLE 8.1 $(k = 1)$.

$$
\begin{aligned}
\mathbb{P}^1 &= \{x_0 : x_1 \mid \text{not zero simultaneously}\} \\
&= \{x_0 : x_1 \mid x_1 \neq 0\} \cup \{1 : 0\}.
\end{aligned}
$$

The former summand is isomorphic to an affine line K through the map

$$
\mathbb{P}^1 - \{1 : 0\} \ni x_0 : x_1 \mapsto x_0/x_1 \in K.
$$

We call the point $1 : 0$ the point at infinity; you must bear in mind that the point at infinity is not a special point in $\mathbb{P}^1$, because if you take another coordinate, any point might have a danger to be called the point at infinity. I shall come back to this point later.

If one uses an inhomogeneous coordinate $y = x_0/x_1$, then the action of the group is expressed by linear fractional transformations:

$$
y \mapsto \frac{ay + b}{cy + d}, \quad \begin{pmatrix} a & b \\ c & d \end{pmatrix} \in GL(2).
$$

EXAMPLE 8.2 $(k = 2)$.

$$
\begin{aligned}
\mathbb{P}^2 &= \{x_0 : x_1 : x_2 \mid \text{not zero simultaneously}\} \\
&= \{x_0 : x_1 : x_2 \mid x_2 \neq 0\} \cup \{x_0 : x_1 : 0 \mid \text{not zero simultaneously}\} \\
&= K^2 \cup \mathbb{P}^1 \\
&= \{x_0 : x_1 : x_2 \mid x_2 \neq 0\} \cup \{x_0 : x_1 : 0 \mid x_1 \neq 0\} \cup \{1 : 0 : 0\} \\
&= K^2 \cup K \cup \{\text{a point}\}.
\end{aligned}
$$

Now you can easily imagine that

$$
\begin{aligned}
\mathbb{P}^k &= K^k \cup \mathbb{P}^{k-1} \\
&= K^k \cup K^{k-1} \cup \cdots \cup K \cup \{\text{a point}\}.
\end{aligned}
$$

If one uses inhomogeneous coordinates $y_1 = x_1/x_0, \ldots, y_k = x_k/x_0$, then again the action of the group $PGL(k+1)$ can be expressed in terms of linear fractional transformations:

$$
y_j \to \frac{a_{j0}y_0 + \cdots + a_{jk}y_k}{a_{00}y_0 + \cdots + a_{0k}y_k},
$$

where $y_0 = 1$.

9. Configuration Space of 4 Points on the Projective Line

I mentioned above that the point at infinity is not a special point. This can be expressed as follows: any point on the projective line can be transformed by an element of the group $GL(2)$, or if you like $PGL(2)$, to any other point on the line. Indeed for any non-zero column 2-vectors x and y you can find a non-singular matrix P such that

$$Px = y.$$

This property is often called the transitivity of the group $GL(2)$ acting on $\mathbb{P}^1$.

Let us play a game. Given two sets each consisting of two distinct points, say $\{x_1, x_2\}$ and $\{y_1, y_2\}$, is there an element of $GL(2)$ sending x_1 to y_1 and x_2 to y_2? The answer is "yes". Indeed, letting

$$u_{11} : u_{21}, \quad u_{12} : u_{22}, \quad v_{11} : v_{21}, \quad v_{12} : v_{22}$$

be homogeneous coordinates of x_1, x_2, y_1, and y_2, respectively, the condition that x_1 and x_2 are distinct and y_1 and y_2 are distinct stipulates that the matrices

$$\begin{pmatrix} u_{11} & u_{12} \\ u_{21} & u_{22} \end{pmatrix} \quad \text{and} \quad \begin{pmatrix} v_{11} & v_{12} \\ v_{21} & v_{22} \end{pmatrix}$$

are non-singular. Now, you can find a non-singular matrix P such that

$$P \begin{pmatrix} u_{11} & u_{12} \\ u_{21} & u_{22} \end{pmatrix} = \begin{pmatrix} v_{11} & v_{12} \\ v_{21} & v_{22} \end{pmatrix}.$$

Next, given two sets of three distinct points, say $\{x_1, x_2, x_3\}$ and $\{y_1, y_2, y_3\}$, ask the same question. Again the answer is "yes". Indeed, I can show that any three distinct points can be transformed to $\infty, 0$, and 1. Let us arrange three homogeneous coordinates of the three x's in columns:

$$\begin{pmatrix} u_{11} & u_{12} & u_{13} \\ u_{21} & u_{22} & u_{23} \end{pmatrix}.$$

Since the matrix consisting of the first and second columns is non-singular, I multiply its inverse matrix from the left by the 2×3-matrix above, to get a matrix of the following form:

$$\begin{pmatrix} 1 & 0 & u \\ 0 & 1 & v \end{pmatrix},$$

for some u and v. Since the three points expressed by this matrix are distinct, u and v are not zero. Since $u : v$ is a homogeneous coordinate, I can assume $u = 1$. If I multiply by the matrix

$$\begin{pmatrix} 1 & 0 \\ 0 & 1/v \end{pmatrix}.$$

from the left, then I have

$$\begin{pmatrix} 1 & 0 & 1 \\ 0 & 1/v & 1 \end{pmatrix}.$$

Since $0 : 1/v = 0 : 1$, we can eventually transform the matrix u_{ij} into

$$\begin{pmatrix} 1 & 0 & 1 \\ 0 & 1 & 1 \end{pmatrix},$$

which represents homogeneous coordinates of the points $\infty, 0$, and 1.

We have therefore proved the well known fact that any three distinct points on the projective line can be transformed to $0, 1$ and ∞. Finally, we ask the same question for the case of four distinct points. This time the answer is "no". Of course the answer is also "no" for any number of distinct points greater than four.

This, however, is not the end of our game, but the starting point. I want to know why it is no. Two systems of four points cannot be transformed to each other; O.K., then when are they so? Let us make the question more precise. The set of systems of four (ordered or labeled) points can be thought of as the following set

$$\mathfrak{X} = \mathbb{P}^1 \times \mathbb{P}^1 \times \mathbb{P}^1 \times \mathbb{P}^1 - \Delta$$

where Δ consists of all systems of four points for which at least two are non-distinct. On $\mathfrak{X}$ the group $GL(2)$ acts as follows:

$$g(x_1, \ldots, x_4) := (gx_1, \ldots, gx_4).$$

The negative answer to the above question for four points implies that the quotient space of $\mathfrak{X}$ with respect to $GL(2)$ is not a point. I want to know what the quotient space looks like.

To make a computation, let us rewrite the space $\mathfrak{X}$ in terms of homogeneous coordinates:

$$\mathfrak{X} = \left\{ x = \begin{pmatrix} x_{11} & x_{12} & x_{13} & x_{14} \\ x_{21} & x_{22} & x_{23} & x_{24} \end{pmatrix} \mid D_x(ij) \neq 0 \text{ for all } i \neq j \right\} / (K^\times)^4,$$

where $D_x(ij)$ is the (i,j)-minor of the matrix x, i.e., the determinant of the matrix consisting of the i-th and j-th columns. The group $(K^\times)^4$ acts on 2×4-matrices from the right as diagonal non-singular matrices. The quotient space of $\mathfrak{X}$ by the group $GL(2)$ will be denoted by $X(4) = X_K(4) = X_K(2,4)$. When I want to recall its definition, since the action of the group is on the left, I shall write

$$X_K(4) = GL(2)\backslash M^*(2,4)/(K^\times)^4,$$

where $M(2,4) = M_K(2,4)$ is the space of 2×4-matrices over K, and $M^*(2,4) = M_K^*(2,4)$ is the subspace of $M(2,4)$ consisting of matrices for which no 2-minor vanishes.

Before we begin a computation to realize $X(4)$, I would like to give you a familiar example in elementary geometry, which is in principle similar to our problem. We work on the Euclidean plane E, on which parallel displacements and rotations can act. In other words, we consider the action of the group G of rigid motions. In this case, any two points can be transformed into each other. Can one system of two points $\{x_1, x_2\}$ be transformed into another $\{y_1, y_2\}$? The answer is "yes" if and only if $d(x_1, x_2) = d(y_1, y_2)$, where d represents distance. d is a function defined on $E \times E$ which is invariant under G, and gives a realization

$$d : G\backslash E \times E \xrightarrow{\sim} [0, +\infty).$$

Next, consider two systems of three distinct, non-collinear points, often called triangles. In order to further understand *invariants*, let us think as follows. You and I are separated by a curtain, you can hear me but you cannot see me, and vice versa. You draw a triangle on a sheet of paper, and I do the same. If I visit you with my paper, we can check by displacement and rotation whether we drew the same triangle. However, I want to know if they are the same only by oral communication, without visiting each other. We know how to do this. You measure the sides of your triangle, I measure mine, and we tell each other three (ordered) numbers. The length of a side is an invariant, but it alone is not enough for our purpose. We need three numbers. This set of numbers is called a *complete set of invariants*. It should be noted that the lengths of two sides and the angle between them works equally well. That is, a complete set of invariants is not necessarily unique. There are many possibilities.

Let us state the above fact in terms of quotient spaces and realizations.

The space of triangles can be expressed as the quotient space

$$G\backslash(E \times E \times E - \Delta),$$

where Δ is the set of all systems of three collinear points. The map

$$E \times E \times E \ni (x_1, x_2, x_3) \mapsto (d(x_2, x_3), d(x_3, x_1), d(x_1, x_2)) \in \mathbb{R}^3$$

satisfies the conditions (1) and (2) in the definition of realizations stated in §3. The image of $E \times E \times E - \Delta$ under this mapping is

$$\{(a, b, c) \in \mathbb{R}^3 \mid a, b, c > 0, \ a + b > c, \ c + a > b, \ b + c > a\}.$$

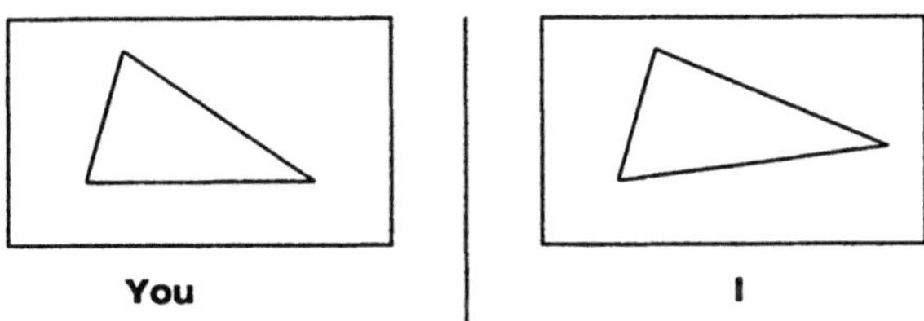

FIGURE 9.1. Two triangles drawn by you and I, separated by a curtain

Let us return to our problem. I will formulate the problem again. The *configuration space $X(n)$ of n distinct points on the projective line* is the quotient space of the set of n distinct, ordered points on $\mathbb{P}^1$ by the group $GL(2)$, i.e., $(x_1, \ldots, x_n)$ is equivalent to $(y_1, \ldots, y_n)$ if and only if there is a projective transformation which takes x_j to y_j for $j = 1, \ldots, n$. The following expression hopefully comes to mind:

$$X(n) = GL(2)\backslash M^*(2, n)/(K^\times)^n.$$

Our task is to find a map defined on $M^*(2, n)$, which is invariant under the group action by $GL(2) \times (K^\times)^n$, and separates the orbits under the group. We have shown that for $n = 1, 2, 3$, the space $X(n)$ is a point. Let us now attack the case $n = 4$. As I mentioned previously, a realization is not necessarily unique. In the present case, there are at least two important realizations.

10. An Easy-going Realization (Cross Ratio)

Though the following realization is easy-going, everyone must know it.

Slogan: *Fix a normal form. Transform the rest into the normal form.*

We found above that any three points can be transformed into, say $\infty, 0, 1$, so let's do it. Since only under the identity are three distinct points kept (pointwise) fixed, the image of the fourth point here should be an invariant. In homogeneous coordinates, every 2×4-matrix with non-vanishing minors can be transformed as follows:

$$\begin{pmatrix} \bullet & \bullet & \bullet & \bullet \\ \bullet & \bullet & \bullet & \bullet \end{pmatrix} \sim \begin{pmatrix} 1 & 0 & \bullet & \bullet \\ 0 & 1 & \bullet & \bullet \end{pmatrix} \sim \begin{pmatrix} 1 & 0 & 1 & 1 \\ 0 & 1 & 1 & \lambda \end{pmatrix}.$$

I choose the third matrix as the normal form. The number λ should be an invariant. Note that there is no particular reason to take this as the normal form. But in any case, let us carry out the computation. Writing $11, 12, \ldots$ in place of $x_{11}, x_{12}, \ldots$, we obtain:

$$\begin{pmatrix} 11 & 12 & 13 & 14 \\ 21 & 22 & 23 & 24 \end{pmatrix}$$

$$\sim \begin{pmatrix} 1 & 0 & 22 \cdot 13 - 12 \cdot 23 & 22 \cdot 14 - 12 \cdot 24 \\ 0 & 1 & -21 \cdot 13 + 11 \cdot 23 & -21 \cdot 14 + 11 \cdot 24 \end{pmatrix}$$

$$\sim \begin{pmatrix} 1 & 0 & 1 & (22 \cdot 14 - 12 \cdot 24)/(22 \cdot 13 - 12 \cdot 23) \\ 0 & 1 & 1 & (-21 \cdot 14 + 11 \cdot 24)/(-21 \cdot 13 + 11 \cdot 23) \end{pmatrix}$$

Therefore we have

$$\lambda(x) = \frac{(-21 \cdot 14 + 11 \cdot 24)(22 \cdot 13 - 12 \cdot 23)}{(-21 \cdot 13 + 11 \cdot 23)(22 \cdot 14 - 12 \cdot 24)}.$$

If we introduce inhomogeneous coordinates:

$$z_1 = x_{11}/x_{21}, \ldots, z_4 = x_{14}/x_{24},$$

the last complicated expression becomes

$$\lambda(x) = \frac{(z_1 - z_4)(z_2 - z_3)}{(z_1 - z_3)(z_2 - z_4)},$$

which is called a *cross ratio* of $(z_1, \ldots, z_4)$. You might find different definitions of the cross ratio in other text books. They are defined by using other numberings of the four points. We will come back to this point later. Anyway, we have proved that the map

$$M(2,4) \ni x = (x_{ij}) \mapsto \lambda(x)$$

is invariant under the left action of $GL(2)$ and the right action of $(K^\times)^4$ and gives a realization of the quotient space $X(4)$ onto $K - \{0,1\}$.

PROPOSITION 10.1. *The map*

$$X(4) \xrightarrow{\sim} K - \{0,1\}$$

given by $M(2,4) \ni x \mapsto \lambda(x)$ *is a realization of the quotient space* $X(4)$.

11. A Democratic Realization

Slogan: *Every point must be treated equally.*

There is no reason for a point x_1 to be sent to 0. It is not fair that only the fourth point is allowed to move freely. There is no reason to choose the normal form above. You should not be satisfied by this realization. I want a democratic one, but how? Above I considered "a function" on $M(2,4)$ to be invariant under the group action. This function, the cross ratio, is a rational function, not a polynomial. Although this function has no poles in the domain we are considering, I prefer polynomials. There is a standard way to answer these demands.

Slogan: *For a quotient space X/G, find a vector valued function f (a system of N functions f_i) on X which is not invariant under the action of G but almost invariant (quasi-invariant), i.e.,*

$$f_i(gx) = a(g)f_i(x), \quad a(g) : independent \ of \ i, \quad g \in G,$$

and projectivize it, i.e.,

$$x \mapsto f_1(x) : f_2(x) : \cdots : f_N(x) \in \mathbb{P}^{N-1}.$$

Let us do this. Since we are looking for polynomials which are quasi-invariant under $GL(2)$, they must be polynomials in $D_x(ij)$. They must also be quasi-invariant under $(K^\times)^4$, and therefore the polynomials must also be column-homogeneous. Linear forms of the $D_x(ij)$ do not work, since they cannot be column-homogeneous. Let us try quadratic polynomials; if these work, we are done. That is, we need not consider polynomials of higher degree. Consider the polynomials

$$D_x(12)D_x(34), \quad D_x(13)D_x(24), \quad D_x(14)D_x(23).$$

These polynomials are quasi-invariants. Indeed for $\{i, j, k, l\} = \{1, 2, 3, 4\}$, we have

$$D_y(ij)D_y(kl) = (\det A)^2(\det a)D_x(ij)D_x(kl), \quad y = Axa,$$

where a is a non-singular diagonal matrix. The following is an answer:

$$f : x \longmapsto D_x(12)D_x(34) : D_x(13)D_x(24) : D_x(14)D_x(23) \in \mathbb{P}^2.$$

You can check directly that $y_1 : y_2 : y_3 \in \mathbb{P}^2$ is in the image of f if and only if the y_j are all non-zero, and satisfy the (so-called Plücker) relation

$$y_1 - y_2 + y_3 = 0.$$

Indeed, in terms of the normal form above, the map f can be given as follows:

$$y_1 : y_2 : y_3 = 1 \cdot (\lambda - 1) : 1 \cdot (-1) : \lambda \cdot (-1) = \lambda - 1 : -1 : -\lambda,$$

which implies the above relation. Let us define a (projective) line in the projective plane as

$$Y := \{y_1 : y_2 : y_3 \in \mathbb{P}^2 \mid y_1 - y_2 + y_3 = 0.\}$$

THEOREM 11.1 (DEMOCRATIC REALIZATION OF $X(4)$). *The map f : $M^*(2, 4) \to \mathbb{P}^2$ induces the isomorphism $\overline{f} : X(4) \to Y_0$, where $Y_0 = \{y_1 : y_2 : y_3 \in Y \mid y_1 y_2 y_3 \neq 0\}$.*

It is very important to realize that the variety Y is *the* compactification of $X(4)$, and to know what kind of degenerate configurations are added to the configuration space $X(4)$ of four *distinct* points to make it compact. Now think about it.

Let us enlarge the domain of definition of f. If three points among the four are equal, then the image would be $0 : 0 : 0$, which is not permitted. We consider degenerate configurations $x = (x_1, x_2, x_3, x_4)$ such that $x_1 = x_2$ or $x_3 = x_4$. The image under f in this case is the same as that of a configuration $x_1 = x_2 \neq x_3 = x_4$, and it is given by $0 : 1 : 1$. Let us code by $(12; 34)$ the configuration $x_1 = x_2 \neq x_3 = x_4$, which is unique up to the group action. The two other configurations

$$(13; 24) : x_1 = x_3 \neq x_2 = x_4,$$
$$(14; 23) : x_1 = x_4 \neq x_2 = x_3$$

are sent to $1 : 0 : -1$ and $1 : 1 : 0$, respectively.

PROPOSITION 11.2. *The isomorphism $\overline{f}$ can be extended to the isomorphism*

$$\overline{X}(4) := X(4) \cup (12; 34) \cup (13; 24) \cup (14; 23) \longrightarrow Y.$$

REMARK 11.1. Recall that, in the easy-going realization of the last section, the three points added were called $\{0, 1, \infty\}$, which give you no idea what kind of degenerate arrangements they represent.

Let us visualize $X_{\mathbb{R}}(4)$, $X_{\mathbb{C}}(4)$, and their compactifications. Since $\mathbb{P}^1_{\mathbb{R}}$ is a circle S^1, the space $X_{\mathbb{R}}(4)$ is the disjoint union of the three parts:

$1 < 2 < 3 < 4 = \{4$ points on S^1 arranged in the order $x_1, x_2, x_3, x_4\}$,

$1 < 2 < 4 < 3 = \{4$ points on S^1 arranged in the order $x_1, x_2, x_4, x_3\}$,

$1 < 3 < 2 < 4 = \{4$ points on S^1 arranged in the order $x_1, x_3, x_2, x_4\}$.

You can find a more kind explanation of this point on the first few pages of Chapter V.

	$\lambda = 1$		$\lambda = 0$		$\lambda = \infty$	
	$(14; 23)$		$(12; 34)$		$(13; 24)$	
$1<4<2<3$		$1<2<3<4$		$1<2<4<3$		$1<4<2<3$

$X_{\mathbb{R}} \subset X_{\mathbb{C}}$

12. Configuration Space of n Points on Projective Spaces

Here I would like to present you a problem. The configuration space of n distinct points in general position in $\mathbb{P}^{k-1}$ is by definition the following quotient space

$$X(k, n) := GL(k) \backslash M^*(k, n) / H_n,$$

where $M^*(k, n)$ is the space of $k \times n$-matrices for which no k-minor vanishes, and $H_n \cong (K^\times)^n$ is the subgroup of $GL(n)$ consisting of diagonal matrices. Each $x \in M^*(k, n)$ represents n distinct points on $\mathbb{P}^{k-1}$; the j-th column of x is regarded as a homogeneous coordinate of the j-th point. The group $GL(k)$ acts on $M^*(k, n)$ from the left as the group of projective transformations on $\mathbb{P}^{k-1}$, and H_n acts on $M^*(k, n)$ from the right lifting the ambiguity of homogeneous coordinates. Note that we have written $X(n)$ in place of $X(2, n)$.

PROBLEM 12.1. Study the geometry of $X(k, n)$.

REMARK 12.1. This is by no means an easy problem; it has been a central problem of algebraic geometry for many years (the so-called *invariant theory*). In this book, in addition to $X(2,4)$, we study $X(2,5)$ and $X(3,6)$.

13. The Grassmann Isomorphism $X(k,n)$ and $X(n-k,n)$

I introduce in this section the *Grassmann isomorphism*

$$* : X(k,n) \xrightarrow{\sim} X(n-k,n),$$

classically called the *association*. Let $x \in M^*(k,n)$ be a $k \times n$-matrix representing a point of $X(k,n)$. We can regard x as a linear map from an n-dimensional linear space U with a fixed basis u_j to a k-dimensional linear space with a fixed basis. Since x is of rank k, its kernel $W \subset U$ is of dimension $n-k$. Choose a basis $w_1,\ldots,w_{n-k}$ of W and express them in terms of u_j as

$$w_i = \sum_{j=1}^{n} x'_{ij} u_j, \quad i = 1,\ldots,n-k.$$

In this way, we obtain an $(n-k) \times n$-matrix $x' = (x'_{ij})$. Please check that $x' \in M^*(n-k,n)$, that is, no $(n-k)$-minor of x' vanishes. Choosing another basis of W, we have $g'x'$ for some $g' \in GL(n-k)$. Thus by this procedure, $x \in M^*(k,n)$ determines the class $GL(n-k)x'$. Since

$$x \, {}^t x' = 0 \quad \text{if and only if} \quad (gxh)^t(g'x'h^{-1}) = 0$$

for $g \in GL(k)$, $g' \in GL(n-k)$ and $h \in H_n$, if we start from $gxh \in M^*(k,n)$, we eventually obtain the class $GL(n-k)xh^{-1}$. Therefore this correspondence leads to the well-defined map $* : X(k,n) \to X(n-k,n)$. Using the space $X(n-k,n)$ instead of $X(k,n)$, we obtain the inverse map of $*$. This implies that $*$ is an isomorphism.

REMARK 13.1. In the case $(k,n) = (2,4)$, the isomorphism $*$ is the identity on $X(2,4)$. In the cases $(k,n) = (k,2k)$ for $k \geq 3$, it defines an involution (an automorphism of order 2) on $X(k,2k)$.

14. Configuration Spaces of Unlabeled Point Sets

Up to now we have been concerned with labeled (ordered) points. Although I have not mentioned this explicitly, we have distinguished two point sets consisting of the same points with different labellings. For example, we have distinguished the system

$$x_1 = 1, \quad x_2 = 2, \quad x_3 = 3, \quad x_4 = 4 \quad \in \mathbb{P}^1$$

from the system

$$x_1 = 2, \quad x_2 = 1, \quad x_3 = 3, \quad x_4 = 4 \quad \in \mathbb{P}^1.$$

Let us now identify them. The symmetric group S_n of degree n acts on $X(k,n)$ as permutations of n columns of $M(k,n)$. We define the configuration space $X\{k,n\}$ of n unlabeled points on $\mathbb{P}^{k-1}$ simply as the quotient space

$$X\{k,n\} := X(k,n)/S_n.$$

As before, we write $X\{n\}$ in place of $X\{2,n\}$. Let us study $X\{4\} = X(4)/S_4$ in the two realizations discussed above.

14.1. $X\{4\}$ in Terms of a Cross Ratio. Let us consider the action of S_4 on $X(4)$. The permutation of the i-th and j-th columns will be denoted by (ij). First note that

$$(12)(34) \begin{pmatrix} 1 & 0 & 1 & 1 \\ 0 & 1 & 1 & \lambda \end{pmatrix} = \begin{pmatrix} 0 & 1 & 1 & 1 \\ 1 & 0 & \lambda & 1 \end{pmatrix} \sim \begin{pmatrix} 1 & 0 & \lambda & 1 \\ 0 & 1 & 1 & 1 \end{pmatrix}$$
$$\sim \begin{pmatrix} 1 & 0 & 1 & 1 \\ 0 & 1 & 1/\lambda & 1 \end{pmatrix} \sim \begin{pmatrix} 1 & 0 & 1 & 1 \\ 0 & \lambda & 1 & \lambda \end{pmatrix} \sim \begin{pmatrix} 1 & 0 & 1 & 1 \\ 0 & 1 & 1 & \lambda \end{pmatrix}.$$

Thus in this way, permutations of the form

$$(ij)(pq), \quad \{i,j,p,q\} = \{1,\ldots,4\}$$

operate trivially on $X(4)$; such permutations form a normal subgroup N of order 4. The quotient group $S_4/N \cong S_3$ acts on $X(4)$ in the following way.

$$(12) \quad \begin{pmatrix} 1 & 0 & 1 & 1 \\ 0 & 1 & 1 & \lambda \end{pmatrix} = \begin{pmatrix} 0 & 1 & 1 & 1 \\ 1 & 0 & 1 & \lambda \end{pmatrix} \sim \begin{pmatrix} 1 & 0 & 1 & \lambda \\ 0 & 1 & 1 & 1 \end{pmatrix}$$

$$\sim \begin{pmatrix} 1 & 0 & 1 & 1 \\ 0 & 1 & 1 & 1/\lambda \end{pmatrix},$$

$$(13) \quad \begin{pmatrix} 1 & 0 & 1 & 1 \\ 0 & 1 & 1 & \lambda \end{pmatrix} = \begin{pmatrix} 1 & 0 & 1 & 1 \\ 1 & 1 & 0 & \lambda \end{pmatrix} \sim \begin{pmatrix} 1 & 0 & 1 & 1 \\ 0 & 1 & -1 & \lambda - 1 \end{pmatrix}$$

$$\sim \begin{pmatrix} 1 & 0 & 1 & 1 \\ 0 & -1 & 1 & 1 - \lambda \end{pmatrix} \sim \begin{pmatrix} 1 & 0 & 1 & 1 \\ 0 & 1 & 1 & 1 - \lambda \end{pmatrix},$$

and so on. In this way, we get six transforms:

$$\lambda, \quad \frac{1}{\lambda}, \quad 1 - \lambda, \quad \frac{1}{1 - \lambda}, \quad \frac{\lambda}{\lambda - 1}, \quad \frac{\lambda - 1}{\lambda},$$

forming a group isomorphic to S_3. Notice that this group can be generated by two transformations $\lambda \mapsto 1 - \lambda$ and $\lambda \mapsto 1/\lambda$. I want a rational function $j(\lambda)$ of degree 6 ($=$ order of S_3) which is invariant under this group. Let us study the fixed points of the transformations which are distinct from the identity:

transform	order	equation	fixed points
$\dfrac{1}{\lambda}$	2	$\lambda^2 - 1 = 0$	$1, \quad -1$
$1 - \lambda$	2	$2\lambda - 1 = 0$	$\dfrac{1}{2}, \quad \infty$
$\dfrac{1}{1-\lambda}, \quad \dfrac{\lambda-1}{\lambda}$	3	$\lambda^2 - \lambda + 1 = 0$	$\rho, \quad \dfrac{1}{\rho}$
$\dfrac{\lambda}{\lambda - 1}$	2	$\lambda^2 - 2\lambda = 0$	$0, \quad 2$

Here, ρ is a primitive sixth root of unity. Since the three points $\{0, 1, \infty\}$ are transitive under the group, these three points are mapped by j to the same point, say ∞. No other point is mapped to ∞. Further, since these three points are fixed points of order 2, the rational function j must have the following form:

$$\frac{\text{(a polynomial of degree 6)}}{\lambda^2(\lambda - 1)^2}.$$

Similarly, the two points ρ and $1/\rho$ are mapped to the same point, say 0. No other point must be mapped to 0. These two points are fixed points of degree 3, so the numerator of the above expression is a constant multiple of $(\lambda^2 - \lambda + 1)^3$. Then, points $1/2$ and 2 must be mapped to the same point, say 1. In this way, we are led to

$$j(\lambda) = \frac{4}{27}\frac{(\lambda^2 - \lambda + 1)^3}{\lambda^2(\lambda - 1)^2}.$$

You can easily check by using generators $1/\lambda$ and $1 - \lambda$ that $j(\lambda)$ is invariant under the group; since it is of order 6, any invariant is a rational function of this invariant. Note that j maps the fixed points as follows

$$
\begin{array}{ccccc}
\lambda & & & & j \\
\\
0, & 1, & \infty & \mapsto & \infty \\
-1, & 2, & \dfrac{1}{2} & \mapsto & 1 \\
\rho, & \dfrac{1}{\rho} & & \mapsto & 0
\end{array}
$$

PROPOSITION 14.1. *The correspondence $\lambda \mapsto j(\lambda) \in \mathbb{C}$ gives a realization of $X\{4\} = X(4)/S_4 (\cong \mathbb{C})$.*

REMARK 14.1. Keep in mind that the action of S_4 on $X(4)$ has fixed points.

14.2. $X\{4\}$ in Terms of the Democratic Realization. Recall that $X(4)$ can be democratically realized by

$$x \mapsto D_x(12)D_x(34) : D_x(13)D_x(24) : D_x(14)D_x(23) \in \mathbb{P}^2.$$

I want to make a rational function in

$$D_x(ij)D_x(pq), \quad \{i,j,p,q\} = \{1,2,3,4\}$$

that is S_4-invariant. This time, since the realization has a democratic shape, I have only to make a symmetric rational function of degree six. Here is one:

$$\frac{[\{D_x(12)D_x(34)\}^2 + \{D_x(13)D_x(24)\}^2 + \{D_x(14)D_x(23)\}^2]^3}{\{D_x(12)D_x(34)\}^2\{D_x(13)D_x(24)\}^2\{D_x(14)D_x(23)\}^2}.$$

By the normal form above, you can easily check that this is $\dfrac{27}{4}2^3$ times j.

CHAPTER II

Elliptic Curves

The theory of elliptic curves is one of the most beautiful and important theories in mathematics. There is no question about this. I know several ways to introduce elliptic curves, as you can find in many books and each method has advantages and disadvantages:

(1) If I follow the history of mathematics, when I first attempt to compute the arc length of an ellipse, I find that this integral, which is called an elliptic integral, cannot be evaluated by trigonometric functions, and so on, as Gauss did. This method is good if you would like to know how discoveries were made, but it takes time and you may feel it is difficult.

(2) I can present the theory of 1-dimensional complex analytic varieties, called Riemann surfaces, define an invariant called the genus, and introduce elliptic curves as compact Riemann surfaces of genus one. Such a presentation is good for learning modern theory of Riemann surfaces, but again it takes too much time.

(3) I can define an elliptic curve as a double cover of the projective line branching at four distinct points. I shall come back to this point of view later.

(4) I can define an elliptic curve as a non-singular cubic plane curve, but this it is too geometry-oriented for this book.

(5) I can define an elliptic curve as the quotient variety (quotient group) of the 1-dimensional complex vector space by a lattice group. While it is difficult for students to understand the notion of a quotient space, this is what I explained in the preceding sections. Using this method, it is a priori obvious that an elliptic curve admits the structure of an abelian group, which may not be easy to see with the other methods. This is the

method we would like to follow.

1. Lattices in $\mathbb{C}$

Let ω_1 and ω_2 be complex numbers satisfying

$$\Im(\omega_1/\omega_2) > 0,$$

and

$$L = L(\omega_1, \omega_2) := \mathbb{Z}\omega_1 + \mathbb{Z}\omega_2$$

be a *lattice* in the additive group $\mathbb{C}$. Notice that $L(\omega_1, \omega_2) = L(\omega_1', \omega_2')$ if and only if ω_1 and ω_2 can be written as $\mathbb{Z}$-linear combinations of ω_1' and ω_2', i.e.

$$\omega_1 = a\omega_1' + b\omega_2',$$
$$\omega_2 = c\omega_1' + d\omega_2',$$

where $a, b, c,$ and d are integers, and ω_1' and ω_2' can be written similarly in terms of ω_1 and ω_2. This implies that the integral matrix

$$A = \begin{pmatrix} a & b \\ c & d \end{pmatrix}$$

has an integral inverse matrix, which is possible if and only if

$$\det A = ad - bc = 1 \text{ or } -1.$$

Since I require $\Im(\omega_1'/\omega_2') > 0$, the determinant must be 1; in other words,

$$A \in SL(2, \mathbb{Z}).$$

2. Elliptic Curves as Quotients of $\mathbb{C}$ by Lattices

Let us consider the quotient space of the additive group $\mathbb{C}$ by the lattice subgroup $L(\omega_1, \omega_2)$:

$$E = E(\omega_1, \omega_2) = \mathbb{C}/L(\omega_1, \omega_2).$$

This quotient space is called an *elliptic curve* and is homeomorphic to a torus, i.e., the direct product of two circles. The parallelogram F with vertices $0, \omega_1, \omega_2, \omega_1 + \omega_2$, as well as any displacement of F (see Figure 2.1) is called a *fundamental parallelogram*. Any point in $\mathbb{C}$ is equivalent to a point in F, and no two points in F are equivalent.

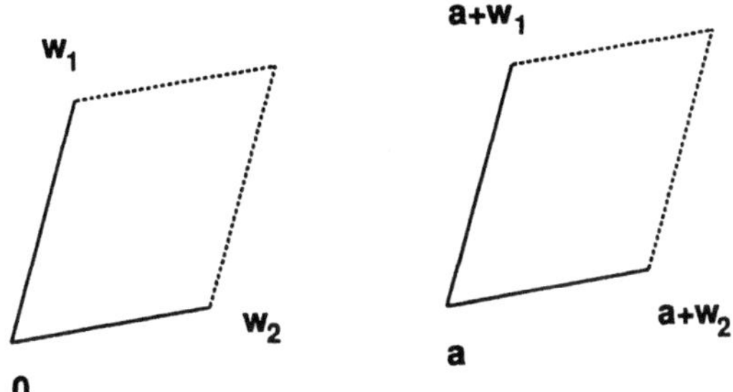

FIGURE 2.1. Fundamental parallelograms

3. Isomorphism Classes of Elliptic Curves

Let α be a non-zero complex number. Then

$$x \to \alpha x$$

gives an automorphism of the additive group $\mathbb{C}$. Therefore, two elliptic curves $E(\omega_1, \omega_2)$ and $E(\alpha\omega_1, \alpha\omega_2)$ are isomorphic. If you would like, by choosing α to be $1/\omega_2$, you can express elliptic curves in the form

$$E(\tau) := E(\tau, 1), \quad \text{where} \quad \Im(\tau) > 0.$$

Two elliptic curves $E(\omega_1, \omega_2)$ and $E(\omega_1', \omega_2')$ are isomorphic if and only if $\omega_1 : \omega_2$ and $\omega_1' : \omega_2'$ are equivalent under $PSL(2, \mathbb{Z})$, or, if you like, two elliptic curves $E(\tau)$ and $E(\tau')$ are isomorphic if and only if τ' can be expressed in terms of τ as follows:

$$\tau' = \frac{a\tau + b}{c\tau + d}, \quad a, b, c, d \in \mathbb{Z}, \ ad - bc = 1.$$

In other words,

$$\tau' = g\tau \quad \text{for some} \quad g \in PSL(2, \mathbb{Z}).$$

Let $\mathbb{H}$ be the upper half space

$$\mathbb{H} = \{\omega_1 : \omega_2 \in \mathbb{P}^1 \mid \Im(\omega_1/\omega_2) > 0\}$$
$$= \{\tau \in \mathbb{C} \mid \Im(\tau) > 0\}.$$

The group $PSL(2, \mathbb{R})$ acts on $\mathbb{H}$; indeed, we have

$$\Im\tau' = \Im\frac{(a\tau + b)(c\bar{\tau} + d)}{|c\tau + d|^2}$$
$$= \frac{(ad - bc)\Im\tau}{|c\tau + d|^2} = \frac{\Im\tau}{|c\tau + d|^2}.$$

Then you can think of the isomorphism classes of elliptic curves as being parameterized by the quotient space $\mathbb{H}/PSL(2,\mathbb{Z})$. However, one must be careful to make sure that this action has fixed points (see Exercise 4).

On an elliptic curve $E(\omega_1,\omega_2)$, there are three points of order 2 (as elements of the abelian group E); they are represented by *half periods*:

$$\frac{\omega_1}{2}, \quad \frac{\omega_2}{2}, \quad \frac{\omega_1+\omega_2}{2} \quad \text{mod } L.$$

When two elliptic curves $E = E(\omega_1,\omega_2)$ and $E' = E(\omega_1',\omega_2')$ are isomorphic under $g \in SL(2,\mathbb{Z})$, the points of order two of E correspond to those of E', of course. But if you require the three points represented by $\omega_1/2, \omega_2/2, \omega_1/2 + \omega_2/2$ to be mapped to those represented by $\omega_1'/2, \omega_2'/2, \omega_1'/2 + \omega_2'/2$ in this order, i.e.,

$$\frac{\omega_1}{2} \mapsto \frac{\omega_1'}{2}, \quad \frac{\omega_2}{2} \mapsto \frac{\omega_2'}{2}, \quad \text{mod } L$$

then g must assume the following form:

$$g = \begin{pmatrix} a & b \\ c & d \end{pmatrix}, \quad a,d : \text{odd}, \quad b,c : \text{even}.$$

This is also a sufficient condition. Such matrices form a group called the *principal congruence subgroup* of level 2. It is denoted by $\Gamma(2)$:

$$\Gamma(2) = \{g \in SL(2,\mathbb{Z}) \mid g \equiv I_2 \mod 2\},$$

where I_r is the identity matrix of size r. If we identify only such pairs of isomorphic elliptic curves, they are parametrized by $\mathbb{H}/\Gamma(2)$.

We note that $\Gamma(2)$ is a normal subgroup of $SL(2,\mathbb{Z})$, that the quotient group $SL(2,\mathbb{Z})/\Gamma(2)$ is isomorphic to the symmetric group S_3, and that the action of $\Gamma(2)$ on $\mathbb{H}$ has no fixed points.

EXERCISE 2. Prove the above statements.

EXERCISE 3. Show that the cosets $SL(2,\mathbb{Z})/\Gamma(2)$ can be represented by the following elements:

$$\begin{pmatrix} 1 & 0 \\ 0 & 1 \end{pmatrix}, \begin{pmatrix} 0 & 1 \\ -1 & 0 \end{pmatrix}, \begin{pmatrix} 1 & 1 \\ -1 & 0 \end{pmatrix}, \begin{pmatrix} 1 & 1 \\ 0 & 1 \end{pmatrix}, \begin{pmatrix} 1 & 0 \\ 1 & 1 \end{pmatrix}, \begin{pmatrix} 0 & 1 \\ -1 & 1 \end{pmatrix}.$$

In the previous section, I showed you a picture of a fundamental parallelogram. Here I show you in Figure 3.1 a set in $\mathbb{H}$ with the properties that any point in $\mathbb{H}$ is equivalent under $SL(2,\mathbb{Z})$ to a point in this set,

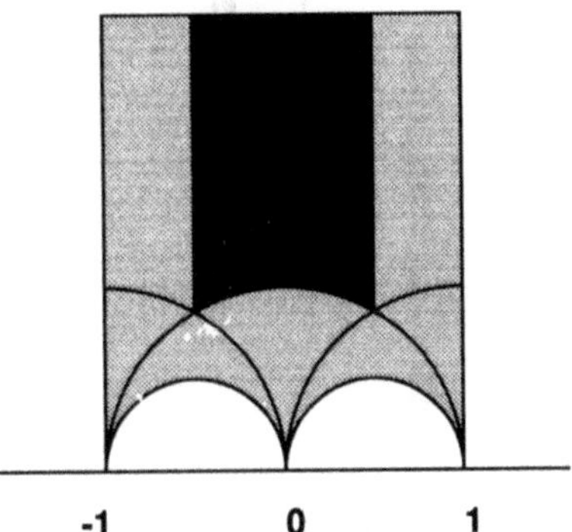

FIGURE 3.1. Six fundamental domains for $SL(2,\mathbb{Z})$

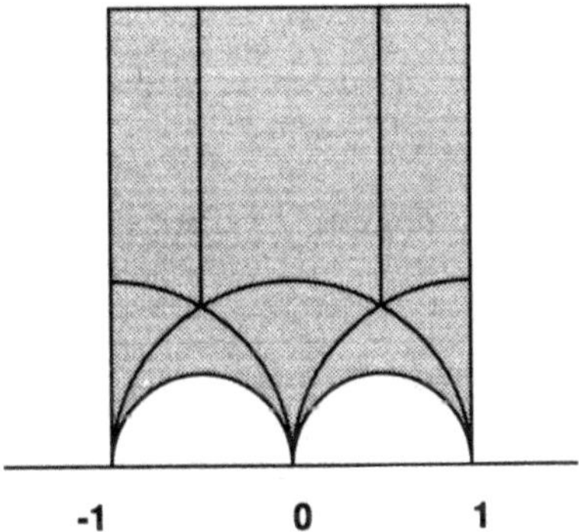

FIGURE 3.2. A fundamental domain for $\Gamma(2)$

and no two points in the set are equivalent. Such a set is called a *funda-mental domain* of the group $SL(2,\mathbb{Z})$. Figure 3.2 shows a fundamental domain for the group $\Gamma(2)$. Notice that halves of the fundamental domains of $SL(2,\mathbb{Z})$ are obtained by the barycentric subdivision of the corresponding halves of $\Gamma(2)$. (See Figure 3.3.)

EXERCISE 4. Check that these domains are fundamental domains.

Solving these exercises is an intelligent way to spend time in boresome meetings. I am sure that they are better than cross-word puzzles.

EXERCISE 5. Show that the group $SL(2,\mathbb{Z})$ can be generated by

$$\begin{pmatrix} 1 & 1 \\ 0 & 1 \end{pmatrix} \quad \text{and} \quad \begin{pmatrix} 0 & 1 \\ -1 & 0 \end{pmatrix}$$

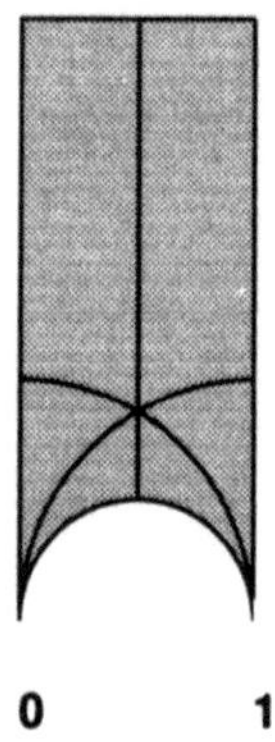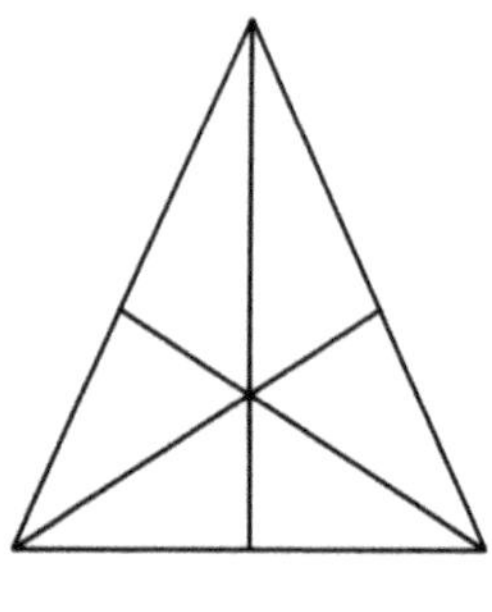

FIGURE 3.3. A barycentric subdivision

or by

$$\begin{pmatrix} 1 & 1 \\ -1 & 0 \end{pmatrix} \quad \text{and} \quad \begin{pmatrix} 0 & 1 \\ -1 & 0 \end{pmatrix};$$

show also that the last two elements have fixed points in $\mathbb{H}$.

Hint: Note that

$$\begin{pmatrix} 1 & 1 \\ 0 & 1 \end{pmatrix}^{\pm 1} \begin{pmatrix} a \\ b \end{pmatrix} = \begin{pmatrix} a \pm b \\ b \end{pmatrix},$$

$$\begin{pmatrix} 0 & 1 \\ -1 & 0 \end{pmatrix} \begin{pmatrix} a \\ b \end{pmatrix} = \begin{pmatrix} b \\ -a \end{pmatrix}.$$

If $b \geq a \geq 1$, then $|b-a| < |b|$. Invoke induction with respect to $|a|^2 + |b|^2$.

EXERCISE 6. Show that the group $\Gamma(2)$ can be generated by

$$\begin{pmatrix} 1 & 2 \\ 0 & 1 \end{pmatrix}, \quad \begin{pmatrix} 1 & 0 \\ 2 & 1 \end{pmatrix} \quad \text{and} \quad -I_2.$$

Hint: If $b > a \geq 1$, then $|b - 2a| < |b|$. Invoke induction with respect to $|a|^2 + |b|^2$.

4. Realization in Terms of the Weierstrass $\wp$ Function

Let us realize the elliptic curve $E = E(\omega_1, \omega_2)$, which is given as the quotient space $\mathbb{C}/L(\omega_1, \omega_2)$. There are at least two important realizations. The first of these which I am going to present now, is due to Weierstrass. The second, which I shall present later, is due to Jacobi, I think, although I am not sure; I am not interested in 'who-did-first'.

We are asked (see Example 3.2 in Chapter I) to make functions on $\mathbb{C}$ which are invariant under parallel displacement by each element of $L = L(\omega_1, \omega_2)$.

4.1. Elliptic Functions in General.

DEFINITION 4.1. A meromorphic function on $\mathbb{C}$ invariant under the action of elements of a lattice L, i.e., a function f satisfying

$$f(u + \omega) = f(u), \quad \omega \in L$$

is called an *elliptic function*.

REMARK 4.1. Instead of the phrase "being invariant under the action of L," I often use "being periodic with periods ω_1 and ω_2," or "having periods ω_1 and ω_2."

REMARK 4.2. A holomorphic elliptic function is a constant.

To prove this statement, we use two well known facts:

(1) A continuous real-valued function defined on a compact set achieves a maximal value on this set.

(2) (Maximum principle) If a holomorphic function achieves a maximal absolute value at a point of an open set on which the function is defined, then it is a constant.

We need only apply these facts, since elliptic functions are by definition defined on the compact set E.

Comment: The maximum principle for holomorphic functions can be proved as follows. Cauchy's estimate

$$|f(a)| \leq \frac{1}{2\pi} \int_0^{2\pi} |f(a + r \exp it)| dt,$$

which can be derived from Cauchy's integral formula

$$f(a) = \frac{1}{2\pi i} \int_{|u-a|=r} \frac{f(u)}{u - a} du,$$

tells us that the absolute value at a certain point a is not greater than the average of absolute values along any curve surrounding the point.

This brings to mind the following scene: I am sitting in an elementary school class after an examination, and my teacher tells me that my score does not exceed the average, but still it is the top in the class. I beat my confused childlike brain and conclude that my classmates and I all got the same score.

We can make the previous remark somewhat stronger.

REMARK 4.3. An elliptic function with at most one simple pole on $\mathbb{C}/L$ is a constant.

PROOF. We show that such a function must be holomorphic. Take a fundamental parallelogram F such that the boundary ∂F does not pass through the poles of f. Integrate the function along ∂F:

$$\int_{\partial F} f(u)\,du;$$

since there is only one simple pole in F, this gives the residue at the point. On the other hand, since the function is periodic with periods ω_1 and ω_2, we have

$$f(u + \omega_1) = f(u), \quad f(u + \omega_2) = f(u).$$

Thus the integrals along the right side and left side of F, as well as those along the bottom and top of F cancel; the integral is zero. A simple pole with zero residue implies no pole. $\square$

REMARK 4.4. The number of zeros (counting multiplicity) and the number of poles (counting multiplicity) in the fundamental parallelogram of an elliptic function are equal.

PROOF. Take a fundamental parallelogram F such that the boundary ∂F does not pass through the zeros and poles of f. Consider the integral

$$\int_{\partial F} \frac{f'(u)}{f(u)}\,du.$$

Since f'/f is an elliptic function, the integral is zero, as we have shown above. On the other hand, at a zero point of order r of f, the function f'/f has a simple pole of residue r, and at a pole of order r of f, the function f'/f has a simple pole of residue $-r$. Thus the integral subtracts the number of poles from the number of zeros in F. $\square$

4.2. The Weierstrass $\wp$ Function. Consider the expression

$$f(u) = \sum_{\omega \in L} \frac{1}{(u - \omega)^3}.$$

You can check that for any point u not in L, there is a small neighborhood around u such that this series converges absolutely. Thus $f(u)$ is a meromorphic function on $\mathbb{C}$ with poles at each point in L. Since convergence is absolute, you can sum up to any order; this means that f is invariant under L. Note that f is an odd function, as is obvious from the definition and the fact that $-L = L$.

The function f has a pole of order 3 at the origin, implying that the map $E \ni u \mapsto f(u) \in \mathbb{P}^1$ is locally 3 to 1. Thus f alone cannot give a realization of E. We must find another elliptic function.

Following Weierstrass, we consider the expression (the *Weierstrass $\wp$ function*)

$$\wp(u) = \frac{1}{u^2} + \sum_{\omega \in L - \{0\}} \left(\frac{1}{(u - \omega)^2} - \frac{1}{\omega^2} \right).$$

Thanks to the term $-1/\omega^2$, we have

$$\frac{1}{(u - \omega)^2} - \frac{1}{\omega^2} = \frac{2u\omega - u^2}{(u - \omega)^2 \omega^2}.$$

Thus the denominator is of degree 4 in ω, while the numerator is of order 1, so that the expression converges, as in the case of $f(u)$, and defines a meromorphic function on $\mathbb{C}$. It is obvious by definition that

$$\wp'(u) = -2f(u),$$

and that $\wp$ is an even function. Let us check that this is an elliptic function, i.e., is invariant under L. Since f is an elliptic function and $\wp$ is a primitive function of f, we have

$$\wp(u + \omega) = \wp(u) + C(\omega), \quad \omega \in L,$$

where $C(\omega)$ is a constant (independent of u). Put $u = -\omega/2$ and recall that $\wp$ is an even function. Then we know that $C(\omega) = 0$. Therefore $\wp$ is also an elliptic function.

Since $\wp$ is an elliptic function with double poles only at points in L, it should have two zeros in a fundamental parallelogram. I do not know in general where they are.

EXERCISE 7. Find the zeros of $\wp$ when

$$\omega_1/\omega_2 = \sqrt{-1}, \quad \omega_1/\omega_2 = \exp \frac{2\pi\sqrt{-1}}{6}.$$

Since $\wp'$ is an elliptic function with triple poles only at points in L, it should have three zeros in a fundamental parallelogram; this time I do know where they are. Since $\wp'$ is an odd function, we have

$$-\wp'(\frac{\omega_1}{2}) = \wp'(-\frac{\omega_1}{2})$$
$$= \wp'(-\frac{\omega_1}{2} + \omega_1) = \wp'(\frac{\omega_1}{2}).$$

So we have found one zero; do the same for $\omega_2/2$ and $\omega_1/2 + \omega_2/2$ and you find two more zeros. These three zeros

$$\frac{\omega_1}{2}, \quad \frac{\omega_2}{2}, \quad \frac{\omega_1}{2} + \frac{\omega_2}{2},$$

are obviously not equivalent modulo L. Thus you know you have found them all.

4.3. The Algebraic Relation between $\wp$ and $\wp'$. There is an algebraic relation between $\wp$ and $\wp'$; let us find it. We make Laurent expansions around the origin. Set

$$g_2 := 60 \sum_{\omega \in L - \{0\}} \frac{1}{\omega^4},$$

$$g_3 := 140 \sum_{\omega \in L - \{0\}} \frac{1}{\omega^6}.$$

Then we have

$$\wp(u) = \frac{1}{u^2} + \frac{g_2}{20}u^2 + \frac{g_3}{28}u^4 + \cdots,$$

$$\wp'(u) = -2\frac{1}{u^3} + \frac{g_2}{10}u + \frac{g_3}{7}u^3 + \cdots.$$

Consider the expression

$$4\wp^3 - (\wp')^2 - g_2\wp - g_3,$$

which is constructed to have no poles at the origin nor a constant term. Since any rational function of elliptic functions is also an elliptic function, this is an elliptic function which takes the value zero at the origin and has no poles. It must therefore be identically zero. Thus we have

PROPOSITION 4.1.

$$(\wp')^2 = 4\wp^3 - g_2\wp - g_3.$$

REMARK 4.5. If you put $z = \wp(u)$, the identity above implies that $\wp(u)$ satisfies the non-linear differential equation

$$\left(\frac{dz}{du}\right)^2 = 4z^3 - g_2 z - g_3;$$

the inverse function $u(z)$ satisfies

$$\left(\frac{du}{dz}\right)^2 = \frac{1}{4z^3 - g_2 z - g_3}.$$

This can be easily integrated, if you are not seriously worried by integrating a 2-valued function. We have

$$u(z) = \int^z \frac{dz}{\sqrt{4z^3 - g_2 z - g_3}}.$$

This is one of the so-called elliptic integrals.

4.4. A Realization. Let us define an affine plane curve $C' = C'(\omega_1, \omega_2)$, a plane curve in $\mathbb{C}^2 \ni (z, w)$, by

$$C' : w^2 = 4z^3 - g_2 z - g_3,$$

and consider the map

$$\varphi' : \mathbb{C} - L \ni u \mapsto (z, w) = (\wp(u), \wp'(u)) \in C' \subset \mathbb{C}^2.$$

PROPOSITION 4.2. *The map* $\varphi' : u \mapsto (\wp(u), \wp'(u))$ *gives an isomorphism between* $E - \{0\}$ *and* C'.

PROOF. We show only that this map is one-to-one. Assume for some u_1 and u_2 that

$$\wp(u_1) = \wp(u_2) \quad \text{and} \quad \wp'(u_1) = \wp'(u_2).$$

Applying Remark 4.4 to the elliptic function $\wp - a$ $(a = \wp(u_1))$, we know that the function $\wp$ takes the same value in any fundamental parallelogram exactly twice. If you remember that $\wp$ is an even function, you know we must have $u_1 \equiv -u_2$ modulo L. Since $\wp'$ is an odd function, u_1 (as well as u_2) must be a zero of $\wp'$. We know the zeros of $\wp'$; they are half periods. But a half period is by definition a point which satisfies $u \equiv -u$ modulo L. Thus $u_1 \equiv u_2$ modulo L. $\square$

When u approaches the origin, both $\wp(u)$ and $\wp'(u)$ tend to infinity, because both functions have poles at the origin. But if you think of (z, w) inhomogeneous coordinates of the projective plane:

$$z : w : 1 = z_1 : z_2 : z_3 \in \mathbb{P}^2,$$

then we have

$$\wp(u) : \wp'(u) : 1 = u^3 \wp(u) : u^3 \wp'(u) : u^3.$$

As u tends to 0, the first and third terms tend to zero, while the second term tends to a non-zero number. Therefore the map φ' can be extended everywhere on $\mathbb{C}$ and thus defines the map

$$\varphi : \mathbb{C} \ni u \mapsto \wp(u) : \wp'(u) : 1 \in \mathbb{P}^2.$$

The image under φ is the curve

$$C := C' \cup \{0 : 1 : 0\} \subset \mathbb{P}^2.$$

PROPOSITION 4.3. *The map $\varphi : u \mapsto \wp(u) : \wp'(u) : 1$ from $\mathbb{C}$ into $\mathbb{P}^2$ gives an isomorphism between E and a curve C; it is a realization of E. The origin is mapped to the point $0 : 1 : 0$.*

Terminology: A realization into a projective space is often called a *projective embedding*. A curve in the projective plane is called a *plane curve*.

4.5. Cubic Plane Curves. Let $z_1 : z_2 : z_3$ be homogeneous coordinates of the projective plane. A curve is defined by a homogeneous polynomial in z_1, z_2, and z_3. If the polynomial is linear, i.e., of degree 1, it defines a line, and it can be parametrized by linear functions, as done in highschool mathematics. If the polynomial is of degree 2, say describing a circle, it can be parametrized by trigonometric functions; recall that

$$(\sin u)^2 + (\cos u)^2 = 1.$$

(Incidentally, the circle also admits the rational parametrization $(2t)^2 + (1 - t^2)^2 = (1 + t^2)^2$.) At any rate, the proposition above means that if the polynomial is of degree 3, then it can be parametrized by elliptic functions.

Let us study the cubic plane curve C defined by

$$z_2^2 z_3 = 4z_1^3 - g_2 z_1 z_3^2 - g_3 z_3^3$$

for a while.

REMARK 4.6. Any non-singular cubic curve can be projectively transformed into this form, the so-called *Weierstrass normal form*.

If we introduce inhomogeneous coordinates,

$$z = \frac{z_1}{z_3}, \quad w = \frac{z_2}{z_3},$$

the equation turns out to define the affine curve C',

$$w^2 = 4z^3 - g_2 z - g_3.$$

The line at infinity $\{z_3 = 0\}$ meets (is tangent to) the curve C at one point, $0 : 1 : 0$.

4.6. Elliptic Curves as Double Covers of the Line. Let us consider the projection

$$\pi' : C' \ni (z, w) \mapsto z \in \mathbb{C}$$

and its extension

$$\pi : C \to \mathbb{P}^1$$

by sending the point $0 : 1 : 0$ to the point at infinity. The projection is 2-to-1 except at $0 : 1 : 0$ and at the three roots of the polynomial equation

$$4z^3 - g_2 z - g_3 = 0.$$

In other words, a cubic plane curve is a double cover of the projective line branching at four points. Note that the double cover is completely determined by the four branching points. We know that the three roots of the cubic equation above are given by

$$e_1 := \wp(\frac{\omega_1}{2}), \quad e_2 := \wp(\frac{\omega_2}{2}), \quad e_3 := \wp(\frac{\omega_1}{2} + \frac{\omega_2}{2}).$$

These are distinct as shown by the argument above; the function $\wp$ takes the same value at u and $-u$ modulo L, nowhere else.

4.7. The Lambda Function – A Realization of $\mathbb{H}/\Gamma(2)$. Now we know that our cubic curve is the double cover of the line branching at e_1, e_2, e_3 and ∞. A cross ratio of these branching points is given by

$$\lambda = \lambda(\omega_1, \omega_2) = \frac{e_1 - e_2}{e_1 - e_3}.$$

There are $6 = 3!$ such cross ratios $\lambda, 1 - \lambda, 1/\lambda, \ldots$ (Do you remember? If not, see §13.1 of Chapter I.) The term *lambda function* refers to any one of these, and any one gives the realization (isomorphism)

$$\lambda : \mathbb{H}/\Gamma(2) \xrightarrow{\sim} \mathbb{C} - \{0, 1\} \cong X(4).$$

This induces the correspondence

$$\mathbb{C} \supset \overline{\mathbb{H}} \ni \quad \begin{matrix} 0 & \longmapsto 1 \\ 1 & \longmapsto 0 \\ +\infty i & \longmapsto \infty \end{matrix} \quad \in \overline{X}(4).$$

4.8. The J-invariant – A Realization of $\mathbb{H}/SL(2, \mathbb{Z})$. Let us compute the invariant j introduced at the end of Chapter I:

$$\begin{aligned}
j(\lambda) &= \frac{4}{27} \frac{(\lambda^2 - \lambda + 1)^3}{\lambda^2 (1 - \lambda)^2} \\
&= \frac{4}{27 \cdot 2^3} \frac{\{(e_1 - e_2)^2 + (e_2 - e_3)^2 + (e_3 - e_1)^2\}^3}{\{(e_1 - e_2)(e_2 - e_3)(e_3 - e_1)\}^2}.
\end{aligned}$$

Since e_1, e_2 and e_3 are zeros of the polynomial $4z^3 - g_2 z - g_3$, we have

$$e_1 + e_2 + e_3 = 0, \quad e_1 e_2 + e_2 e_3 + e_3 e_1 = -\frac{g_2}{4}, \quad e_1 e_2 e_3 = \frac{g_3}{4}.$$

Using these relations, we get

$$J(\omega_1 : \omega_2) := j(\lambda) = \frac{g_2^3}{g_2^3 - 27 g_3^2}.$$

The function J gives the realization (isomorphism)

$$J : \mathbb{H}/SL(2, \mathbb{Z}) \xrightarrow{\sim} \mathbb{C} \cong X\{4\}.$$

REMARK 4.7. By a straightforward computation, you can show that two cubic curves written in the Weierstrass normal form

$$C' : w^2 = 4z^3 - g_2 z - g_3$$

are projectively equivalent if and only if the quantity

$$J = \frac{g_2^3}{g_2^3 - 27 g_3^2}$$

is the same for both. This is called the *J-invariant* of a cubic curve.

Resumé of this section. We found a projective embedding of an elliptic curve whose image is a cubic plane curve, and we now know that a cubic curve is the double cover of the projective line branching at four distinct points. The story is now fairly complicated, so I repeat once more. We started from $\omega_1 : \omega_2 = \tau \in \mathbb{H}$, defined an elliptic curve $E(\omega_1 : \omega_2)$, obtained its projective embedding as a cubic plane curve, and found that this curve is the double cover of the line branching at four distinct points. Since two points in $\mathbb{H}$ related by $SL(2, \mathbb{Z})$ give isomorphic elliptic curves, we get the isomorphism

$$J : \mathbb{H}/SL(2, \mathbb{Z}) \xrightarrow{\sim} X\{4\}.$$

If we consider two isomorphic elliptic curves $E(\omega_1 : \omega_2)$ and $E(\omega_1' : \omega_2')$ to be equivalent only when the half periods $\omega_1/2, \omega_2/2$ correspond to $\omega_1'/2, \omega_2'/2$ (in this order), we obtain the isomorphism

$$\lambda : \mathbb{H}/\Gamma(2) \xrightarrow{\sim} X(4).$$

5. A Realization in Terms of the Theta Functions

In the realization of elliptic curves given above, we used meromorphic functions $\wp$ and $\wp'$. In this section, I would like to make a realization using *holomorphic* functions. But you already know that only constants are holomorphic functions invariant under a lattice L. Recall the technique we used when we made the democratic realization of the configuration space $X(4)$ in Chapter I §11. There, instead of an invariant rational function, we used three almost invariant polynomials. I will follow this line.

We are going to find holomorphic functions $f_1, f_2, \ldots$ such that

$$f_j(u + \omega) = a(\omega)f_j(u), \quad \omega \in L, \quad a(\omega) \neq 0, \quad j = 1, 2, \ldots;$$

then the map $u \mapsto f_1(u) : f_2(u) : \cdots$ will give a realization. It will really be a hard job, so let us relax for a while and play with highschool mathematics.

5.1. Coffee Break? – Pencils of Quadrics.

PROBLEM 5.1. Find the line passing through the two intersection points of the following two circles in the real (x, y)-plane:

$$f_1 = x^2 + y^2 - 9 = 0,$$
$$f_2 = (x - 2)^2 + (y - 3)^2 - 4 = 0.$$

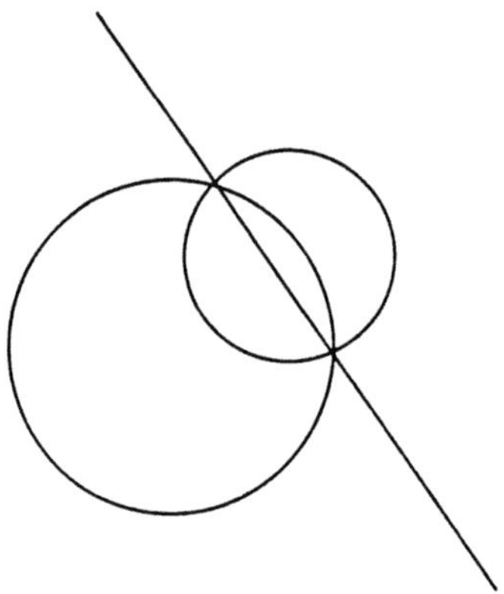

FIGURE 5.1. Two circles

An honest but stupid method: Express y in terms of x by solving $f_1 = 0$. Put this expression into f_2, and solve the resulting quartic equation. Then choose the suitable two out of the four roots, ...

A clever method: If $f_1(a, b) = f_2(a, b) = 0$ at a point (a, b), then for arbitrary functions g_1 and g_2, we have

$$g_1(a, b)f_1(a, b) + g_2(a, b)f_2(a, b) = 0,$$

of course. This means that the curve given by the equation

$$g_1 f_1 + f_2 g_2 = 0$$

passes through the intersection points of the two curves defined by $f_1 = 0$ and $f_2 = 0$. If you are aware of this fact, the answer is immediate:

$$\text{Answer:} \quad f_1 - f_2 = 0.$$

PROBLEM 5.2. Find all the lines which pass through two intersection points of the following circle and ellipse:

$$f_1 = x^2 + y^2 - 4 = 0,$$
$$f_2 = x^2 + 9y^2 - 9 = 0.$$

If you draw a picture, you can easily find six lines.

Now reflect on these two problems. Both problems treat intersections of two quadric curves, but we found one line in Problem 5.1 and six lines in Problem 5.2. I do not like this disagreement. It is due to the fact that the real number field is not algebraically closed and that we worked in the affine plane, not in the projective plane. So, let us reformulate the two

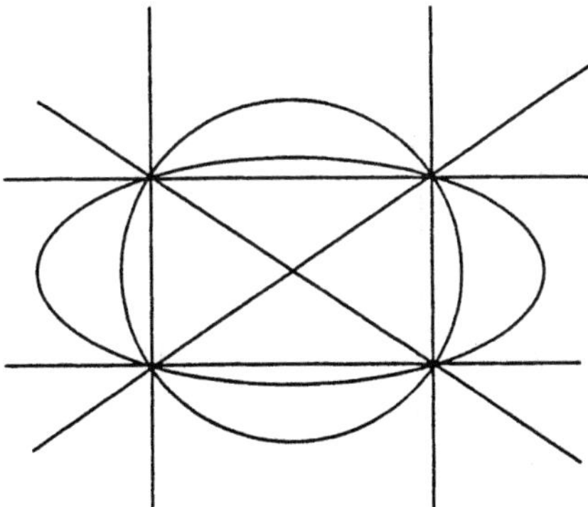

FIGURE 5.2. A circle and an ellipse

problems in the complex projective plane with homogeneous coordinates $x : y : z$.

PROBLEM 5.3 (THE PROJECTIVE VERSION OF PROBLEM 5.1). Find all the lines passing through two intersection points of the two quadrics:

$$f_1 = x^2 + y^2 - 9z^2 = 0,$$
$$f_2 = (x - 2z)^2 + (y - 3z)^2 - 4z^2 = 0.$$

PROBLEM 5.4 (THE PROJECTIVE VERSION OF PROBLEM 5.2). Find all the lines passing through two intersection points of the two quadrics:

$$f_1 = x^2 + y^2 - 4z^2 = 0,$$
$$f_2 = x^2 + 9y^2 - 9z^2 = 0.$$

A quadric f can be written as follows by making use of a symmetric matrix A:

$$f = (x, y, z) \begin{pmatrix} a_{11} & a_{12} & a_{13} \\ a_{21} & a_{22} & a_{23} \\ a_{31} & a_{32} & a_{33} \end{pmatrix} \begin{pmatrix} x \\ y \\ z \end{pmatrix}, \quad a_{ij} = a_{ji}.$$

For example, in Problem 5.3, the corresponding symmetric matrices A_1 and A_2 of f_1 and f_2 are given as

$$A_1 = \begin{pmatrix} 1 & 0 & 0 \\ 0 & 1 & 0 \\ 0 & 0 & -9 \end{pmatrix}, \quad A_2 = \begin{pmatrix} 1 & 0 & -2 \\ 0 & 1 & -3 \\ -2 & -3 & 9 \end{pmatrix}.$$

Consider two quadrics, $f_1 = 0$ and $f_2 = 0$. For any complex numbers t_1 and t_2, which are not simultaneously zero, we consider the quadric

$$\begin{aligned} f(t_1, t_2) : &= t_1 f_1 + t_2 f_2 \\ &= (x, y, z)(t_1 A_1 + t_2 A_2)^t(x, y, z) = 0. \end{aligned}$$

Such a curve passes through all the intersection points of the two given quadrics. The 1-dimensional family of curves defined by this equation is called the *pencil* generated by the two conics. The quadric defined by $f(t_1, t_2) = 0$ reduces to a pair of lines if and only if the rank of the matrix

$$t_1 A_1 + t_2 A_2$$

is 2, to a double line when the rank is 1. In any case, we can determine all the degenerate quadrics passing through all the intersection points of the two quadrics by solving the homogeneous cubic equation

$$\det(t_1 A_1 + t_2 A_2) = 0.$$

Generically there are three roots, i.e., three points on the projective line with homogeneous coordinates $t_1 : t_2$. For each root $t_1 : t_2$, the equation $f(t_1, t_2) = 0$ defines two lines (let us think of a double line as two lines). In summing, we proved that there are three pairs of lines passing through four intersection points of the two quadrics.

I am not interested in solving the above cubic equation for the following reason. Put

$$f_1' = af_1 + bf_2, \quad f_2' = cf_1 + df_2, \quad ad - bc \neq 0,$$
$$A_1' = aA_1 + bA_2, \quad A_2' = cA_1 + dA_2,$$

and consider the same problem as above. Note that the set of intersection points of $f_1 = 0$ and $f_2 = 0$ and that of $f_1' = 0$ and $f_2' = 0$ coincide. Following the recipe above, we consider the cubic equation in $t_1' : t_2'$

$$\det(t_1' A_1' + t_2' A_2') = 0.$$

Since this is identical to

$$\det\{(at_1' + ct_2')A_1 + (bt_1' + dt_2')A_2\} = 0,$$

the three roots $t_1' : t_2'$ of this cubic equation and the three roots $t_1 : t_2$ of the cubic equation above are related as follows:

$$\begin{pmatrix} t_1 \\ t_2 \end{pmatrix} = \begin{pmatrix} a & c \\ b & d \end{pmatrix} \begin{pmatrix} t_1' \\ t_2' \end{pmatrix}.$$

So the two sets of three roots are transformed into each other by a projective transformation. Since we know that any three points can be transformed projectively into any three points, explicit values of the three roots have no intrinsic meaning.

Let us work now in the 3-dimensional projective space with homogeneous coordinates $x : y : z : w$. A *quadric surface* is a surface defined by a quadric equation:

$$F(x, y, z, w) = (x, y, z, w)A \, {}^t(x, y, z, w) = 0,$$

where A is a 4×4 non-zero matrix. A quadric surface is said to be *degenerate* if $\det A = 0$; generically, such a surface looks like a cone. (Recall that in the plane curve case, a degenerate quadric is a pair of lines. Since two lines in a plane intersect at a point unless they coincide, it looks like a cone.) We now pose a problem in this 3-dimensional projective space analogous to that in the plane case:

PROBLEM 5.5. Find all the degenerate quadric surfaces passing through the intersection curve B of the two quadric surfaces defined by the equations

$$F_1 = x^2 + y^2 + z^2 - w^2 = 0,$$
$$F_2 = x^2 + y^2 + 16z^2 - 4w^2 = 0.$$

REMARK 5.1. If you work in the real affine space, by putting $w = 1$, these equations correspond to a sphere and an oval (see Figure 5.3).

Let us consider two different quadrics $F_1 = 0$ and $F_2 = 0$ given by symmetric 4×4-matrices A_1 and A_2:

$$F_1(x, y, z, w) = (x, y, z, w)A_1{}^t(x, y, z, w),$$
$$F_2(x, y, z, w) = (x, y, z, w)A_2{}^t(x, y, z, w).$$

Form the pencil of quadric surfaces

$$F(t_1, t_2) = (x, y, z, w)(t_1 A_1 + t_2 A_2)^t(x, y, z, w) = 0, \quad t_1 : t_2 \in \mathbb{P}^1,$$

which pass through the intersection curve B of the two quadrics. A member of this pencil is a degenerate quadric surface if and only if

$$a(t_1, t_2) := \det(t_1 A_1 + t_2 A_2) = 0.$$

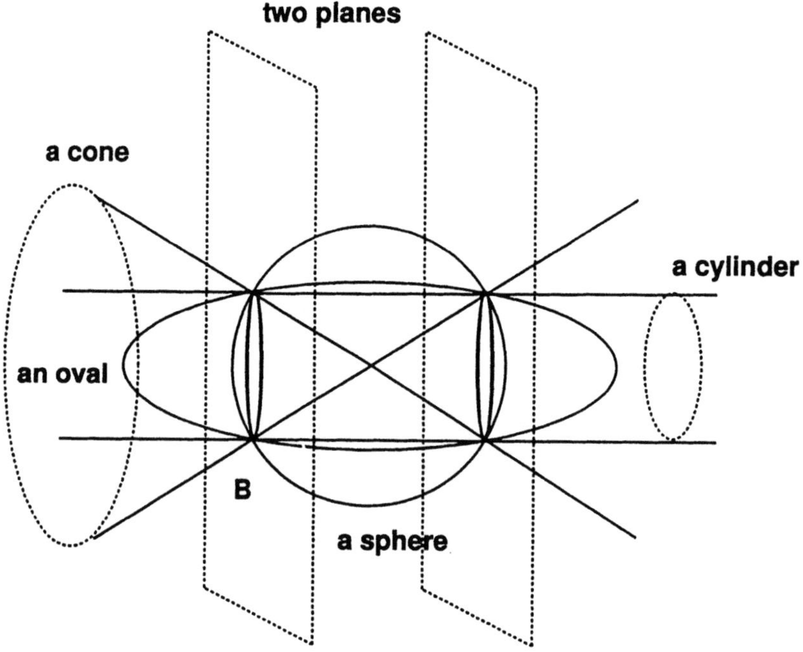

FIGURE 5.3. A sphere and an oval

Assume that the generic member is non-degenerate. Then this equation is quartic in $t_1 : t_2$, and thus it has four roots, i.e., four points on the line. If we put

$$F_1' = aF_1 + bF_2,$$
$$F_2' = cF_1 + dF_2, \quad ad - bc \neq 0,$$

the intersection curve of $F_1' = 0$ and $F_2' = 0$ coincides with the intersection curve B above. By repeating the previous step for F_1' and F_2', we get four points on the line. These points can be different from those obtained above but are sent to them by the transformation

$$\begin{pmatrix} t_1 \\ t_2 \end{pmatrix} = \begin{pmatrix} a & c \\ b & d \end{pmatrix} \begin{pmatrix} t_1' \\ t_2' \end{pmatrix}.$$

I am not interested in the explicit value of each of the four points, but I am interested in the position of the four points as a whole. This set of points can be represented by a point in the configuration space $X\{4\}$. This point is determined only by the intersection curve B (or equivalently, the pencil). That is, it is largely independent of the choice of the two quadrics defining it.

REMARK 5.2. In Problem 5.3, the quartic equation has a double root, which corresponds to the pair of parallel planes in Figure 5.3.

Assume further that the quartic equation $a(t_1, t_2) = 0$ has four distinct roots. Then we obtain a map f from the set $\{B\}$ of such curves to $X\{4\}$. We have

PROPOSITION 5.1. *Let $\{B\}$ be the set of curves which each constitutes the intersection of two quadric surfaces such that the quartic equation $a(t_1 : t_2) = 0$ has four distinct roots, and let $f(B)$ be the set of four roots of the quartic equation. Then two such curves B and B' satisfying $f(B) = f(B')$ are transformed into each other by projective transformations of the space, i.e., by $PGL(4)$, and f gives the isomorphism*

$$\{B\}/PGL(4) \xrightarrow{\sim} X\{4\}.$$

This can be proved by the following lemma, which is an exercise in linear algebra, elementary but delicate.

LEMMA 5.2. *The following conditions for two distinct non-degenerate 4×4-symmetric matrices A_1 and A_2 are equivalent.*

(1) *The equation $a(t_1 : t_2) = 0$ has distinct four roots.*

(2) $B := \{F_1 = F_2 = 0\}$ *is non-singular and does not lie in a plane.*

(3) *The two surfaces $F_1 = 0$ and $F_2 = 0$ are not tangent at any point on B.*

(4) *The two matrices A_1 and A_2 can be diagonalized simultaneously, i.e., there exists $P \in GL(4)$ such that $D_1 := {}^t P A_1 P$ and $D_2 := {}^t P A_2 P$ are diagonal, and $\det(t_1 D_1 + t_2 D_2) = 0$ has distinct four roots.*

(5) *There exist no $v \in \mathbb{C}^4 - \{0\}$ and $\lambda \in \mathbb{C}^\times$ such that $v A_1 {}^t v = v A_2 {}^t v = 0$ and $A_1 v = \lambda A_2 v$.*

(6) *rank $(t_1 A_1 + t_2 A_2) \geq 3$ for all $(t_1, t_2) \neq (0, 0)$.*

A sketch of proof. By definition, $(3) \Leftrightarrow (5)$. Since A_2 is non-degenerate, we can assume A_2 is the identity matrix. Assume (5). Then by a standard method of diagonalizing a symmetric matrix with an orthogonal matrix, we get (4). The remaining part is easy and is left to the reader as an exercise.

5.2. Theta Functions. At the very beginning of this book, we learned that a realization of the quotient space $\mathbb{C}/\mathbb{Z}$ requires the exponential function (§4 of Chapter I). We are now attempting to make a realization of $\mathbb{C}/L$ in terms of holomorphic functions. These functions must be functions of $\exp z$, but you should not expect that they are simple functions. I am very sorry, but I must introduce very suddenly the following function. Please be patient. Your suffering will not last long. For notational simplicity we set $\omega_1 = \tau$, $\omega_2 = 1$.

DEFINITION 5.1 (THE THETA FUNCTION). The theta function is defined as

$$\theta(z, \tau) := \sum_{-\infty}^{\infty} \exp\{\pi i(n^2 \tau + 2nz)\}, \quad z \in \mathbb{C}, \ \tau \in \mathbb{H}.$$

Since

$$|\exp\{\pi i(n^2 \tau + 2nz)\}| = |\exp \Re\{\pi i(n^2 \tau + 2nz)\}|$$
$$= \exp\{-\pi n^2 \Im \tau\} \cdot \exp\{-2\pi n \Im z\}$$
$$= [\exp\{-\pi \Im \tau\}]^{n^2} \cdot [\exp\{-2\pi \Im z\}]^n$$

and $\Im \tau > 0$, the series converges rapidly and defines a holomorphic function in $z \in \mathbb{C}$ and $\tau \in \mathbb{H}$. The theta function has the following properties:

(1) $\theta(z+1,\tau) = \theta(z,\tau)$.
(2) $\theta(-z,\tau) = \theta(z,\tau)$.
(3) $\theta(z+\tau,\tau) = \exp\{-\pi i(\tau+2z)\}\theta(z,\tau)$.

The first and the third properties imply that, as a function of z, θ is nearly L-periodic. Only the third property is not obvious. It can be shown as follows:

$$\theta(z+\tau,\tau) = \sum \exp\{\pi i n^2 \tau + 2\pi i n(z+\tau)\}$$
$$= \sum \exp\{\pi i(n+1)^2\tau - \pi i \tau + 2\pi i(n+1)z - 2\pi i z\}$$
$$= \exp\{-\pi i \tau - 2\pi i z\}\theta(z,\tau).$$

Please remember this technique as *passing your responsibility to the running n*.

Using the property (3) twice, you obtain

$$\theta(z+2\tau,\tau) = \exp\{-4\pi i(\tau+z)\}\theta(z,\tau).$$

5.3. Number of Zeros of Theta Functions.

PROPOSITION 5.3. *For fixed τ the function $\theta(z) = \theta(z,\tau)$ has exactly one simple zero in a fundamental parallelogram F.*

PROOF. I use the same technique here that I used in proving that a meromorphic elliptic function with only one simple pole in F is holomorphic (Remark 4.3). The following integral along the boundary of F counts the zeros of $\theta(z)$ in F:

$$\frac{1}{2\pi i}\int \frac{\theta'(z)}{\theta(z)}dz.$$

On the other hand, we have

$$\frac{\theta'}{\theta}(z+1) = \frac{\theta'}{\theta}(z),$$
$$\frac{\theta'}{\theta}(z+\tau) = -2\pi i + \frac{\theta'}{\theta}(z),$$

which tells us that the integral is 1. $\square$

Surely you want to know where the zero is; you will find it soon. By the same argument as in the proof above, you can prove

PROPOSITION 5.4. *A non-zero holomorphic function f satisfying*

$$f(z+1) = f(z),$$
$$f(z+\tau) = \exp\{-4\pi i(\tau + 2z)\}f(z),$$

has exactly four zeros in a fundamental parallelogram.

5.4. Position of Zeros of Theta Functions.

DEFINITION 5.2. For $j, k = 0, 1$, we define four theta functions:

$$\theta_{jk}(z,\tau) = \sum \exp\{\pi i(n + \frac{j}{2})^2\tau + 2\pi i(n + \frac{j}{2})(z + \frac{k}{2})\}.$$

Explicitly, we have

$$\theta_{00}(z,\tau) = \theta(z,\tau),$$
$$\theta_{01}(z,\tau) = \theta(z + \frac{1}{2},\tau),$$
$$\theta_{10}(z,\tau) = \exp\{\frac{\pi i\tau}{4} + \pi iz\}\theta(z + \frac{\tau}{2},\tau),$$
$$\theta_{11}(z,\tau) = \exp\{\frac{\pi i\tau}{4} + \pi i(z + \frac{1}{2})\}\theta(z + \frac{1}{2} + \frac{\tau}{2},\tau).$$

Note that, up to trivial exponential factors, these are simply parallel displacements (of the variable z) by half periods of the original theta function. Thus each theta function has exactly one zero in F.

PROPOSITION 5.5. $\theta_{11}(z) = \theta_{11}(z,\tau)$ *is odd, and the other three theta functions are even.*

PROOF. The proofs for all four cases are similar, so I consider only the case of $\theta_{11}(z)$. Recalling our slogan, *pass your responsibility to the running n*, we have

$$\theta_{11}(-z) = \sum \exp\{\pi i(n + \frac{1}{2})^2\tau + 2\pi i(n + \frac{1}{2})(-z + \frac{1}{2})\}$$
$$= \sum \exp\{\pi i(-n - \frac{1}{2})^2\tau + 2\pi i(-n - \frac{1}{2})(z - \frac{1}{2})\}$$
$$= \sum \exp\{\pi i(m + \frac{1}{2})^2\tau + 2\pi i(m + \frac{1}{2})(z - 1 + \frac{1}{2})\}$$
$$= \exp\{-2\pi i\frac{1}{2}\}\theta_{11}(z) = -\theta_{11}(z),$$

where I put $-n = m + 1$. $\quad\square$

You know that any odd function has a zero at the origin, do you not? What? The function $\wp'$ is a counterexample? Oh, it is not defined at the origin; if you like, please consider $1/\wp'$. Anyway, $\theta_{11}(z)$ has a zero at the origin, the unique zero in F. Since other thetas are parallel displacements of this one, we know their zeros too.

COROLLARY 5.6. *For any fixed τ, all zeros of thetas (as functions of z) are simple and situated as follows:*

$$\theta_{00}(z): \quad \mathbb{Z} + \tau\mathbb{Z} + \frac{1+\tau}{2},$$

$$\theta_{01}(z): \quad \mathbb{Z} + \tau\mathbb{Z} + \frac{\tau}{2},$$

$$\theta_{10}(z): \quad \mathbb{Z} + \tau\mathbb{Z} + \frac{1}{2},$$

$$\theta_{11}(z): \quad \mathbb{Z} + \tau\mathbb{Z}.$$

5.5. A Projective Embedding.

LEMMA 5.7.

$$\theta_{jk}(z+1,\tau) = (-)^j\theta_{jk}(z,\tau),$$

$$\theta_{jk}(z+\tau,\tau) = (-)^k\exp\{-\pi i(\tau+2z)\}\theta_{jk}(z,\tau).$$

If you still remember our slogan, *pass your responsibility to the running n*, you can prove this by yourself.

COROLLARY 5.8.

$$\theta_{jk}(z+2,\tau) = \theta_{jk}(z,\tau),$$

$$\theta_{jk}(z+2\tau,\tau) = \exp\{-4\pi i(\tau+z)\}\theta_{jk}(z,\tau).$$

Now we have four quasi-invariants. Let us make a projective embedding of the elliptic curve $E(\tau)$. We associate with $z \in E(\tau)$ the following point $\varphi(z)$ in the projective 3-space:

$$\varphi : E(\tau) \ni z \mapsto \theta_{00}(2z,\tau) : \theta_{01}(2z,\tau) : \theta_{10}(2z,\tau) : \theta_{11}(2z,\tau) \in \mathbb{P}^3.$$

THEOREM 5.9. *The map φ gives an embedding of $E(\tau)$.*

We prove only that φ is injective. Assume there are z_1 and $z_2(\neq z_1)$ in a fundamental parallelogram F such that $\varphi(z_1) = \varphi(z_2)$. Take any point $w \in F$ distinct from $z_1, z_2, z_1 + 1/2, z_2 + 1/2$ (if $z_1 \equiv z_2 + 1/2$ then consider the four points $z_1, z_2, z_1 + \tau/2, z_2 + \tau/2$). Let us define a non-zero function

$$f(z) := a_{00}\theta_{00}(2z) + \cdots + a_{11}\theta_{11}(2z)$$

such that $f(z_1) = f(z_1 + 1/2) = f(w) = 0$ (three linear equations in the four unknowns $a_{00}, \ldots, a_{11}$). The assumption $\varphi(z_1) = \varphi(z_2)$ together with Lemma 5.7 implies $\varphi(z_2 + 1/2) = \varphi(z_1 + 1/2)$. Thus we have $f(z_2 + 1/2) = (\text{constant})f(z_1 + 1/2) = 0$, so that our non-zero function f has at least five zeros $\{z_1, z_2, z_1 + 1/2, z_2 + 1/2, w\}$. On the other hand, f satisfies

$$f(z + 1) = f(z)$$
$$f(z + \tau) = \exp\{-4\pi i(\tau + 2z)\}f(z),$$

and we know that such a function has exactly four zeros in F, which is a contradiction.

5.6. How Does the Image Look?

DEFINITION 5.3 (THETA ZERO-VALUES).

$$\theta_{jk} := \theta_{jk}(\tau) := \theta_{jk}(0, \tau), \quad j, k = 0, 1.$$

Note that, since $\theta_{11}(z) = \theta_{11}(z, \tau)$ is an odd function, we have $\theta_{11} = 0$.

THEOREM 5.10 (RIEMANN RELATIONS). *For $\theta_{ij}(z) = \theta_{ij}(z, \tau)$,*

$$\theta_{00}^2\theta_{00}(z)^2 - \theta_{01}^2\theta_{01}(z)^2 - \theta_{10}^2\theta_{10}(z)^2 = 0,$$
$$\theta_{10}^2\theta_{01}(z)^2 - \theta_{01}^2\theta_{10}(z)^2 - \theta_{00}^2\theta_{11}(z)^2 = 0.$$

COROLLARY 5.11 (JACOBI'S IDENTITY).

$$\theta_{00}(\tau)^4 - \theta_{01}(\tau)^4 - \theta_{10}(\tau)^4 = 0.$$

PROOF. To find such relations, you must be talented, or at least you must be lucky. However once one is found, to prove one is a student's exercise. I prove the second one. Consider the expression

$$\frac{\theta_{10}^2\theta_{01}(z)^2 - \theta_{01}^2\theta_{10}(z)^2}{\theta_{00}^2\theta_{11}(z)^2}.$$

By the transformation formulae, it is easy to see that this is an elliptic function. Its only possible poles are in L. Since the numerator is zero at any element of L, this function has at most one simple pole in F. This implies that it is a constant. Put $z = 1/2$. Then since $\theta_{01}(1/2) = \theta_{00}$, $\theta_{10}(1/2) = 0$, and $\theta_{11}(1/2) = \theta_{10}$, you know this constant is 1. $\square$

Let $x : y : z : w$ be coordinates on the projective 3-space, which I consider as the target of our projective embedding, i.e.,

$$x : y : z : w = \theta_{00}(2z, \tau) : \theta_{01}(2z, \tau) : \theta_{10}(2z, \tau) : \theta_{11}(2z, \tau).$$

The Riemann relations imply that the image is on the intersection of the following two quadrics:

$$\begin{aligned}
F_1 &= \theta_{00}^2 x^2 - \theta_{01}^2 y^2 - \theta_{10}^2 z^2 && = 0, \\
F_2 &= \phantom{\theta_{00}^2 x^2 -{}} \theta_{10}^2 y^2 - \theta_{01}^2 z^2 - \theta_{00}^2 w^2 &&= 0.
\end{aligned}$$

5.7. Invariants of the Space Curves in Question. This is the object we discussed in the coffee break! At any rate, let us follow the recipe we found there. Do you remember?

(1) Express the quadrics by matrices:

$$\begin{aligned}
A_1 &= diag\ (\theta_{00}^2,\ -\theta_{01}^2,\ -\theta_{10}^2,\ 0\ \ ), \\
A_2 &= diag\ (\ \ \ 0,\ \ \ \theta_{10}^2,\ -\theta_{01}^2,\ -\theta_{00}^2).
\end{aligned}$$

(2) Introduce homogeneous coordinates $t_1 : t_2$ on a projective line, and form

$$\det(t_1 A_1 + t_2 A_2) = 0.$$

(3) Solve the quartic equation. Since A_1 and A_2 are degenerate, $1 : 0$ and $0 : 1$ are clearly OK. Since A_1 and A_2 are diagonal, two other roots can be found immediately, and we obtain (please rotate this book by an angle of $\pi/2$)

$$\begin{pmatrix} t_1 \\ \cdot\cdot \\ t_2 \end{pmatrix} = \begin{pmatrix} 1 \\ \cdot\cdot \\ 0 \end{pmatrix},\ \begin{pmatrix} 0 \\ \cdot\cdot \\ 1 \end{pmatrix},\ \begin{pmatrix} \theta_{10}^2 \\ \cdot\cdot \\ \theta_{01}^2 \end{pmatrix},\ \begin{pmatrix} \theta_{01}^2 \\ \cdot\cdot \\ -\theta_{10}^2 \end{pmatrix}.$$

(4) Since the above are four points on the projective line, let us make the democratic realization. Do you remember? We compute 2×2-minors $D(ij)D(kl)$ and arrange them in the following way:

$$D(12)D(34) : D(13)D(24) : D(14)D(23).$$

The result is

$$-\theta_{10}^4 - \theta_{01}^4 : -\theta_{01}^4 : \theta_{10}^4 = -\theta_{00}^4(\tau) : -\theta_{01}^4(\tau) : \theta_{10}^4(\tau).$$

Here I used Jacobi's identity to make the formula nicer. The map

$$\lambda : \tau \mapsto y_1 : y_2 : y_3 = -\theta_{00}^4(\tau) : -\theta_{01}^4(\tau) : \theta_{10}^4(\tau)$$

gives the realization (isomorphism)

$$\lambda : \mathbb{H}/\Gamma(2) \xrightarrow{\sim} Y_0 = \{y_1 : y_2 : y_3 \in \mathbb{P}^2 \mid y_1 - y_2 + y_3 = 0, \; y_1 y_2 y_3 \neq 0\}$$
$$(\cong X(4)).$$

This induces the following correspondence of boundary points:

$$\mathbb{\overline{H}} \ni \begin{array}{rcll} 1 & \longmapsto & 0:1:1 & = (12;34) \\ 0 & \longmapsto & 1:0:-1 & = (13;24) \\ +\infty i & \longmapsto & 1:1:0 & = (14;23) \end{array} \in Y.$$

These are proved in the next subsection.

(5) A realization of four unlabeled points is given by

$$\tau \mapsto J(\tau) = (\text{const})\frac{[\theta_{00}^8(\tau) + \theta_{01}^8(\tau) + \theta_{10}^8(\tau)]^3}{\theta_{00}^8(\tau)\theta_{01}^8(\tau)\theta_{10}^8(\tau)}.$$

This gives the realization (isomorphism)

$$J : \mathbb{H}/SL(2,\mathbb{Z}) \xrightarrow{\sim} X\{4\}.$$

5.8. The Values of the Theta Functions at $0, 1$ and ∞. By the definition of the theta functions, we have

$$\theta_{00}(it) = \sum e^{-\pi n^2 t} = 1 + 2(q + q^4 + q^9 + \cdots),$$
$$\theta_{01}(it) = \sum e^{-\pi n^2 t + \pi i n} = 1 - 2(q - q^4 + q^9 - \cdots),$$
$$\theta_{10}(it) = \sum e^{-\pi(n+\frac{1}{2})^2 t} = 2(q^{\frac{1}{4}} + q^{\frac{9}{4}} + \cdots),$$

where $q = e^{-\pi t}$. These imply

$$\theta_{00}(\infty) := \lim_{t \to +\infty} \theta_{00}(it) = 1, \quad \theta_{01}(\infty) = 1, \quad \theta_{10}(\infty) = 0.$$

To know the values

$$\theta_{jk}(0) := \lim_{t \to +0} \theta_{jk}(it),$$

we need some preparation.

Let f be a rapidly decreasing function on $(-\infty, +\infty)$ and $\hat{f}$ its Fourier transformation:

$$\hat{f}(\xi) := \int_{-\infty}^{+\infty} f(x)e^{-2\pi i x \xi}dx.$$

We then have the *Poisson summation formula*

$$\sum_{n \in \mathbb{Z}} f(n) = \sum_{n \in \mathbb{Z}} \hat{f}(n).$$

This can be proved by considering the periodic function

$$F(x) = \sum_{n \in \mathbb{Z}} f(x + n).$$

Its Fourier coefficients are given by

$$\int_0^1 F(x)e^{-2\pi inx}dx = \sum \int_0^1 f(x+n)e^{-2\pi inx}$$
$$= \int_{-\infty}^{+\infty} f(x)e^{-2\pi inx}$$
$$= \hat{f}(n).$$

Since F has the expansion

$$F(x) = \sum \hat{f}(n)e^{2\pi inx},$$

we need only put $x = 0$ to obtain the summation formula.

We apply this formula to

$$f = e^{-\pi tx^2}, \quad \hat{f} = \frac{1}{\sqrt{t}}e^{-\frac{\pi}{t}\xi^2} \quad (t > 0)$$

and

$$f = e^{-\pi tx^2 + \pi ix}, \quad \hat{f} = \frac{1}{\sqrt{t}}e^{-\frac{\pi}{t}(\xi-\frac{1}{2})^2} \quad (t > 0).$$

(In the computation getting these Fourier transformations, the famous formula

$$\int_{-\infty}^{+\infty} e^{-\pi tx^2}dx = \frac{1}{\sqrt{t}}$$

is used.) Then we have

$$\theta_{00}(it) = \frac{1}{\sqrt{t}}\theta_{00}(\frac{i}{t}), \quad \text{and} \quad \theta_{01}(it) = \frac{1}{\sqrt{t}}\theta_{10}(\frac{i}{t}).$$

The second identity implies

$$\theta_{10}(it) = \frac{1}{\sqrt{t}}\theta_{01}(\frac{i}{t}).$$

Thus we have

$$\theta_{00}(0) : \theta_{01}(0) : \theta_{10}(0) = 1 : 0 : 1.$$

The values at 1 can be obtained from those at 0, since we have the transformation formulae:

$$\theta_{00}(\tau + 1) = \theta_{01}(\tau),$$
$$\theta_{01}(\tau + 1) = \theta_{00}(\tau),$$
$$\theta_{10}(\tau + 1) = e^{\frac{\pi i}{4}}\theta_{10}(\tau),$$

which can be readily seen from the definition of thetas.

CHAPTER III

Modular Interpretations of $X(2,4)$

I have a belief that any nice theory must be presented first classically and intuitively in plain language. I therefore make this chapter as classical and intuitive as possible.

1. The Hypergeometric Series

Following Euler, let us define the power series

$$F(a, b, c; x) = \sum_{n=0}^{\infty} \frac{(a, n)(b, n)}{(c, n)(1, n)} x^n,$$

where

$$(a, n) = a(a + 1) \cdots (a + n - 1),$$

and c is not $0, -1, -2, \ldots$ This is called the *hypergeometric series*. Notice that this series is symmetric in a and b, i.e., $F(a, b, c; x) = F(b, a, c; x)$. Let $A(n)$ be the n-th coefficient; then it satisfies

$$\frac{A(n + 1)}{A(n)} = \frac{(a + n)(b + n)}{(c + n)(1 + n)}.$$

Thus the radius of convergence of this series is 1, unless a or b is a non-positive integer, in which case the series is finite (and c is allowed to be a negative integer smaller than the non-positive integer a or b.) In any case, this series defines a function holomorphic in x in (at least) the open unit disc centered at 0 and is holomorphic in (a, b, c) if $c \neq 0, -1, -2, \ldots$

59

2. The Hypergeometric Differential Equation

The series introduced in the previous section satisfies a linear differential equation. It is very uncomfortable work to find this equation if you differentiate this series once, twice, ... , and seek a linear relation with coefficients in $\mathbb{Z}[a,b,c,x]$. Here is a clever way. Let us, as Euler, introduce the differential operator

$$D = x\frac{d}{dx}.$$

This is a simple step, but it is important to recognize that power functions are eigenfunctions of this operator:

$$Dx^n = nx^n,$$

and a power series is the eigenfunction-expansion with respect to this operator. Notice that for a polynomial f with constant coefficients, we have

$$f(D)x^n = f(n)x^n.$$

Now we are ready. Since I am a specialist of this business, I know immediately by glancing at the formula

$$\frac{A(n+1)}{A(n)} = \frac{(a+n)(b+n)}{(c+n)(1+n)}$$

that the hypergeometric series satisfies the following differential equation:

$$E(a,b,c) : [(a+D)(b+D) - (c+D)(1+D)\frac{1}{x}]u = 0.$$

After the following proof, you will also be a specialist.

$$[(a+D)(b+D) - (c+D)(1+D)\frac{1}{x}]\sum A(n)x^n$$
$$= \sum[(a+D)(b+D)A(n)x^n - (c+D)(1+D)A(n)x^{n-1}]$$
$$= \sum[(a+n)(b+n)A(n)x^n - (c+n-1)(1+n-1)A(n)x^{n-1}]$$
$$= \sum[(a+n)(b+n)A(n)x^n - (c+n)(1+n)A(n+1)x^n]$$
$$= 0.$$

I wrote $(a+D)(b+D)$ without defining products of operators, believing that the reader will not misunderstand. If a constant b is regarded as the operator multiplying the constant b, then D and b commute:

$$Db = bD.$$

If x is regarded as the operator multiplying x, then

$$Dx = x + xD.$$

The above differential equation is equivalent to (carrying out the computation is a good exercise to master products of operators)

$$E(a,b,c) : x(1-x)\frac{d^2u}{dx^2} + \{c - (a+b+1)x\}\frac{du}{dx} - abu = 0.$$

This equation is called the *hypergeometric (differential) equation*, and is denoted by $E(a,b,c)$. It is linear, of second order (more than obvious!), and has singularities at $x = 0, 1$ and maybe infinity. Cauchy's fundamental theorem asserts that for any point x_0 not equal to 0 or 1 there are two linearly independent holomorphic solutions around x_0. In other words, the set of solutions around x_0 forms a 2-dimensional linear space over $\mathbb{C}$. Any of these solutions can be analytically continued along any path in $X = \mathbb{C} - \{0, 1\}$. The solutions are not single-valued in general, but they do not change much. Let γ be a loop in X starting and ending at x_0, and let u_1 be a non-zero solution at x_0 and u_2 another solution which is not a constant multiple of u_1. Further, let $\gamma_* u_1$ and $\gamma_* u_2$ be the functions obtained by the analytic continuation of u_1 and u_2 along γ. Since $\gamma_* u_1$ and $\gamma_* u_2$ remain to be linearly independent solutions at x_0, we have

$$\gamma_* \begin{pmatrix} u_1 \\ u_2 \end{pmatrix} = M(\gamma) \begin{pmatrix} u_1 \\ u_2 \end{pmatrix} \quad \text{for some} \quad M(\gamma) \in GL(2, \mathbb{C}).$$

The matrix $M(\gamma)$ is called the *circuit matrix of (u_1, u_2) along γ*. If γ_1

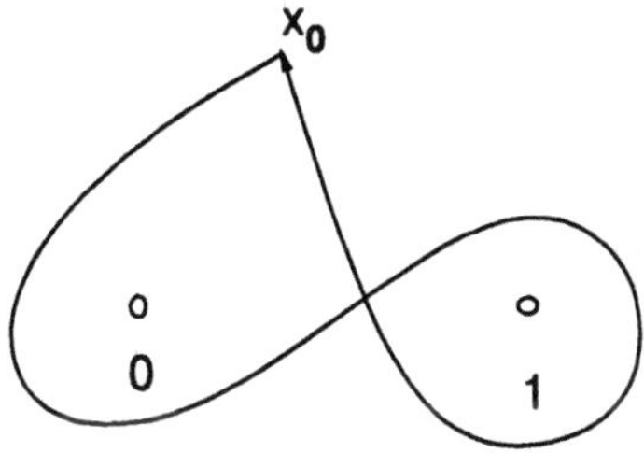

FIGURE 2.1. A loop γ

and γ_2 are curves starting from and ending at x_0, then their composition

$\gamma_1 \cdot \gamma_2$ is defined to be the curve following first γ_1 and then γ_2. We have

$$(\gamma_1 \cdot \gamma_2) \begin{pmatrix} u_1 \\ u_2 \end{pmatrix} = (\gamma_2)_*(\gamma_1)_* \begin{pmatrix} u_1 \\ u_2 \end{pmatrix}$$

$$= (\gamma_2)_* M(\gamma_1) \begin{pmatrix} u_1 \\ u_2 \end{pmatrix} = M(\gamma_1)(\gamma_2)_* \begin{pmatrix} u_1 \\ u_2 \end{pmatrix}$$

$$= M(\gamma_1)M(\gamma_2) \begin{pmatrix} u_1 \\ u_2 \end{pmatrix}$$

and so

$$M(\gamma_1 \cdot \gamma_2) = M(\gamma_1)M(\gamma_2).$$

If γ_1 can be continuously deformed in X (fixing x_0) into γ_2, then we have $M(\gamma_1) = M(\gamma_2)$. Therefore denoting by $\pi_1(X, x_0)$ *the fundamental group of X with base point x_0* (the group of homotopy equivalence classes of curves starting and ending at x_0), the correspondence $\gamma \mapsto M(\gamma)$ induces the homomorphism

$$\pi_1(X, x_0) \longrightarrow GL(2, \mathbb{C}),$$

which is called *the monodromy representation* of the differential equation.

This map depends on the choice of u_1 and u_2 as well as on x_0. If we change u_1 and u_2, the new representation is conjugate to the old one. This is also the case when we change x_0. So the differential equation determines the conjugacy class of a representation $\pi_1(X, x_0) \longrightarrow GL(2, \mathbb{C})$.

The monodromy group of the hypergeometric equation is now defined as the image of the monodromy representation in $GL(2, \mathbb{C})$. *The projective monodromy group* is defined to be the image of the monodromy group under the natural map $GL(2, \mathbb{C}) \to PGL(2, \mathbb{C})$. The conjugacy class of the monodromy group in $GL(2, \mathbb{C})$, as well as the conjugacy class in $PGL(2, \mathbb{C})$ of the projective monodromy group, is determined uniquely by the equation. In §5 of Chapter IV we give the explicit monodromy representation of the hypergeometric equation.

3. Another Solution around the Origin

I am going to study solutions around 0. By virtue of the symmetry of the hypergeometric equation, studied in the next section, the situation around 1 and ∞ will also become clear. You may ask why I study the equation only around the singularities. Well, at a non-singular point, I mean at any point in $X = \mathbb{C} - \{0, 1\}$, this equation contains no information at all. In other words, there are no invariants, if you know what

I mean: Around a non-singular point, I can find a new local coordinate, say ξ, and a new unknown $v = a(\xi)u$ (a is holomorphic and non-vanishing at the point) which take the equation $E(a, b, c)$ into $d^2v/d\xi^2 = 0$.

Although the hypergeometric equation $E(a, b, c)$ is singular at the origin, it admits a holomorphic solution $F(a, b, c; x)$. Do you feel funny? I do not think this is funny, but it is indeed mysterious. I am going to give you another solution which is not holomorphic at the origin. For preparation, let us compute a little – don't worry, only a little:

$$D(x^s u) = sx^s u + x^s Du = x^s(s + D)u, \quad \text{i.e.,} \quad Dx^s = x^s(s + D).$$

That's all (if you set $s = 1$, this proves the product formula $Dx = x + xD$ above). Please remember this formula as "*when x^s passes through D from the right to the left, then D changes into $s + D$*". Watch the operator

$$(a + D)(b + D) - (c + D)(1 + D)\frac{1}{x}$$

carefully and think what happens if you let x^{1-c} pass through this operator from the right to the left. You got it? That's it:

$$[(a + D)(b + D) - (c + D)(1 + D)\frac{1}{x}]x^{1-c}$$

$$= x^{1-c}[(a + 1 - c + D)(b + 1 - c + D) - (1 + D)(2 - c + D)\frac{1}{x}]$$

This means that

$$x^{1-c}F(a + 1 - c, b + 1 - c, 2 - c; x)$$

is also a solution of $E(a, b, c)$. Unfortunately this series is defined only when

$$2 - c \neq 0, -1, -2, \ldots, \quad \text{that is,} \quad c \neq 2, 3, 4, \ldots$$

If $c = 1$ the series is well-defined, but it coincides with $F(a, b, c; x)$. So when c is an integer (recall that if $c = 0, -1, -2, \ldots$, then $F(a, b, c; x)$ is not defined), we must find another solution. You may think that since c is a complex parameter, and the set of integers is a small part of $\mathbb{C}$, we can neglect these accidental cases. I want to agree with you, but I know that many interesting examples are found in accidental cases (this is almost a rule). We will soon see a demonstration of this point for the present case. Since I do not wish to study the cases $c \in \mathbb{Z}$ in general (if you would like, see *Frobenius's method* in [Yos1] or [IKSY]), let us make a deal: I will give another solution only in the exceptional case $c = 1$. If c is near but not equal to 1, all $F(a,b,c;x)$ and $x^{1-c}F(a+1-c,b+1-c,2-c;x)$ are well-defined solutions, and so too is the linear combination

$$\frac{x^{1-c}F(a+1-c,b+1-c,2-c;x)-F(a,b,c;x)}{c-1}.$$

Let c tend to 1. Then we know that the above expression tends to

$$\frac{d}{dc}x^{1-c}F(a+1-c,b+1-c,2-c;x)|_{c=1}.$$

This gives a solution with a logarithmic singularity at the origin. In particular this is independent of the holomorphic solution $F(a,b,c;x)$.

4. Symmetries of the Hypergeometric Equation

Let us transform the coordinate x into

$$1-x \quad \text{and} \quad \frac{1}{x}$$

in order to study the singularities at 1 and ∞, respectively.

Comment: I have a bad habit to say "Transform the coordinate x into $1-x$ (or $1/x$) in the (hypergeometric) equation" instead of "First transform the equation by putting $x=1-\xi$ (or $x=1/\xi$). Then, in the resulting equation, replace the symbol ξ by x." Please get used to this abbreviated expression.

For the transformation $x \mapsto 1-x$, I use the second form of the equation $E(a,b,c)$, given in §2:

$$x(1-x)\frac{d^2u}{dx^2}+\{c-(a+b+1)x\}\frac{du}{dx}-abu=0.$$

The first and the third terms are not changed under the transformation, while the coefficient of the second term is changed into

$$-\{c-(a+b+1)(1-x)\}=a+b+1-c-(a+b+1)x.$$

Thus the transformed equation is again the hypergeometric equation with parameters

$$(a,b,a+b+1-c).$$

Therefore, unless $c-a-b$ is an integer, as in the previous section, you get two solutions around 1:

$$F(a,b,a+b+1-c,1-x), \quad (1-x)^{c-a-b}F(c-a,c-b,c+1-a-b;1-x).$$

When $c - a - b = 0$, the second solution should be replaced by

$$\lim_{\epsilon \to 0} \frac{F(a, b, 1 - \epsilon, 1 - x) - (1 - x)^{\epsilon} F(\epsilon + b, \epsilon + a, \epsilon + 1; 1 - x)}{\epsilon}.$$

For the transformation $x \mapsto 1/x$, I use the first form of the equation $E(a, b, c)$, given in §2:

$$[(a + D)(b + D) - (c + D)(1 + D)\frac{1}{x}]u = 0.$$

Under this transformation the operator D changes into $-D$. (Recall the above comment.) Thus the equation becomes

$$[(a - D)(b - D) - (c - D)(1 - D)x]u = 0.$$

I let x^a pass through the above operator from the right to the left:

$$[(a - D)(b - D) - (c - D)(1 - D)x]x^a$$
$$= (-a + D)(-b + D)x^a - (-c + D)(-1 + D)x^{1+a}$$
$$= -[(-c + D)(-1 + D)x^{1+a} - (-a + D)(-b + D)x^{1+a}\frac{1}{x}]$$
$$= -x^{1+a}[(1 + a - c + D)(a + D) - (1 + D)(a - b + 1 + D)\frac{1}{x}].$$

If you neglect the factor $-x^{1+a}$, this is again the hypergeometric operator with parameters

$$(a, 1 + a - c, a - b + 1).$$

Therefore, unless $a - b$ is an integer, we get two solutions at ∞:

$$t^a F(a, 1 + a - c, a - b + 1; t), \quad t^b F(b, 1 + b - c, b - a + 1; t); \quad t = \frac{1}{x}.$$

I will not repeat what we should do when $a = b$.

5. Time to Pay

In the very beginning of this book, in §5 of Chapter I, I gave a definition of power functions

$$x^a = \exp(a \log C_x),$$

and said that they are not single-valued, and so you must be careful. But so far in this chapter, I have deliberately ignored this difficulty and made computations as if there were nothing to worry about. Now it is time to pay.

I assume the parameters are real, and

$$c, \quad c - a - b, \quad a - b$$

are not integers. I fix branches of the six solutions above by restricting their domains of definition:

$$f_{01} := F(a, b, c; x),$$

$$f_{02} := x^{1-c} F(a + 1 - c, b + 1 - c, 2 - c; x)$$

are defined in $\mathbb{C} - (-\infty, 0] \cup [1, +\infty)$ and are real-valued on $(0, 1)$. (In the figures below, dotted lines represent non-inclusion in the domain of definition.)

FIGURE 5.1. $\mathbb{C} - (-\infty, 0] \cup [1, +\infty)$

$$f_{11} := F(a, b, a + b + 1 - c; 1 - x),$$

$$f_{12} := (1 - x)^{c-a-b} F(c - a, c - b, c + 1 - a - b; 1 - x)$$

are defined in $\mathbb{C} - (-\infty, 1]$ and real-valued on $(1, +\infty)$.

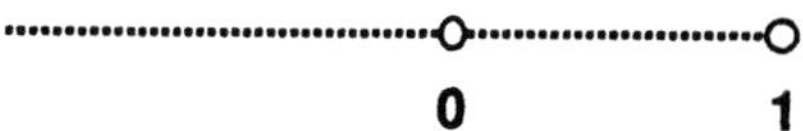

FIGURE 5.2. $\mathbb{C} - (-\infty, 1]$

$$f_{\infty 1} := (-\frac{1}{x})^a F(a, 1 + a - c, a - b + 1; \frac{1}{x}),$$

$$f_{\infty 2} := (-\frac{1}{x})^b F(b, 1 + b - c, b - a + 1; \frac{1}{x})$$

are defined in $\mathbb{C} - [0, +\infty)$ and real-valued on $(-\infty, 0)$.

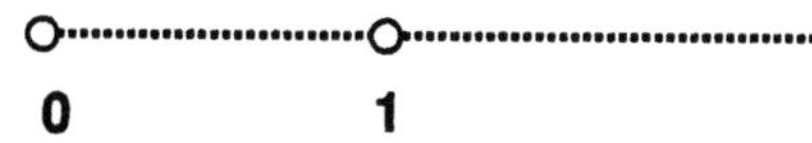

FIGURE 5.3. $\mathbb{C} - [0, +\infty)$

I wish to also consider the cases $1 - c = 0$, $c - a - b = 0$ and $a - b = 0$. As I explained in the previous section, I re-define f_{02}, f_{12} and $f_{\infty 2}$ as follows:

$$f_{02} := \lim_{c \to 1} \frac{f_{01}(a, b, c) - f_{02}(a, b, c)}{1 - c},$$

$$f_{12} := \lim_{c - a - b \to 0} \frac{f_{11}(a, b, c) - f_{12}(a, b, c)}{c - a - b},$$

$$f_{\infty 2} := \lim_{a \to b} \frac{f_{\infty 1}(a, b, c) - f_{\infty 2}(a, b, c)}{a - b}.$$

In this way I have defined, without any cheating, linearly independent solutions in the three simply connected domains. Since the equation is second order and linear, in a domain of common definition, say,

$$X_+ = \{x \in X \mid \Im x > 0\},$$

the three pairs of solutions are linearly related. That is to say, e.g.,

$$(f_{01}, f_{02}) = (f_{11}, f_{12})M_+^{10} \quad \text{for some} \quad M_+^{10} \in GL(2, \mathbb{C}),$$

$$(f_{01}, f_{02}) = (f_{\infty 1}, f_{\infty 2})M_+^{\infty 0} \quad \text{for some} \quad M_+^{\infty 0} \in GL(2, \mathbb{C}),$$

etc. These matrices $M_+^{10}, M_+^{\infty 0}, \ldots$, often called *connection matrices*, are explicitly known (cf. [IKSY]). 1 will give them implicitly in the following sections; this is enough for our purposes.

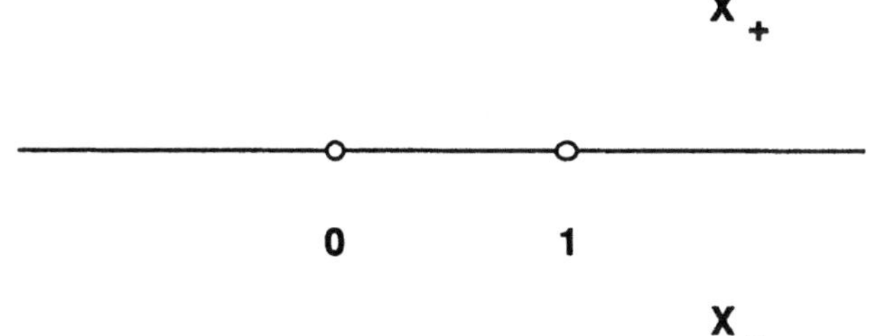

FIGURE 5.4. X_+ and X_-

6. The Schwarz Map and Schwarz Triangles

Let us consider after H.A. Schwarz the following map (referred to as the *Schwarz map*):

$$f = f_0 : X_+ \ni x \longmapsto f_{01}(x) : f_{02}(x) \in \mathbb{P}^1.$$

(Cauchy's fundamental theorem asserts that the linearly independent solutions f_{01} and f_{02} do not vanish simultaneously at any point of $\mathbb{C} - \{0,1\}$.) How does the image $f(X_+)$ look? Since f_0 is real-valued on the interval $(0,1)$, the interval $(f(0), f(1))$ is a part of the boundary of the image. In the same way, by the map

$$f_1 : X_+ \ni x \longmapsto f_{11}(x) : f_{12}(x) \in \mathbb{P}^1,$$

the interval $(1, \infty)$ is sent to the interval $(f_1(1), f_1(\infty))$. A similar statement is also true for

$$f_\infty : X_+ \ni x \longmapsto f_{\infty 1}(x) : f_{\infty 2}(x) \in \mathbb{P}^1.$$

On the other hand, it is well known (if you do not know, please learn) that, by a linear fractional transformation, circles and lines are mapped to circles and lines (lines are circles passing through the point at infinity). Since f, f_1 and f_∞ are related in a linear fractional way, as we saw at the end of the previous section, I conclude that the image under f of the union $(\infty, 0) \cup (0,1) \cup (1, \infty)$ is the boundary of a triangle bounded by three arcs (parts of circles). This triangle is called a *Schwarz triangle*.

Thanks to the local study given above around the three singular points, you can even determine the three angles! They are

$$\begin{array}{lll} \pi|1 - c| & \text{at} & f(0), \\ \pi|c - a - b| & \text{at} & f(1), \\ \pi|a - b| & \text{at} & f(\infty). \end{array}$$

If you admit the conformality of the locally bi-holomorphic map, proof is immediate. An angle zero means the two circles in question are tangent at the point. This is the reason I also defined f_{02}, f_{12} and $f_{\infty 2}$ when $1 - c$, $c - a - b$ and $a - b$ are zero, respectively.

In this way, we obtain

PROPOSITION 6.1. *If*

$$|1 - c|, \ |c - a - b|, \ |a - b| < 1,$$

the Schwarz map f sends the half-plane X_+ bijectively to a Schwarz triangle.

Note that without the condition for angles the statement does not hold.

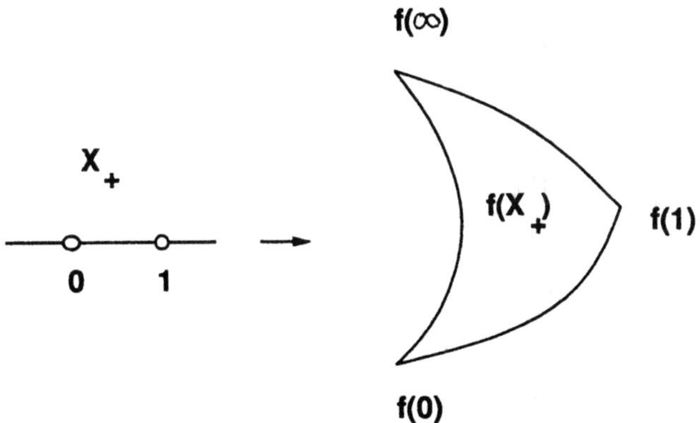

FIGURE 6.1. The Schwarz triangle $f(X_+)$

7. Schwarz's Reflection Principle

The following is *Schwarz's reflection principle.*

PROPOSITION 7.1. *Let f be holomorphic in a domain D whose boundary contains a real interval (a, b). Assume f can be extended to a continuous function on $D \cup (a, b)$ and is real-valued on the interval (a, b). Extend f by*

$$f(x) := \overline{f(\overline{x})}, \quad x \in \overline{D} := \{\overline{\xi} \mid \xi \in D\}$$

to the mirror image $\overline{D}$ of D. Then the extended f is holomorphic on $D \cup (a, b) \cup \overline{D}$, and its image is the union of $f(D)$, its mirror image $\overline{f(D)}$, and the interval $f((a, b))$ bonding them (see Figure 7.1).

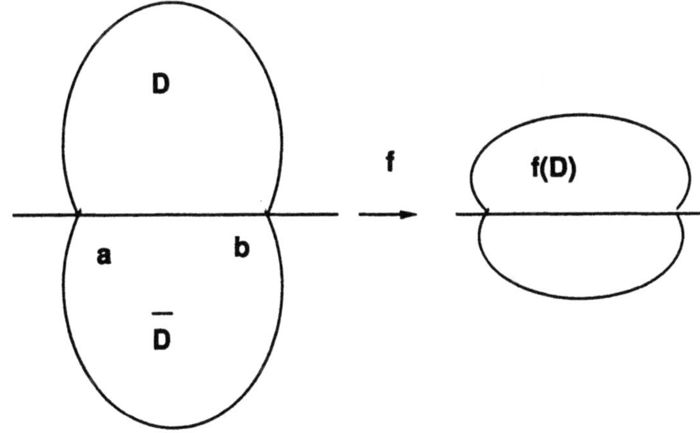

FIGURE 7.1. Schwarz's reflection principle

PROOF. It is clear by the assumption that f is continuous in $D \cup (a,b) \cup \overline{D}$, and is holomorphic in D and $\overline{D}$. We show that the integral of f along any closed curve close enough to the interval (a,b) is zero. We need only cut the curve at the intersection with the real axis and close the two halves along (a,b) (see Figure 7.2). $\square$

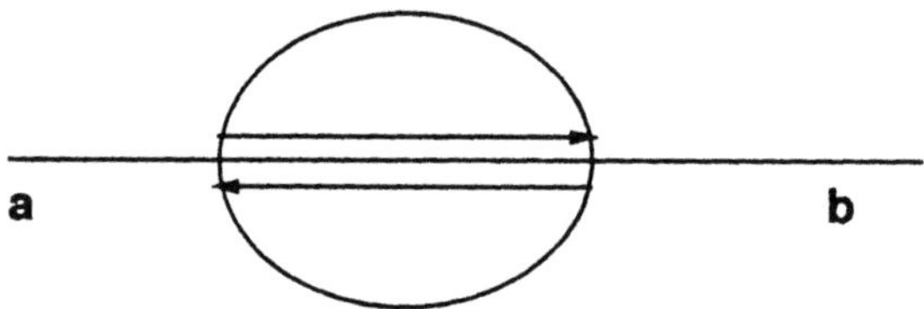

FIGURE 7.2. Dividing the integral into two parts

We shall apply this principle in the following situation.

COROLLARY 7.2. *Let f be holomorphic on $D \cup (a,b) \cup \overline{D}$ and $f((a,b))$ be part of a circle C. Then $f(\overline{D})$ is the mirror image of $f(D)$ with respect to C, i.e., $f(\overline{D}) = g^{-1}(\overline{g \circ f(D)})$, where g is any element of $PGL(2,\mathbb{C})$ sending C into the real axis.*

PROOF. Apply Proposition 7.1 to $g \circ f$, and note that the analytic continuation of $f|_D$ to $D \cup (a,b) \cup \overline{D}$ is unique. $\square$

Let us apply this principle to our map f defined on X_+ and to the intervals

$$(-\infty, 0), \quad (0,1), \quad (1,+\infty).$$

This can be extended to

$$X_- := \{ x \in \mathbb{C} \mid \Im x < 0 \},$$

through any one of these three intervals. Apply the principle again on X_- through any one of the three intervals. For instance, let γ be a path starting at $\overset{\bullet}{x} \in X_+$, passing through the interval $(\infty, 0)$ into X_- and returning through the interval $(0,1)$. Then the image of γ is a path in $\mathbb{P}^1$ starting at $f(\overset{\bullet}{x})$ in the original triangle $T = f(X_+)$, passing through the arc $(f(\infty), f(0))$ of T into the mirror image $T' = f(X_-)$ of T with respect to this arc, and then passing through the arc $(f(0), f(1))$ of T' into the the mirror image T'' of T' with respect to this new arc (see Figure 7.3). The consequent map $\gamma_* f$ by analytic continuation along γ is a linear fractional transformation of f, i.e., there is a matrix $M \in GL(2, \mathbb{C})$ such that

$$\gamma_* f = \frac{m_{11} f + m_{12}}{m_{21} f + m_{22}}, \quad M = (m_{ij}).$$

In this way, you can make analytic continuations from X_+ to X_- and from

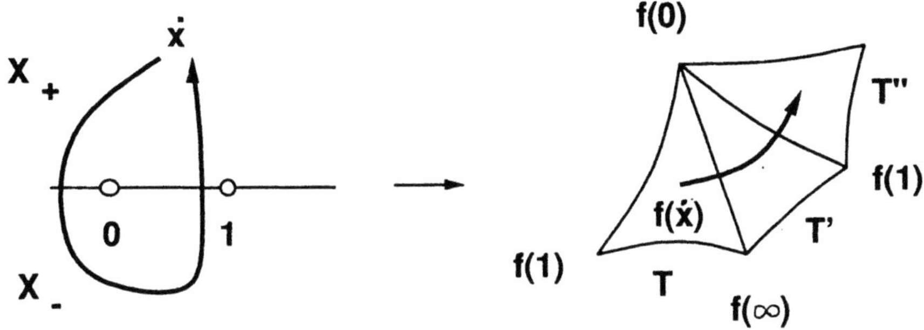

FIGURE 7.3. The image of a loop

X_- to X_+ as many times as you like. Images are conformal reflections of the original Schwarz triangle $f(X_+)$, as is shown in Figure 7.4. If you

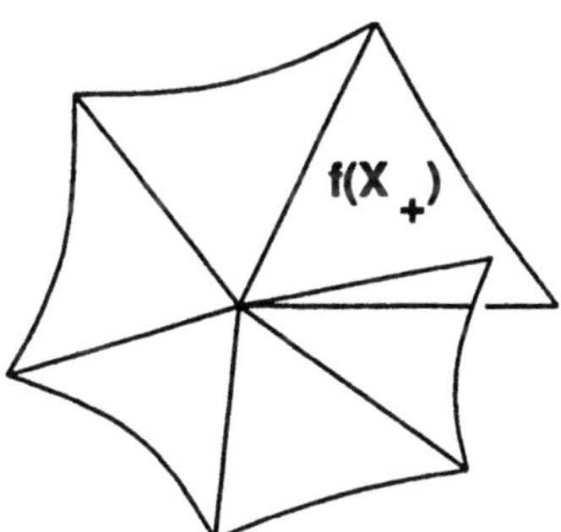

FIGURE 7.4. Conformal reflections of the Schwarz triangle

make an even number of reflections, you get a linear fractional transformation, as above. Such transformations form the projective monodromy group,

$$\Gamma \subset PGL(2, \mathbb{C}).$$

8. Modular Interpretations

I would like to make the above picture (in Figure 7.4) nice; I do not like the triangles overlapping again and again. So let us assume the angles of the triangle to be integral quotients of π:

$$|1 - c| = \frac{1}{p}, \quad |c - a - b| = \frac{1}{q}, \quad |a - b| = \frac{1}{r},$$

where

$$p, q, r \in \{2, 3, \dots\} \cup \{\infty\}.$$

There are three cases. For each case, a picture of a Schwarz triangle and all its possible reflections are shown.

8.1. $1/p + 1/q + 1/r > 1.$ There are only four possibilities:

$$
\begin{array}{ccll}
p & q & r & \text{Nicknames} \\
\\
2 & 2 & n & \text{Dihedral} \quad (n = 2, 3, \dots) \\
2 & 3 & 3 & \text{Tetrahedral} \\
2 & 3 & 4 & \text{Octahedral} \\
2 & 3 & 5 & \text{Icosahedral}
\end{array}
$$

In each case, a finite number of triangles cover the whole sphere $\mathbb{P}^1$. The projective monodromy group $\Gamma(p, q, r)$ is a finite group having the above nicknames. In the tetrahedral case, for instance, starting from a regular tetrahedron, you make the barycentric subdivision of the four faces, and then project this to a sphere from the center of your tetrahedron. You then obtain 24 triangles with angles $\pi/2, \pi/3$ and $\pi/3$, which is the picture of the Schwarz triangles you want. The triangles are colored in checkerboard pattern; white triangles are images of X_+, and black ones are images of X_-. The elements of the group $\Gamma(p, q, r)$ act on the sphere as rotations fixing the vertices of the triangles maintaining the checkerboard pattern. For other polyhedra, following the same steps, you obtain the pictures appearing in Figures 8.1 and 8.2. The Schwarz map induces the isomorphism

$$\mathbb{C} - \{0, 1\} \xrightarrow{\sim} (\mathbb{P}^1 - \{\text{vertices of the triangles}\})/\Gamma(p, q, r).$$

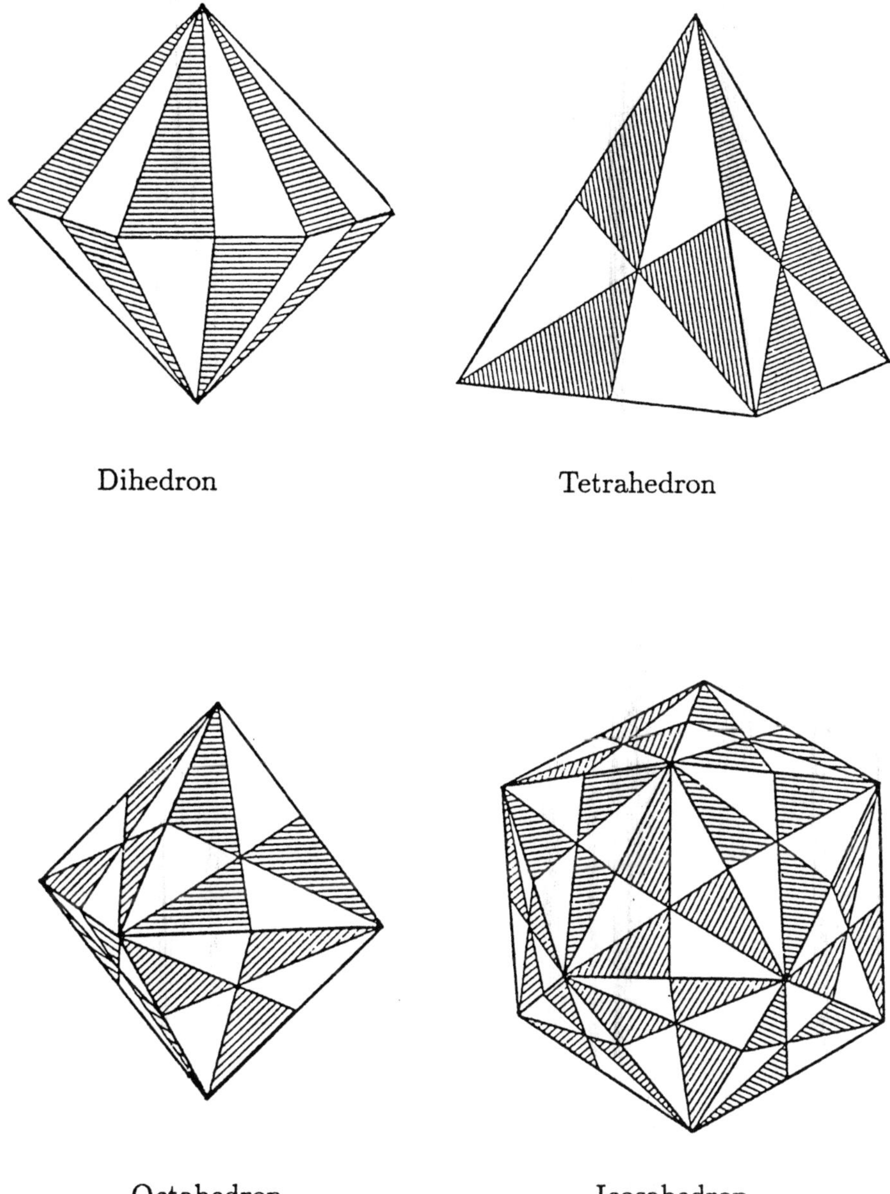

FIGURE 8.1. Barycentric subdivision of the faces of regular polyhedrons

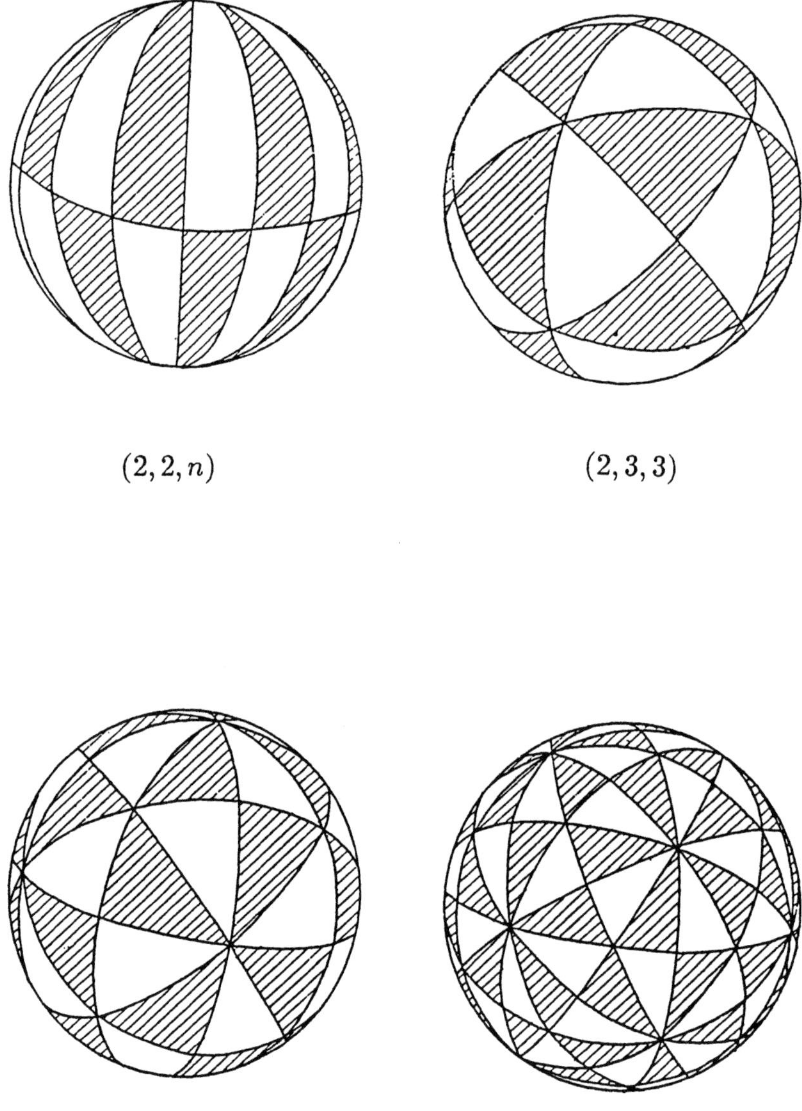

FIGURE 8.2. Schwarz' triangles filling a sphere

8.2. $1/p + 1/q + 1/r = 1$**.** There are only four possibilities:

p	q	r	Nicknames
2	2	∞	Ruler
2	3	6	Hexagonal lattice
2	4	4	Square lattice
3	3	3	Equilateral-triangle lattice

In each case, the triangles cover the plane $\mathbb{C} \subset \mathbb{P}^1$ (see Figure 8.3). The projective monodromy group $\Gamma(p, q, r)$ is a discrete subgroup of the euclidean motion group $\{z \to cz + d \mid c \in \mathbb{C}^\times, d \in \mathbb{C}\}$. Such a group is called a *crystallographic group*. The Schwarz map induces the isomorphism

$$\mathbb{C} - \{0, 1\} \xrightarrow{\sim} (\mathbb{C} - \{\text{vertices of the triangles}\})/\Gamma(p, q, r).$$

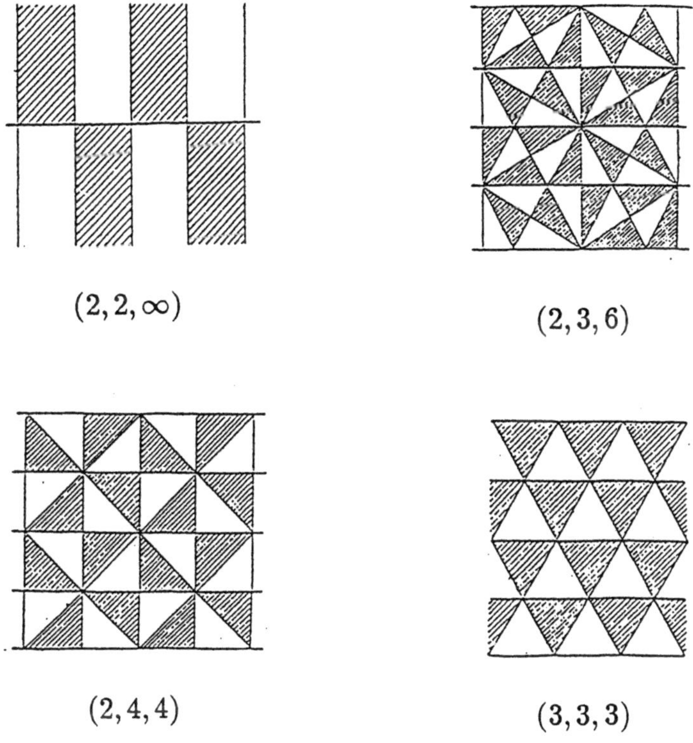

FIGURE 8.3. Crystallography: Schwarz' triangles tilling a plane

8.3. $1/p + 1/q + 1/r < 1$. There are...

...infinitely many possibilities!

In this case, you can make conformal reflections of a Schwarz triangle so that the union of the reflected triangles fills a disc D, and you have the isomorphism

$$\mathbb{C} - \{0,1\} \xrightarrow{\sim} (D - \{\text{vertices of the triangles}\})/\Gamma(p,q,r).$$

Instead of proving this fact, I tabulate the three most well known triangle groups $\Gamma(p,q,r)$ with their pictures (see Figures 8.4 and 8.5). (Hint for a proof: There is a unique circle which intersects perpendicularly the three circles bounding the Schwarz triangle.)

p	q	r	Reason why they are well known
2	3	∞	$PSL(2,\mathbb{Z})$
∞	∞	∞	$\Gamma(2)$
2	3	7	Area of the triangle is minimum

Are you wondering why the groups $\Gamma(2,3,\infty)$ and $\Gamma(\infty,\infty,\infty)$ have something to do with $PGL(2,\mathbb{Z})$ and $\Gamma(2)$?

Something to do with?

Oh! Look at these pictures! They are...

...exactly the same as the pictures I showed you in Chapter II, Figures 3.1 and 3.2, the fundamental domains of $PSL(2,\mathbb{Z})$ and $\Gamma(2)$.

Well, sorry,

not exactly the same. If you perform the transformation

$$\tau \to \frac{\tau - i}{\tau + i}$$

to take the pictures on the upper-half plane $\mathbb{H}$ to those on the unit disc D, then...

...they are the same!

Now you must agree. The monodromy groups $\Gamma(2,3,\infty)$ and $\Gamma(\infty,\infty,\infty)$ are conjugate in $PGL(2,\mathbb{C})$ to $PSL(2,\mathbb{Z})$ and $\Gamma(2)$, respectively. Of course you are probably not satisfied by the reasoning above, that "the two pictures agree." We shall see this again in the next chapter, using an integral representation of the hypergeometric functions.

FIGURE 8.4. Schwarz' triangles of type $(2, 3, \infty)$ filling a disc

FIGURE 8.5. Schwarz' triangles of type (∞, ∞, ∞) filling a disc

For

$$(a, b, c) = (\frac{1}{2}, \frac{1}{2}, 1) \quad \text{and} \quad (\frac{1}{12}, \frac{5}{12}, 1),$$

let us compute the values $1 - c, c - a - b$ and $a - b$:

$$1 - 1 = 0 = \frac{1}{\infty}, \quad 1 - \frac{1}{2} - \frac{1}{2} = 0 = \frac{1}{\infty}, \quad \frac{1}{2} - \frac{1}{2} = 0 = \frac{1}{\infty},$$

$$1 - 1 = 0 = \frac{1}{\infty}, \quad 1 - \frac{1}{12} - \frac{5}{12} = \frac{1}{2}, \quad \frac{1}{12} - \frac{5}{12} = -\frac{1}{3}.$$

Thus by virtue of the agreement of the pictures in Figures 8.4 and 8.5 and those in Chapter II, Figures 3.1 and 3.2, we obtain the following

FIGURE 8.6. Schwarz' triangles of type $(2,3,7)$ filling a disc

propositions.

PROPOSITION 8.1. *The Schwarz map of the hypergeometric equation* $E(1/2, 1/2, 1)$ *gives an isomorphism,*

$$X(4) \cong \mathbb{C} - \{0, 1\} \xrightarrow{\sim} \mathbb{H}/\Gamma(2),$$

which is the inverse map of the lambda function.

PROPOSITION 8.2. *The Schwarz map of the hypergeometric equation* $E(1/12, 5/12, 1)$ *gives an isomorphism,*

$$X\{4\} \cong \mathbb{C} \xrightarrow{\sim} \mathbb{H}/PSL(2),$$

which is the inverse map of the J-invariant function.

Summary

Our preparatory journey has now come to an end. I hope you enjoyed the journey. We started from a study of the configuration space $X(4)$ and its democratic embedding $X(4) \to Y$. We then studied elliptic curves, and with their help we obtained the map $\mathbb{H}/\Gamma(2) \to Y$. Finally, we turned to the hypergeometric differential equation, and with its help, we obtained the inverse map $X(4) \to \mathbb{H}/\Gamma(2)$. These can be summarized in the following commutative diagram:

$$
\begin{array}{ccc}
& \text{hypergeometric functions} & \\
X(4) & \longrightarrow & \mathbb{H}/\Gamma(2) \\
\text{democratic embedding} \searrow & & \swarrow \text{theta functions} \\
& Y \subset \mathbb{P}^2 &
\end{array}
$$

CHAPTER IV

Hypergeometric Integrals and Loaded Cycles

We started our hypergeometric business by studying the series $F(a, b, c; x)$, found the linear differential equation to which this series is a solution, found other solutions of this equation, and made use of these solutions to define the Schwarz map. Another way to introduce hypergeometric functions is to start with the so-called hypergeometric integrals. These are integrals of simple multi-valued functions. In this treatment, we will try to avoid algebro-geometric and homology-algebraic discussions.

After introducing the hypergeometric integrals, I give explicit generators of the monodromy group of the hypergeometric equation, and give a geometric meaning to the story in the last section of the previous chapter. Thanks to this formulation, we can proceed smoothly to the modular interpretations of the configuration spaces $X(2,5), X(2,6), \ldots$, which will be treated in Chapter VI.

1. Hypergeometric Integrals

The hypergeometric differential equation $E(a, b, c)$ and its solutions (*the hypergeometric functions*) are defined on $\mathbb{C} - \{0, 1\}$. The configuration space $X(4)$ is also known to be isomorphic to $\mathbb{C} - \{0, 1\}$. This is not a coincidence. I am going to convince you that the hypergeometric functions are naturally defined on this configuration space, and that the symmetry appearing in §4 of Chapter III will be understood as the symmetry of $X(4)$.

For four disjoint points $x_1, \ldots, x_4 \in \mathbb{P}^1$ (which are temporarily assumed to be distinct from the point at infinity) and four complex numbers μ_j satisfying

$$\mu_1 + \cdots + \mu_4 = 2,$$

the following *formal* integral is called a *hypergeometric integral*:

$$\int_C \prod_{j=1}^{4} (t - x_j)^{-\mu_j} \, dt.$$

The meaning of the word 'formal' here is that no path of integration C has yet been specified, no branch of the multi-valued form has been assigned, and convergence has not been discussed; let us worry about these points later. The condition for the μ_j is such that the integrand does not branch at $t = \infty$.

I am sure that you are anxious to know whether the hypergeometric integral is a function defined on the configuration space $X(4)$. Unfortunately, it is not quite well-defined on $X(4)$; a little modification is necessary. In order to see this, let us change notation a little by making use of honest homogeneous coordinates, and consider, for $x = (x_{ij}) \in M^*(2,4)$ and $1 \leq p \neq q \leq 4$, the integral

$$I(x) := \int_C \prod_{j=1}^{4} (x_{1j}t_1 + x_{2j}t_2)^{\beta_j} d\log \frac{x_{1p}t_1 + x_{2p}t_2}{x_{1q}t_1 + x_{2q}t_2},$$

where $t_1 : t_2$ are homogeneous coordinates of t-space, and

$$\beta_1 + \cdots + \beta_4 = 0.$$

Note that the hypergeometric integral above can be recovered by normalizing the integral $I(x)$, by putting

$$x_{1j} = 1, \quad x_{2j} = x_j, \quad t_1 = t, \quad t_2 = -1,$$
$$\mu_j = -\beta_j \ (j \neq p, q), \quad \mu_p = -\beta_p + 1, \quad \mu_q = -\beta_q + 1.$$

As a function of $x \in M^*(2,4)$, $I(x)$ is invariant under the left action of $GL(2)$. In fact, the transformation $x \mapsto gx \ (g \in GL(2))$ is equivalent to the coordinate transformation $(t_1, t_2) \mapsto (t_1, t_2)g^{-1}$. However, $I(x)$ is not invariant under the right action of H_4. In fact, we have

$$I(xh) = I(x) \prod h_j^{\beta_j}, \quad h = \mathrm{diag}(h_1, \ldots, h_4).$$

To make $I(x)$ invariant also under the right action of H_4, we multiply (following K. Matsumoto) $I(x)$ by the inverse of

$$(D_x^\beta)^{1/(n-2)}, \quad \text{where} \quad D_x^\beta = \prod_{i \neq j} D_x(ij)^{\beta_i}, \quad n = 4.$$

Since we have $D^\beta_{gx} = D^\beta_x$,

$$D^\beta_{xh} = \prod_{i\neq j} h_i^{\beta_i} h_j^{\beta_i} D_x(ij)^{\beta_i} \quad \text{and}$$

$$\prod_{i\neq j} h_i^{\beta_i} h_j^{\beta_i} = \prod_{i,j} h_i^{\beta_i} h_j^{\beta_i} (\prod_i h_i^{\beta_i} h_i^{\beta_i})^{-1}$$

$$= (\prod_i h_i^{\beta_i})^n \prod_j h_j^{\sum_i \beta_i} (\prod_i h_i^{\beta_i})^{-2}$$

$$= (\prod_i h_i^{\beta_i})^{n-2},$$

the product

$$(D^\beta_x)^{-1/2} I(x)$$

is invariant under the left action of $GL(2)$ and the right action of H_4.
Hence this product is defined on $X(4) = GL(2)\backslash M^*(2,4)/H_4$.

By the above argument, we find that though a hypergeometric integral
itself is not defined on $X(4)$, the 'ratio' of hypergeometric integrals with
different paths of integration is defined on $X(4)$. Since, in the following,
we mainly treat the ratio of hypergeometric integrals, we study these
rather than this complicated expression $(D^\beta_x)^{-1/2} I(x)$.

Note that when $x_4 = \infty$ the hypergeometric integral should be understood as

$$\int_C \prod_{j=1}^3 (t - x_j)^{-\mu_j} dt.$$

If we put $\mu_j = 1/2$, then the integral takes the form

$$\int \frac{dt}{\sqrt{(t - x_1)(t - x_2)(t - x_3)}},$$

which is an elliptic integral.

2. Paths of Integration

Let us now specify the paths along which we would like to integrate.
We shall characterize in §4 all possible paths. In this section, however,
we specify only two particularly useful and well-known ones, a compact
path and a non-compact path. To make the notation simple, let us write
the integral as follows:

$$\int_C t^{p-1}(1 - t)^{q-1} f(t) dt,$$

where two of the x_j have been normalized to 0 and 1, and the remaining part is denoted by $f(t)$. The function f is holomorphic around the closed interval $[0, 1]$.

You probably wonder why integrals such as these are called *hypergeometric integrals*; the following proposition gives an answer.

2.1. The Segment $(0, 1)$. If $\Re(p) > 0$ and $\Re(q) > 0$, then the integral along the segment $(0, 1)$ converges, and this segment can serve as a path of integration.

PROPOSITION 2.1. *Suppose* $\Re(c) > \Re(a) > 0$. *Then we have*

$$F(a, b, c; x) = \frac{\Gamma(c)}{\Gamma(a)\Gamma(c - a)} \int_0^1 t^{a-1}(1 - t)^{c-a-1}(1 - tx)^{-b}dt$$

for $|x| < 1$, *where the branch of the integrand is determined by the following assignment of the arguments*

$$\arg t = 0, \quad \arg(1 - t) = 0, \quad |\arg(1 - xt)| < \frac{\pi}{2}, \quad (0 < t < 1).$$

This proposition can be proved by substituting the *binomial formula*

$$(1 - tx)^{-b} = \sum_{n=0}^{\infty} \frac{(b, n)}{(1, n)}(tx)^n, \quad |tx| < 1$$

into the integrand, and by using the *beta function formula*:

$$(B(p, q) :=) \int_0^1 t^{p-1}(1 - t)^{q-1}dt = \frac{\Gamma(p)\Gamma(q)}{\Gamma(p + q)}, \quad \Re(p),\ \Re(q) > 0,$$

which can be proved by a famous, tricky method. Here the *gamma function* $\Gamma(p)$ is defined by the integral

$$\Gamma(p) = \int_0^{\infty} t^{p-1}e^{-t}dt.$$

Comment: Since $(1 + \varepsilon)^{1/\varepsilon}$ tends to e as $\varepsilon \to 0$, we have

$$\Gamma(p) = \lim_{\varepsilon \to 0} \frac{1}{\varepsilon^p} B(p, \frac{1}{\varepsilon}).$$

This formula and the beta formula above pose the following question: which is more elementary, the beta function or the gamma function?

Of course you would like to get rid of the restriction $\Re(c) > \Re(a) > 0$. This can be seen by considering the following loop (a path that ends where it starts).

2.2. A Double Contour Loop around 0 and 1. Let γ be a closed curve starting from a point T in the open interval $(0,1)$, making a loop in the counterclockwise direction around 0 and 1, and then making a second loop around 0 and 1 but this time in the clockwise direction, and ending at T, as shown in Figure 2.1. (The four oriented segments are labeled by $s_1, \ldots, s_4$, and the four oriented circles by C_0, C_0', C_1, C_1'). This is called a *double contour loop* or a *Pochhammer loop* around 0 and 1.

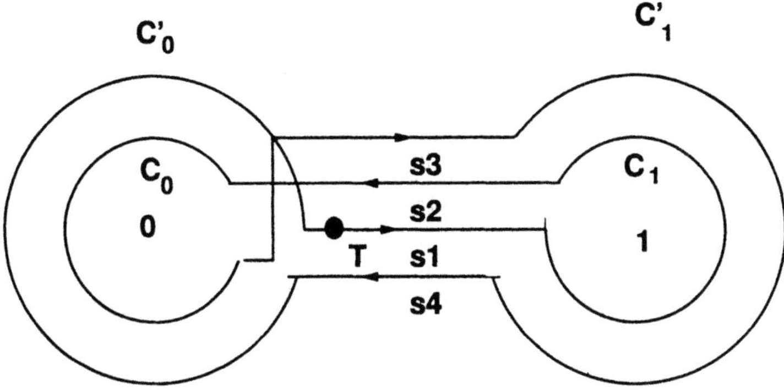

FIGURE 2.1. γ: A double contour loop around 0 and 1

For any given complex numbers p and q, let us fix a branch of the multivalued function

$$u = t^{p-1}(1 - t)^{q-1}f(t)$$

at the starting point T. If $f(t)$ is a function holomorphic in some neighborhood U of the closed interval $[0,1]$ in $\mathbb{C}$, and if γ is inside U, we can continue the function u analytically along γ, and we find that the branch at the end point of γ is the same as that at the starting point. (This fact can be checked by tracing the change of the arguments of t and $1 - t$.) This implies in particular that the integrand is continuous on γ. Thus the integral

$$\int_{\gamma} u\,dt$$

is well defined in the sense that it is independent of the origin T and even of the shape of the loop, provided that you deform it smoothly in $U - \{0,1\}$, and provided of course that it remains a loop during the deformation. It should be pointed out that this does depend on the choice of a branch of the multi-valued function $u = t^{p-1}(1 - t)^{q-1}f(t)$. We shall come back to this point later.

Since γ is compact, the integral

$$F = \int_\gamma t^{a-1}(1-t)^{c-a-1}(1-tx)^{-b}dt$$

converges uniformly in x, so that one can differentiate under the integration symbol. We shall show that F is a solution of the hypergeometric equation in two different ways in the following sections. (Another elegant explanation is given in [IKSY].) In addition to γ, a double contour loop around any two of the four points $\{0,1,1/x,\infty\}$ works as the path of integration as well. A picture of a double contour loop around the two points $1/x$ and ∞ is given in Figure 2.2.

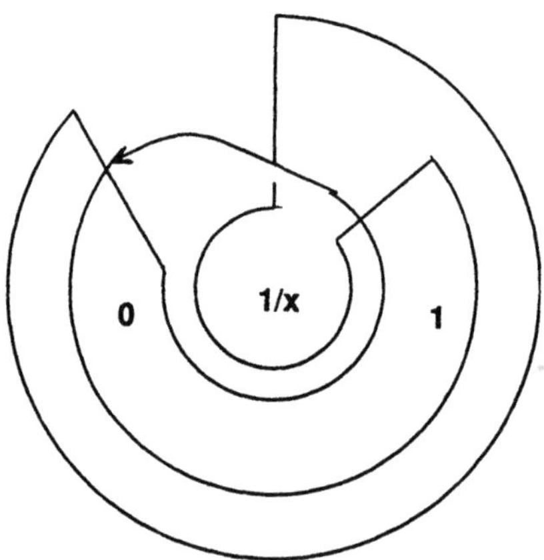

FIGURE 2.2. Double contour loop around $1/x$ and ∞

2.3. The Euler Transformation. For any complex number λ, consider the Euler kernel

$$K(\lambda) = K(\lambda,t,x) := (1-tx)^{-\lambda}.$$

The Euler transformation $F(x)$ of the function $G(t) = t^{a-1}(1-t)^{c-a-1}$ is given by the integral

$$F(x) = \int_\gamma G(t)K(\lambda,t,x)dt,$$

where γ is a double contour loop around any two of the four points $\{0,1,1/x,\infty\}$. We have

$$PF = \int G(t)PK(\lambda)dt,$$

where P is the *hypergeometric operator*

$$P = P(x, D) = (a + D)(b + D) - (c + D)(1 + D)\frac{1}{x}.$$

We are going to find a linear differential operator of first order

$$Q = Q(t, \theta), \quad \text{where} \quad \theta = t\frac{d}{dt},$$

and a function $L = L(x, t)$ such that

$$PK(\lambda) = QL.$$

With such Q and L we have, using integration by parts several times,

$$\begin{aligned}
PF &= \int G(t)QL\,dt \\
&= \int Q^*G(t)L\,dt,
\end{aligned}$$

where Q^* is the formal adjoint operator of Q. By definition the formal adjoint operator Q^* of any linear differential operator

$$Q = \sum a_j(t)\frac{d^j}{dt^j}$$

is

$$Q^* = \sum (-1)^j \frac{d^j}{dt^j} \cdot a_j(t).$$

Notice that for any $\lambda \in \mathbb{C}$, we have

$$\begin{aligned}
DK(\lambda) &= \theta K(\lambda), \\
(b + D)K(\lambda) &= \lambda K(\lambda + 1) + (b - \lambda)K(\lambda), \\
(1 + D)\frac{1}{x}K(\lambda) &= \lambda t K(\lambda + 1).
\end{aligned}$$

By setting $\lambda = b$ we get

$$\begin{aligned}
PK(b) &= (a + D)bK(b + 1) - (c + D)btK(b + 1) \\
&= b\{(a + \theta) - t(c + \theta)\}K(b + 1).
\end{aligned}$$

This computation suggests that we set

$$\begin{aligned}
Q &= a + \theta - t(c + \theta), \\
L &= K(b + 1).
\end{aligned}$$

Now we have

$$Q = a - ct + t\frac{d}{dt} - t^2\frac{d}{dt},$$

$$Q^* = a - ct - \frac{d}{dt}t + \frac{d}{dt}t^2$$

$$= a - ct - t\frac{d}{dt} - 1 + 2t + t^2\frac{d}{dt}$$

$$= t(t-1)\{\frac{d}{dt} - \frac{a-1}{t} - \frac{c-a-1}{t-1}\}.$$

Up to multiplicative constant, the equation $Q^*w = 0$ has the unique solution

$$G = t^{a-1}(1-t)^{c-a-1}.$$

Therefore

$$F = \int_\gamma t^{a-1}(1-t)^{c-a-1}(1-tx)^{-b}dt$$

is a solution of the hypergeometric equation.

2.4. The Derivation ∇ Acting on Rational Forms. For

$$u = t^{a-1}(1-t)^{c-a-1}(1-tx)^{-b},$$

we consider

$$f = \int_\gamma u\,dt, \quad \frac{df}{dx} = \int_\gamma ug_1\,dt, \quad \frac{d^2f}{dx^2} = \int_\gamma ug_2\,dt, \quad \ldots,$$

where the g_j are rational functions in t, over the rational function field $\mathbb{C}(x)$, given as follows:

$$g_1 = \frac{bt}{1-tx}, \quad g_2 = b(b+1)\frac{t^2}{(1-tx)^2}, \ldots.$$

We are looking for a linear relation among f, df/dx, $d^2f/dx^2,\ldots$ over $\mathbb{C}(x)$.

By the way, if h is smooth on the double contour loop γ — in particular, if h is rational with poles only at $t = 0, 1, 1/x$ — then by the *Stokes theorem* we have

$$\int_\gamma d(uh) = 0,$$

where d is the exterior derivation in t. The integrand can be computed by the Leibniz rule, and we obtain an important formula:

$$d(uh) = u\nabla h.$$

Here ∇ is a derivation (often called a *connection*), sending a function to a 1-form, defined by

$$\nabla h := dh + \omega h,$$

where ω is the rational 1-form (often called the *connection form*):

$$\omega = \frac{du}{u} \quad \left(= (\frac{a-1}{t} - \frac{c-a-1}{1-t} + \frac{bx}{1-tx})dt \right).$$

We have the following fact: Define

$$Z := \{\text{rational 1-forms with poles only at } 0, 1, 1/x\},$$
$$B := \nabla\{\text{rational functions with poles only at } 0, 1, 1/x\}.$$

Then Z modulo B is 2-dimensional over $\mathbb{C}(x)$. Since this fact is not crucial in the overall argument, we do not bother to give its proof (which is not difficult). If you accept this fact, you will realize that the three forms

$$1 \cdot dt, \quad g_1 dt, \quad g_2 dt$$

modulo B must have a $\mathbb{C}(x)$-linear relation. This implies a $\mathbb{C}(x)$-linear relation among f, df/dx and $d^2 f/dx^2$, yielding a second-order linear differential equation.

Actual computation can be carried out (without using the above fact) by calculating $P(x, D)u$ as follows:

$$P(x, D)u = ub \left(\frac{a}{1-tx} - \frac{ct}{1-tx} + (b+1)\frac{tx}{(1-tx)^2} - (b+1)\frac{t^2x}{(1-tx)^2} \right).$$

Since we have

$$\nabla \frac{t}{1-tx} = \frac{dt}{1-tx} + (\frac{a-1}{t} - \frac{c-a-1}{1-t})\frac{tdt}{1-tx} + (b+1)\frac{txdt}{(1-tx)^2},$$
$$\nabla \frac{t^2}{1-tx} = \frac{2tdt}{1-tx} + (\frac{a-1}{t} - \frac{c-a-1}{1-t})\frac{t^2dt}{1-tx} + (b+1)\frac{t^2xdt}{(1-tx)^2},$$

we conclude that

$$P(x, D)udt = ub\nabla \left(\frac{t-t^2}{1-tx} \right).$$

This implies that $P(x, D)f = 0$.

2.5. The Relation between the Two Kinds of Paths. We now study the relation between the two kinds of paths appearing in §2.1 and §2.2. Let f be a function holomorphic near the segment $[0,1]$ and γ be the double contour loop around 0 and 1. We fix the arguments of t and $1 - t$ on the open interval $(0,1)$ as

$$\arg t = 0, \quad \arg(1 - t) = 0, \quad (0 < t < 1),$$

and at the starting point (=the ending point) of γ we assume these arguments. When $\Re(p) > 0$ and $\Re(q) > 0$, the contributions along the four circles around 0 and 1 can be made as small as desired. Thus by considering the variation of the argument along γ, we see that

$$\int_\gamma t^{p-1}(1-t)^{q-1}f(t)dt = \{1-e^{2\pi iq}+e^{2\pi i(p+q)}-e^{2\pi ip}\} \int_0^1 t^{p-1}(1-t)^{q-1}f(t)dt.$$

In the factor outside the integral on the right-hand side, the term 1 represents the contribution of the path s_1 from 0 to 1 passing through the base T. (See Figure 2.1.) After a positive turn around the point 1, the value of the function $t^p(1 - t)^q$ changes by $e^{2\pi iq}$, and thus the second term $-e^{2\pi iq}$ corresponds to the path s_2 from 1 to 0. Now you know why the third and fourth terms correspond to the paths s_3 and s_4. If neither p nor q is an integer, since we have

$$1 - e^{2\pi iq} + e^{2\pi i(p+q)} - e^{2\pi ip} = (1 - e^{2\pi ip})(1 - e^{2\pi iq}),$$

the above equality provides us with a method of converting an integral on the segment $(0,1)$ into an integral on the double contour loop around 0 and 1.

It is very troublesome to assign a branch of the multi-valued function $t^p(1-t)^q$ on a portion s_1 of the contour loop γ and to continue analytically along γ, determining the branches for s_2, s_3 and s_4. It is convenient to consider the path (we assume $p, q \notin \mathbb{Z}$, of course)

$$\gamma' := \frac{\gamma}{(1 - e^{2\pi ip})(1 - e^{2\pi iq})}$$

instead of γ. This is a one-way path on the segment $(0,1)$ (see Figure 2.3). If you carefully trace the value of $t^p(1 - t)^q$ along the four circles in γ around 0 and 1, you find

$$\gamma' = -\frac{C_\varepsilon(0)}{1 - e^{2\pi ip}} + \overrightarrow{(\varepsilon, 1 - \varepsilon)} + \frac{C_\varepsilon(1)}{1 - e^{2\pi iq}},$$

where $C_\varepsilon(0)$ is a positively oriented circle starting and ending at $t = \varepsilon$ with radius ε and center 0, and $C_\varepsilon(1)$ a positively oriented circle starting and ending at $t = 1 - \varepsilon$ with radius ε and with center 1. (Along the segment $(\varepsilon, 1 - \varepsilon)$ and on the two starting points of the circles we assign the given branch of $t^p(1 - t)^q$ and continue along the circles.)

Indeed, the two circles around 1 in γ are C_1 and C_1' (see Figure 2.1), and we have

$$C_1 = C_\varepsilon(1),$$

of course. The branches at the starting and ending points of C_1' are $e^{2\pi i(p+q)}$ times and $e^{2\pi ip}$ times the original branch, respectively. Since C_1' is oriented in the clockwise direction, we have

$$C_1' = -e^{2\pi ip}C_\varepsilon(1),$$

so that

$$\frac{C_1 + C_1'}{(1 - e^{2\pi ip})(1 - e^{2\pi iq})} = \frac{(1 - e^{2\pi ip})C_\varepsilon(1)}{(1 - e^{2\pi ip})(1 - e^{2\pi iq})} = \frac{C_\varepsilon(1)}{(1 - e^{2\pi iq})}.$$

FIGURE 2.3. The double contour loop divided by $(1 - e^{2\pi ip})(1 - e^{2\pi iq})$

The situation around 0 is similar to that around 1.

Therefore, under the conditions $\Re(p) > 0$ and $\Re(q) > 0$, taking branches as above, we have

$$\int_{\gamma'} t^{p-1}(1 - t)^{q-1}f(t)\,dt = \int_0^1 t^{p-1}(1 - t)^{q-1}f(t)\,dt.$$

3. Loaded Paths

To assign a branch along paths is still a tedious task, but since the function $t^p(1 - t)^q$ is not single-valued, we cannot get rid of it. Up to this point we have been thinking in terms of integrating the continuous function $t^{p-1}(1 - t)^{q-1}f(t)$ along a topological path γ. What if we think of values of the function

$$u = t^p(1 - t)^q f(t)$$

as being assigned along the path (let us call it a *loaded path*) and that we are integrating the rational form

$$\frac{dt}{t(1-t)}$$

along this path? You might say: "But nothing has changed other than the words you use!" And of course you would be right. But you will see that this is a useful way of thinking.

A loaded p-chain ($p = 0, 1, 2$) is a formal sum of loaded p-simplices (points, paths, curvilinear triangles). A loaded p-simplex is a topological simplex on which a branch of u is assigned. The boundary is naturally defined. For example, the boundary of a loaded path should be given by

(ending point loaded with the value of the function there)

$-$(starting point loaded with the value of the function there).

Let us, for instance, compute the boundary of the loaded path γ' (in the previous subsection, the loaded function $t^p(1 - t)^q f(t)$ was discussed). Let u denote a branch of the function $u = t^p(1 - t)^q f(t)$ on the interval $(0, 1)$; it does not matter which arguments are used. The same letter u will also denote the functions on the circles $C_\varepsilon(0)$ and $C_\varepsilon(1)$, starting from ε and $1 - \varepsilon$, respectively. To indicate with what the path γ' is loaded, we denote the loaded path by $\gamma' \otimes u$, and a point x associated with the "load" $u(x)$ (a loaded point) will be denoted by

$$[x] \otimes u(x).$$

Then we have

$$\partial\{C_\varepsilon(0) \otimes u\} = [\varepsilon] \otimes e^{2\pi i p}u(\varepsilon) - [\varepsilon] \otimes u(\varepsilon),$$
$$\partial\{\overrightarrow{(\varepsilon, 1 - \varepsilon)} \otimes u\} = [1 - \varepsilon] \otimes u(1 - \varepsilon) - [\varepsilon] \otimes u(\varepsilon),$$
$$\partial\{C_\varepsilon(1) \otimes u\} = [1 - \varepsilon] \otimes e^{2\pi i q}u(1 - \varepsilon) - [1 - \varepsilon] \otimes u(1 - \varepsilon),$$

and therefore $\partial\gamma' = 0$. Since we have the habit of calling an object with zero boundary a cycle, we call a loaded path with zero boundary a *loaded cycle*. Two loaded paths γ_1 and γ_2 are called homologous if the difference $\gamma_1 - \gamma_2$ is the boundary of a loaded 2-chain. In this case we write $\gamma_1 \sim \gamma_2$.

Now, what about the loaded path $\overrightarrow{(0, 1)}$? Well, since it has no boundary in $X = \mathbb{C} - \{0, 1\}$, we must also call it a cycle. Thus far we defined two kinds of loaded cycles, compact and non-compact. The correspondence

$$\text{reg} : \overrightarrow{(0,1)} \mapsto \gamma'$$

is called the *regularization*, and γ' will be denoted by $\text{reg}\overrightarrow{(0,1)}$.

REMARK 3.1. Many people (perhaps everyone except me) call loaded cycles *twisted cycles*. I do not like this terminology, since there is nothing to *twist*.

4. Relations among Loaded Cycles

In order to make the story appear more systematic, we change notation. Let $x_1, \ldots, x_n$ be distinct real points on $\mathbb{P}^1$ satisfying

$$x_1 < \cdots < x_n.$$

Then put

$$T(x) = \mathbb{P}^1 - \{x_1, \ldots, x_n\} \cup \{x_0 = x_{n+1} = \infty\},$$

and consider the multi-valued function

$$u = \prod_1^n (t - x_j)^{\alpha_j}.$$

On each oriented interval $\overrightarrow{(x_p, x_{p+1})}$, we load a branch of the function u determined by

$$\arg(t - x_j) = \begin{cases} 0 & j \leq p \\ -\pi & p + 1 \leq j, \end{cases}$$

and call this loaded path I_p. Although this assignment is by no means complicated, whenever you see formulae like this, it takes some time to recognize the meaning. Note that if you analytically continue the branch of u corresponding to some loaded path I_j through the *lower* half part of the t-plane $T(x)$, then you get the branches of u corresponding to other loaded paths I_k. Bear in mind that if you pass through the *upper* part, you do not get the correct branches. Therefore, instead of assigning the arguments of $t - x_j$ by the formula above, we often describe this assignment by saying simply

u is defined on the lower half plane,

without mentioning the exact values of the arguments, which are rarely used.

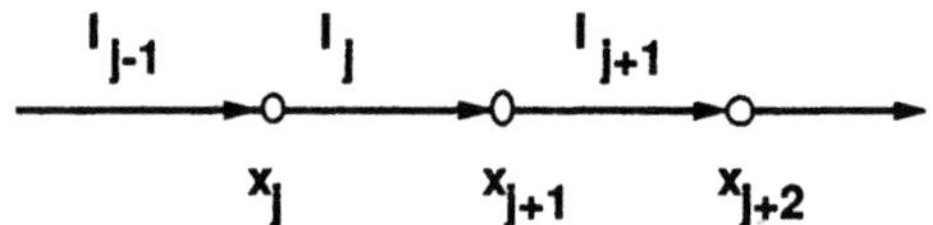

FIGURE 4.1. Loaded paths I_j on $T(x)$

Considering the boundaries of the two loaded 2-chains, one in the lower and one in the upper half of the t-space $T(x)$, as shown in Figure 4.2, we have

PROPOSITION 4.1. *The loaded paths $I_0, I_1, \ldots, I_n$, loaded with the function $u = \prod_1^n (t-x_j)^{\alpha_j}$ defined on the lower half plane, are related according to the relations*

$$I_0 + I_1 + \cdots + I_n \sim 0,$$

and

$$I_0 + e(-\alpha_1)I_1 + e(-\alpha_1 - \alpha_2)I_2 + \cdots + e(-\alpha_1 - \cdots - \alpha_n)I_n \sim 0,$$

where $e(\cdot) = \exp(2\pi i \cdot)$.

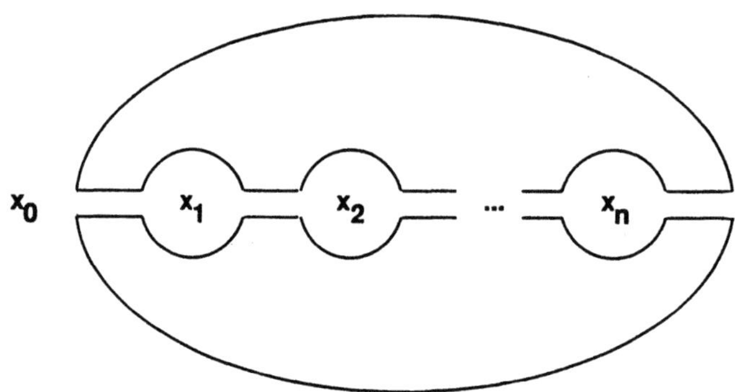

FIGURE 4.2. Two loaded 2-chains and their boundaries

Under the condition

$$\alpha_j \notin \mathbb{Z}, \quad \alpha_0 := -\sum_{j=1}^n \alpha_j \notin \mathbb{Z},$$

the above proposition can be paraphrased by the correspondence

$$I_j \longrightarrow \mathrm{reg} I_j.$$

Under the same condition, there are just $n-1$ linearly independent loaded cycles. This will be shown in §7.

5. Monodromy of Loaded Cycles and of Hypergeometric Functions

Now we are going to see what happens if we let a point x_{j+1} travel, making a loop around other points. In particular, let us consider the situation in which x_{j+1} moves in the counterclockwise direction around the adjacent point, x_j. It will be useful later to know what happens when the loop is halfway completed, and x_{j+1} and x_j have exchanged positions. This situation is depicted in Figure 5.1. Define $x'_0, \ldots, x'_n$ by

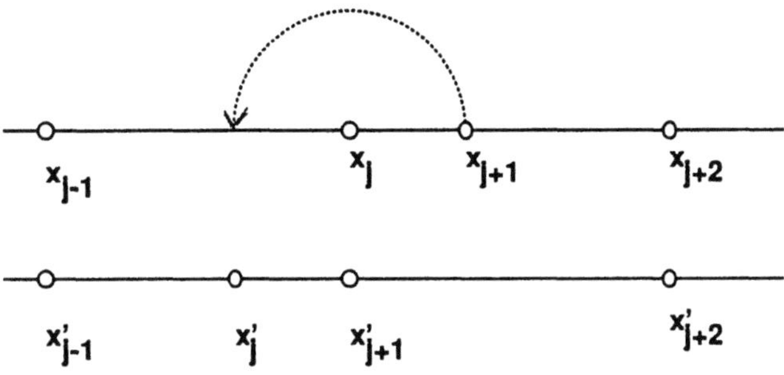

FIGURE 5.1. A half-turn: exchanging x_{j+1} and x_j

$$x'_j = x_{j+1}, \quad x'_{j+1} = x_j, \quad x'_k = x_k, \quad k \neq j,\, j+1,$$

and the loaded cycles I'_i by the oriented intervals $\overrightarrow{(x'_i, x'_{i+1})}$ loaded with the function

$$u' = \prod_{k \leq j-1} (t - x'_k)^{\alpha_k} \cdot (t - x'_j)^{\alpha_{j+1}} (t - x'_{j+1})^{\alpha_j} \prod_{j+2 \leq k} (t - x'_k)^{\alpha_k},$$

defined on the lower half plane. In Figure 5.2, the changes suffered by I_{j-1}, I_j and I_{j+1} are illustrated. Carefully tracing how $\arg(t - x_j)$ changes on I_j, we obtain the *half-turn formula*:

$$I_{j-1} \longrightarrow I'_{j-1} + I'_j,$$
$$I_j \longrightarrow -c_j I'_j,$$
$$I_{j+1} \longrightarrow I'_{j+1} + c_j I'_j,$$
$$I_k \longrightarrow I'_k, \quad |k - j| \geq 2,$$

where

$$c_j = e(\alpha_j) = \exp 2\pi i \alpha_j.$$

Note that the result is the same if you let the point x_j travel halfway around x_{j+1} in the counterclockwise direction. If x_{j+1} continues, traveling

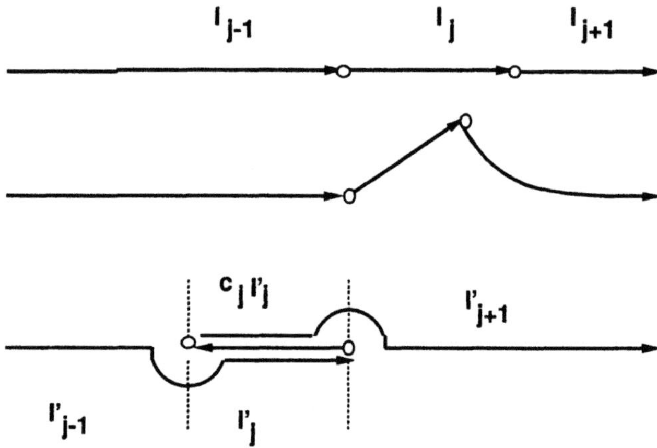

FIGURE 5.2. An animation of the deformation of the half-turn of x_{j+1} around x_j

back to the original position, by applying the transformation above (do not forget to exchange α_j and α_{j+1}) again, we get

$$I'_{j-1} + I'_j \longrightarrow \{I_{j-1} + I_j\} - c_{j+1}I_j,$$
$$-c_j I'_j \longrightarrow -c_j(-c_{j+1})I_j,$$
$$I'_{j+1} + c_j I'_j \longrightarrow \{I_{j+1} + c_{j+1}I_j\} + c_j(-c_{j+1})I_j,$$
$$I'_k \longrightarrow I_k, \quad |k - j| \geq 2.$$

We thus obtain the *full-turn formula*:

$$I_{j-1} \longrightarrow I_{j-1} + (1 - c_{j+1})I_j,$$
$$I_j \longrightarrow c_j c_{j+1} I_j,$$
$$I_{j+1} \longrightarrow I_{j+1} + c_{j+1}(1 - c_j)I_j,$$
$$I_k \longrightarrow I_k, \quad |k - j| \geq 2.$$

If you do not rest at the half-turn but make the full-turn at once, we get the same formula, of course (see Figure 5.3).

In general, when x_j travels around x_k along a loop γ_{jk} as in Figure 5.4, we first apply the half-turn formula the necessary number of times, then the full-turn formula, and finally apply the half-turn formula the necessary number of times. In so doing, we obtain a transformation $M(\gamma_{jk}, \alpha)$ of the loaded cycles. The possible moves of the points x_j can be described by the *fundamental group* $\pi_1(X(n + 1), \overset{\bullet}{x})$, where $\overset{\bullet}{x}$ is the

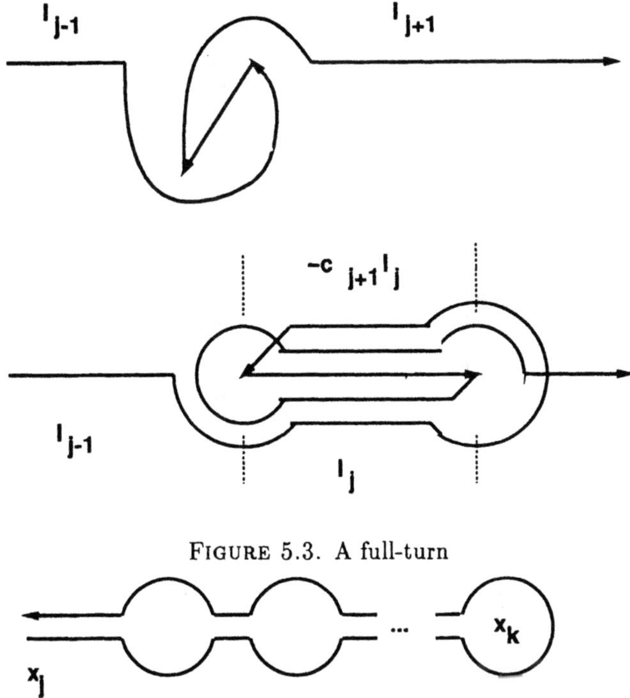

FIGURE 5.3. A full-turn

FIGURE 5.4. The loop γ_{jk}

configuration

$$x_1 < \cdots < x_n, \quad x_{n+1} = \infty,$$

which serves as a base point in $X(n+1)$. The matrices $M(\gamma_{jk}, \alpha)$ generate the monodromy group $\Gamma(\alpha)$. Let u_j be the hypergeometric integral of u along the loaded cycle I_j. The ratio $u_1 : \cdots : u_{n-1}$ is defined on $X(n+1)$ and gives a (multi-valued) map: $X(n+1) \longrightarrow \mathbb{P}^{n-2}$, with the projective monodromy group $\Gamma(\alpha) \mod \mathbb{C}^{\times}$.

When $n = 3$, we know the space well: $X(2,4) \cong \mathbb{P}^1 - \{\text{three points}\}$. (Do you remember?) Here is the picture we saw in Chapter I:

$(14;23)$	$\overset{\bullet}{x}$	$(12;34)$	$(13;24)$
$1 < 4 < 2 < 3$	$1 < 2 < 3 < 4$	$1 < 2 < 4 < 3$	

The fundamental group can be generated freely by the loops γ_{12} and γ_{23} directed in the counterclockwise direction around the holes labelled

$(12; 34)$ and $(14; 23)$, respectively. The loop γ_{12} corresponds to the move of x_1 around x_2 described above, and the loop γ_{23} to the move of x_2 around x_3. Thus the circuit matrices with respect to a basis of the loaded cycles I_1 and I_2 are given by

$$\gamma_{12} : \begin{pmatrix} I_1 \\ I_2 \end{pmatrix} \longrightarrow \begin{pmatrix} c_1 c_2 & 0 \\ c_2(1 - c_1) & 1 \end{pmatrix} \begin{pmatrix} I_1 \\ I_2 \end{pmatrix},$$

$$\gamma_{23} : \begin{pmatrix} I_1 \\ I_2 \end{pmatrix} \longrightarrow \begin{pmatrix} 1 & 1 - c_3 \\ 0 & c_2 c_3 \end{pmatrix} \begin{pmatrix} I_1 \\ I_2 \end{pmatrix}.$$

Let us denote the two matrices above by $M(12)$ and $M(23)$. These two matrices generate the monodromy group. For example, if $\alpha_j \in 1/2 + \mathbb{Z}$, they become

$$\begin{pmatrix} 1 & 0 \\ -2 & 1 \end{pmatrix} \quad \text{and} \quad \begin{pmatrix} 1 & 2 \\ 0 & 1 \end{pmatrix},$$

which generate $\Gamma(2)$ up to ± 1.

When $n = 4$, the corresponding space $X(2, 5)$ will be studied in Chapter V. The fundamental group in this case is generated by five loops corresponding to the move of x_j around x_{j+1} for $j = 1, \ldots, 5$. This time these generators are not free; relations are given in §7 of Chapter V.

When $n \geq 5$, the fundamental group of the corresponding space $X(2, n+1)$ possesses generators in addition to the loops described by points x_j moving around the adjacent points.

6. Invariant Hermitian Forms

Let us assume for simplicity that all the α_j are non-integral and real.

PROPOSITION 6.1. *There is a unique hermitian form H (up to a multiplicative real constant) invariant under the monodromy group:*

$$^t\overline{M} H M = H \quad \text{for all} \quad M \in \Gamma(\alpha).$$

Since we know generators of the monodromy group, we can find this H. The computation is straightforward but very uncomfortable (you shall know if you carry it out). When $n = 3$, since the matrices are 2×2, your suffering is not so great, and you shall find (up to a multiplicative real constant)

$$H = \frac{1}{\sqrt{c_1 c_2 c_3}} \begin{pmatrix} d_1 d_{23} & d_1 d_3 \\ d_1 c_2 d_3 & d_{12} d_3 \end{pmatrix},$$

where

$$c_j = \exp 2\pi i \alpha_j, \quad d_j = c_j - 1, \quad c_{jk} = c_j c_k, \quad d_{jk} = c_{jk} - 1.$$

Incidentally, the inverse of H is given by

$$H^{-1} = \frac{\sqrt{c_1 c_2 c_3}}{d_{123}} \begin{pmatrix} d_{12}/d_1 d_2 & -1/d_2 \\ -c_2/d_2 & d_{23}/d_2 d_3 \end{pmatrix},$$

where $d_{123} = c_1 c_2 c_3 - 1$. For example, when $\alpha_j \in 1/2 + \mathbb{Z}$,

$$H = (\text{real constant}) \begin{pmatrix} 0 & i \\ -i & 0 \end{pmatrix}.$$

There is a tricky way to find the invariant form which will be presented in the next section.

When $n = 3$, the invariant hermitian form H has the following geometric meaning. If H is non-degenerate and indefinite, and if a point of the image $\varphi(x_0)$(a set of points) of x_0 under the Schwarz map

$$\varphi : X(4) \ni x \mapsto u_1(x) : u_2(x) \in \mathbb{P}^1$$

is inside (or outside) the set

$$\{u_1 : u_2 \in \mathbb{P}^1 \mid (\overline{u}_1, \overline{u}_2) H^t(u_1, u_2) > 0\},$$

then is also the whole $\varphi(x_0)$. Because we have

$$\begin{aligned}
(\overline{u}_1', \overline{u}_2') H^t(u_1', u_2') &= (\overline{u}_1, \overline{u}_2)^t \overline{M} H M^t(u_1, u_2) \\
&= (\overline{u}_1, \overline{u}_2) H^t(u_1, u_2),
\end{aligned}$$

where

$$^t(u_1', u_2') = M^t(u_1, u_2), \quad M \in \Gamma(\alpha).$$

For example, for

$$H = \begin{pmatrix} 0 & i \\ -i & 0 \end{pmatrix},$$

we have

$$\begin{aligned}
(\overline{u}_1, \overline{u}_2) H^t(u_1, u_2) &= i(\overline{u}_1 u_2 - u_1 \overline{u}_2) \\
&= 2|u_2|^2 \Im(u_1/u_2),
\end{aligned}$$

so that

$$\{u_1 : u_2 \in \mathbb{P}^1 \mid (\overline{u}_1, \overline{u}_2) H^t(u_1, u_2) > 0\} = \mathbb{H}.$$

7. Intersections of Loaded Cycles

If nothing is loaded onto a cycle, it is an ordinary topological cycle. (Of course, you can think in this case that the constant function 1 is loaded, if you wish to consider these also as loaded cycles.) You can define the intersection number of two such cycles: after you make a deformation causing two cycles to meet transversally, you count – do not forget the orientation! – how many times they meet. Note that this number is invariant under continuous moves of the cycles.

Now think of what is the natural definition of the *intersection numbers* of loaded cycles. If C_1 and C_2 are loaded with the holomorphic functions u_1 and u_2, respectively, and if they meet at a point t in a positive way (see Figure 7.1), then the only possible definition of the intersection number

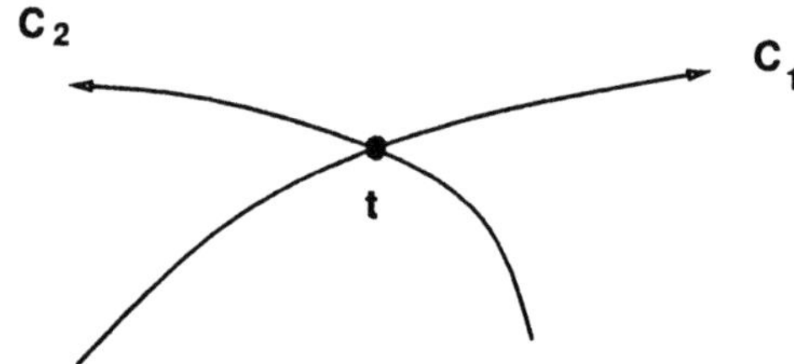

FIGURE 7.1. A positive intersection of C_1 and C_2 at t

of C_1 and C_2 at the point t would be

$$(C_1 \cdot C_2)_t = u_1(t)u_2(t);$$

the intersection number $C_1 \cdot C_2$ of C_1 and C_2 would be the *sum* of the intersection numbers at all the intersection points. If you let the cycles move slightly, the value of the right-hand side changes! This is no good. Is there any escape? Yes. If the function u_2 is a constant times $1/u_1$, then the right-hand side is a constant! There would be no other escape.

Let us evaluate the intersection number of loaded cycles I_j loaded with a branch u_0 of

$$u = \prod (t - x_j)^{\alpha_j}$$

defined on the lower half plane and $\check{I}_j$ loaded with $1/u_0$. How do we count intersections of I_j and $\check{I}_j$? Of course you can slightly deform them so that they do not meet in $T(x)$. Thus should we conclude that the intersection number is zero? If this were the case, all intersection numbers would be zero. This too is no good. Note that since these are non-compact cycles, even in the usual non-loaded case we cannot define intersection numbers for them. We must make at least one of them compact. This is the regularization we defined in §3. With this in mind, we are ready to begin. Let us evaluate the intersection number of $\mathrm{reg}I_j$ and $\check{I}_j$. Considering the orientation in Figure 7.2, we have

$$(\mathrm{reg}I_j) \cdot \check{I}_j = -\frac{1}{d_j} - 1 - \frac{1}{d_{j+1}} = -\frac{d_{j,j+1}}{d_j d_{j+1}}.$$

Here I deformed $\check{I}_j$ to have the shape of a sine curve. You can check that

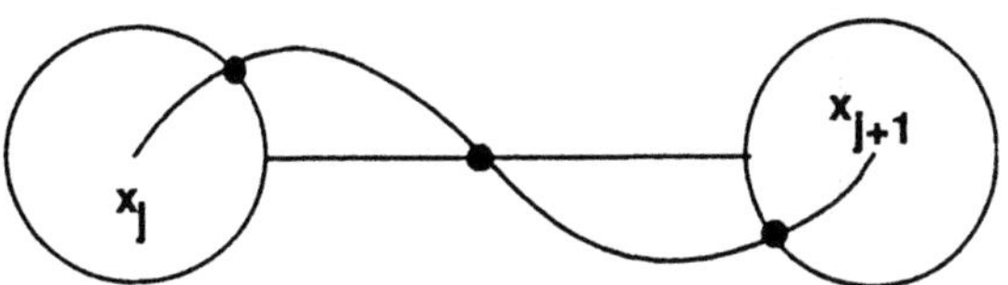

FIGURE 7.2. Intersection of $\mathrm{reg}I_j$ and I_j

the intersection number does not depend on the manner of deformation. If $|j - k| \geq 2$, then I_j and $\check{I}_k$ have no topological intersection. In this case the intersection number would be zero. The remaining cases can be seen in Figure 7.3. We have

$$(\mathrm{reg}I_j) \cdot \check{I}_{j-1} = \frac{c_j}{d_j},$$
$$(\mathrm{reg}I_{j-1}) \cdot \check{I}_j = \frac{1}{d_j}.$$

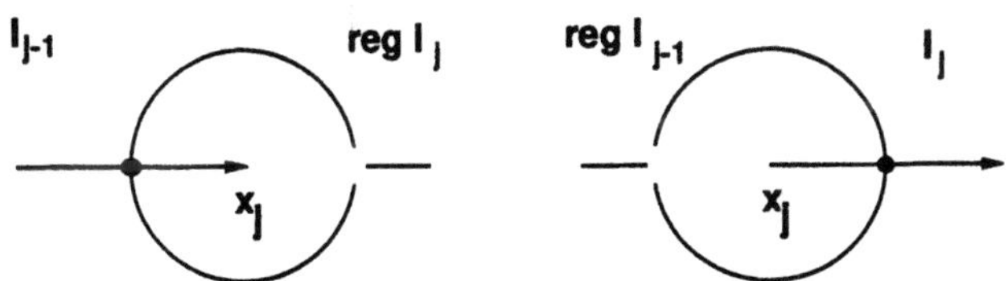

FIGURE 7.3. Intersection of $\mathrm{reg}I_j$ and I_{j-1}, and that of $\mathrm{reg}I_{j-1}$ and I_j,

Therefore the *intersection matrix*

$$Int(n+1,\alpha) = \begin{pmatrix} reg I_1 \\ \vdots \\ reg I_{n-1} \end{pmatrix} \cdot (\check{I}_1, \ldots, \check{I}_{n-1})$$

$$= \begin{pmatrix} reg I_1 \cdot \check{I}_1 & \cdots & reg I_1 \cdot \check{I}_{n-1} \\ \vdots & & \vdots \\ reg I_{n-1} \cdot \check{I}_1 & \cdots & reg I_{n-1} \cdot \check{I}_{n-1} \end{pmatrix}$$

is given by the following tri-diagonal matrix:

$$-\begin{pmatrix} d_{12}/d_1 d_2 & -1/d_2 & 0 & \cdots & & 0 & & 0 \\ -c_2/d_2 & d_{23}/d_2 d_3 & \cdots & & & 0 & & 0 \\ 0 & \vdots & & & & & & \vdots \\ \vdots & & & & \vdots & & & 0 \\ 0 & 0 & \cdots & d_{n-2,n-1}/d_{n-2}d_{n-1} & & -1/d_{n-1} \\ 0 & 0 & \cdots & 0 & -c_{n-1}/d_{n-1} & & d_{n-1,n}/d_{n-1}d_n \end{pmatrix}.$$

We have defined intersection numbers so that they are invariant under continuous deformations of the loaded cycles. Let's investigate the nature of $Int(n+1,\alpha)$ by considering what happens to the loaded cycles when the points $x_1, \ldots, x_n$ travel around and return, that is, the point $\overset{\bullet}{x} = (x_0, \ldots, x_n) \in X(n+1)$ travels along a loop γ in the configuration space $X(n+1)$. Accordingly, the cycles loaded with u change as follows:

$$\begin{pmatrix} reg I_1 \\ \vdots \\ reg I_{n-1} \end{pmatrix} \longrightarrow M(\gamma,\alpha) \begin{pmatrix} reg I_1 \\ \vdots \\ reg I_{n-1} \end{pmatrix},$$

where $M(\gamma,\alpha)$ is the circuit matrix along γ, and the cycles loaded with u^{-1} change in the following way:

$$(\check{I}_1, \ldots, \check{I}_{n-1}) \longrightarrow (\check{I}_1, \ldots, \check{I}_{n-1})^t M(\gamma, -\alpha).$$

During the deformation of cycles, the intersection numbers do not change, and we therefore have

$$M(\gamma,\alpha) Int(n+1,\alpha)^t M(\gamma,-\alpha) = Int(n+1,\alpha),$$

for all $\gamma \in \pi_1(X(n+1), \overset{\bullet}{x})$, which is equivalent to

$$^t M(\gamma,-\alpha) Int(n+1,\alpha)^{-1} M(\gamma,\alpha) = Int(n+1,\alpha)^{-1},$$

for all $\gamma \in \pi_1(X(n+1), \overset{\bullet}{x})$. If all the α_j are real, then since

$$M(\gamma, -\alpha) = \overline{M(\gamma, \alpha)} \quad \text{(complex conjugate)},$$

the matrix $Int(n+1, \alpha)^{-1}$ should be a constant multiple of the invariant hermitian form we discussed in the previous section. Let us check.

By induction, one can prove the following:

$$\det Int(n+1; \alpha) = (-1)^{n-1} \frac{d_{1\cdots n}}{d_1 \cdots d_n},$$

where

$$c_{jk\cdots} = c_j c_k \cdots, \quad d_{jk\cdots} = c_{jk\cdots} - 1.$$

This shows that the loaded cycles $I_1, \ldots, I_{n-1}$ are linearly independent provided that

$$\alpha_j \notin \mathbb{Z}, \quad \alpha_0 := -\sum_{j=1}^{n} \alpha_j \notin \mathbb{Z}.$$

The inverse matrix of $Int(n+1; \alpha)$ is $(-1/d_{1\cdots n})D$, where

$$D = \begin{pmatrix} d_1 d_{2\cdots n} & d_1 d_{3\cdots n} & d_1 d_{4\cdots n} & \cdots & d_1 d_n \\ d_1 c_2 d_{3\cdots n} & d_{12} d_{3\cdots n} & d_{12} d_{4\cdots n} & \cdots & d_{12} d_n \\ d_1 c_{23} d_{4\cdots n} & d_{12} c_3 d_{4\cdots n} & d_{123} d_{4\cdots n} & \cdots & d_{123} d_n \\ \vdots & \vdots & \vdots & & \vdots \\ d_1 c_{2\cdots n-1} d_n & d_{12} c_{3\cdots n-1} d_n & d_{123} c_{4\cdots n-1} d_n & \cdots & d_{1\cdots n-1} d_n \end{pmatrix}.$$

Note that the upper triangular part is obtained by transposing the lower part and then deleting the $c_{jk\cdots}$ and that the determinant of this matrix is

$$d_1 \cdots d_n (d_{1\cdots n})^{n-2}.$$

Note also if all the α_j are real, i.e., $|c_j| = 1$, then $(c_{12\cdots n})^{-1/2}D$ is hermitian. In any case, when $n = 3$, this matrix is identical to the invariant hermitian form given in the previous section.

8. A Review of the Modular Interpretation of $X(4)$

For four points $x_1, \ldots, x_4 \in \mathbb{P}^1$ and four parameters μ_j satisfying

$$\sum_{j=1}^{4} \mu_j = 2,$$

we consider the hypergeometric integrals

$$u_k(x) = \int_{I_k} \prod_{j=1}^{4} (t - x_j)^{-\mu_j} dt, \quad k = 1, 2.$$

Recall that if one normalizes as

$$x_1 = 0, \quad x_2 = 1, \quad x_3 = 1/x, \quad x_4 = \infty,$$

then these integrals are solutions of the hypergeometric equation $E(a, b, c)$, where

$$-\mu_1 = a - 1, \quad -\mu_2 = c - a - 1, \quad -\mu_3 = -b, \quad -\mu_4 = b - c.$$

I hope you remember that, using the above normalization, the three angles of the Schwarz triangle at the images (under the Schwarz map) of the points

$$x = 0 \quad \text{(in which case } x_1 = x_2 \neq x_3 = x_4\text{)},$$
$$x = 1 \quad \text{(in which case } x_1 = x_4 \neq x_2 = x_3\text{)},$$
$$x = \infty \quad \text{(in which case } x_1 = x_3 \neq x_2 = x_4\text{)}$$

are π times

$$|1 - c|, \quad |c - a - b|, \quad |a - b|,$$

which are equal to

$$|\mu_1 + \mu_2 - 1| = |\mu_3 + \mu_4 - 1|, \quad |\mu_1 + \mu_4 - 1| = |\mu_2 + \mu_3 - 1|,$$
$$|\mu_1 + \mu_3 - 1| = |\mu_2 + \mu_4 - 1|,$$

respectively. So we know how to treat the four points equally.

Assume the μ_j are rational numbers and there are integers $n_{ij} = n_{ji}$ $(i \neq j)$ greater than 1 or equal to ∞ such that

$$|\mu_i + \mu_j - 1| = \frac{1}{n_{ij}} \quad (i \neq j).$$

Note that $n_{ij} = n_{kl}$ if $\{i, j, k, l\} = \{1, 2, 3, 4\}$ and the number

$$s(\mu) := \sum_{j=1,\ j \neq i}^{4} \frac{1}{n_{ij}}$$

is independent of $i = 1, \ldots, 4$. Also note that for any three integers

$$2 \leq n_{12} = n_{34}, \ n_{23} = n_{14}, \ n_{31} = n_{24} \leq \infty, \quad n_{ij} = n_{ji} \ (i \neq j),$$

there are rational numbers $\mu_1, \ldots, \mu_4$ (not unique) satisfying the above equalities.

While the integrals $u_1(x)$ and $u_2(x)$ are not defined on the configuration space $X(4)$, the ratio $u_1(x) : u_2(x)$ defines a map $X(4) \to \mathbb{P}^1$. This is the Schwarz map.

Let

$$D = \mathbb{P}^1 \quad \text{when} \quad s(\mu) > 1,$$
$$D = \mathbb{C} \subset \mathbb{P}^1 \quad \text{when} \quad s(\mu) = 1,$$
$$D = H \subset \mathbb{P}^1 \quad \text{when} \quad s(\mu) < 1.$$

The Schwarz triangle in D is bounded by three geodesics (arcs and segments) with angles π/n_{12}, π/n_{23}, π/n_{31}. Let $\Gamma(\mu) \subset Aut(D)$ be the Schwarz triangle group (the doubled triangle is its fundamental domain). This is the projective monodromy group of the corresponding hypergeometric equation. The triple (n_{12}, n_{23}, n_{31}) is called the *characteristic* of the triangle group. Then the period map

$$x \mapsto u_1(x) : u_2(x) \in \mathbb{P}^1$$

gives the injection

$$X(4) \longrightarrow D/\Gamma(\mu)$$

and the isomorphism

$$\overline{X(4)} \overset{\sim}{\longrightarrow} \overline{D/\Gamma(\mu)},$$

where $\overline{D/\Gamma(\mu)}$ is the space added to $D/\Gamma(\mu)$ by cusps (the $\Gamma(\mu)$-orbits of the vertices of the Schwarz triangles on ∂D).

You see, this section is a paraphrase of Chapter III.

9. The Relation between $s(\mu)$ and the Hermitian form $H(\alpha)$

You would naturally guess that, when the $\mu_j = -\alpha_j$ are real, the quantity $s(\mu)$ and the invariant Hermitian form $H(\alpha)$ must be related. Of course your guess is right. We show in this section the following

PROPOSITION 9.1. *Under the assumption*

$$|\mu_j - \mu_k| \le 1 \quad 1 \le j \ne k \le 4,$$

the Hermitian form $H(\alpha)$ is definite, degenerate, or indefinite if and only if $s(\mu) > 1, = 1$, or < 1, respectively. (Note that this assumption holds if

$$|\mu_i + \mu_j - 1| \le \frac{1}{2} \quad (i \ne j),$$

which is assumed in the previous section.)

Put

$$a_1 = \mu_2 + \mu_3 - 1, \quad a_2 = \mu_3 + \mu_1 - 1, \quad a_3 = \mu_1 + \mu_2 - 1.$$

The function

$$S(a) = |a_1| + |a_2| + |a_3| \quad (= s(\mu))$$

of $a = (a_1, a_2, a_3)$ takes the value 1 exactly on the boundary of the octahedron in the a-space bounded by the eight planes

$$\pm a_1 \pm a_2 \pm a_3 = 1.$$

One can readily see that

$$S(a) > 1, \quad = 1, \quad \text{or} \quad < 1$$

if a is outside, on, or inside this octahedron. Since we have

$$|a_1 - a_2| = |\mu_1 - \mu_2|,$$

$$|a_1 + a_2| = |\mu_3 - \mu_4|,$$

the assumption in the proposition defines a *rhombic dodecahedron* inscribing this octahedron. The twelve faces of this semi-regular dodecahedron are rhombi meeting four by four at six of its fourteen vertices and three by three at the eight remaining ones. (See Figure 9.1.) The long diagonals of the rhombic faces are the edges of this octahedron.

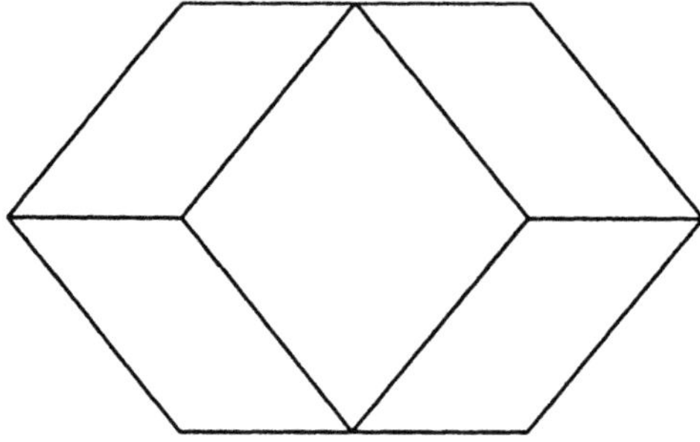

FIGURE 9.1. Top view of the rhombic dodecahedron sitting on one of its twelve faces

On the other hand, the real valued function in α_j (mod 1)

$$\det H(\alpha) = \det \frac{1}{\sqrt{c_1 c_2 c_3}} \begin{pmatrix} d_{12} d_3 & -d_1 c_2 d_3 \\ -d_1 d_3 & d_1 d_{23} \end{pmatrix}$$

$$= \frac{d_1 d_2 d_3}{c_1 c_2 c_3} d_{123} = -d_1 d_2 d_3 d_4$$

(where $c_j = \exp 2\pi i \alpha_j$ and $d_{jk...}$ are defined in §§6 and 7) vanishes only when $d_j = 0$ for some $j = 1, \ldots, 4$. We consider, without loss in generality, the case $d_4 = 0$. This identity is equivalent to

$$c_1 c_2 c_3 = \exp 2\pi i(\mu_1 + \mu_2 + \mu_3) = 1,$$

and to

$$\mu_1 + \mu_2 + \mu_3 = \frac{1}{2}\{a_1 + a_2 + a_3\} + \frac{3}{2} \in \mathbb{Z}.$$

Thus under the assumption in the proposition, this is equivalent to

$$a_1 + a_2 + a_3 = \pm 1.$$

Thus H, as a function of a in the said semi-regular dodecahedron, vanishes exactly on the boundary of the octahedron and takes negative (resp. positive) values inside (resp. outside) the octahedron. This proves the proposition.

10. Periods of Curves

The Schwarz map was defined as the ratio of two linearly independent solutions of the hypergeometric differential equation, which can also be presented as the ratio of two hypergeometric integrals. There is another interpretation that smells more algebro-geometric.

Let us give a typical example. For four distinct points $x_1, \ldots, x_4$ on $\mathbb{P}^1$, there is a unique double cover S_x of $\mathbb{P}^1$ branching at the four points:

$$S_x : s^2 = \prod_{j=1}^{4}(t - x_j).$$

Consider the 1-form

$$\eta(x) := \frac{dt}{\sqrt{\prod_{j=1}^{4}(t - x_j)}}$$

with parameter x. This (precisely speaking, its pull-back under the projection $S_x \ni (s,t) \mapsto t \in \mathbb{P}^1$) can be seen as a holomorphic 1-form on S_x. Since S_x is homeomorphic to the torus $S^1 \times S^1$, there are two linearly independent cycles, say $c_1 = c_1(x)$ and $c_2 = c_2(x)$. The integrals

$$u_k(x) = \int_{c_k(x)} \eta(x)$$

are called periods of S_x. If we normalize x as

$$x_1 = 0, \quad x_2 = 1, \quad x_3 = \infty, \quad x_4 = x,$$

then these integrals are linear independent solutions of the hypergeometric equation $E(1/2, 1/2, 1)$. The Schwarz map in this case is defined by the ratio of two (linearly independent) periods. In this sense one can call the Schwarz map a *period map*.

Now we turn to more general cases. Consider the family of curves

$$S_x : s^d = \prod_{j=1}^{4} (t - x_j)^{d\mu_j},$$

where d is the smallest common denominator of the rational numbers μ_j and the following two periods of S_x:

$$u_k(x) = \int_{c_k(x)} \prod_{j=1}^{4} (t - x_j)^{-\mu_j} dt \quad (k = 1, 2),$$

where the $c_k(x)$ are suitable cycles of the curve S_x, for example, the lifts of the double contour loop around (x_1, x_2) and (x_2, x_3). Notice that the genus of the curve can be very high and that linear independence (over $\mathbb{Z}$) of the two cycles is not enough to produce a non-constant period map. At any rate, $u_1(x)$ and $u_2(x)$ are periods of the curve S_x, and the Schwarz map is defined by these periods of the family of the curves.

11. Excuse for My Using Many Kinds of Parameters

For the hypergeometric functions, series, integrals and differential equations, I use several kinds of parameters such as α_j, β_j, μ_j and $a, b, c, \ldots$ You may blame me for this disorganization. I would like to explain, in this section, why I use different parameters and then I make clear the relation between these. As you will see, historical usage and notational convenience of the object in question require various parameters.

(1) When I speak of the hypergeometric series, I would like to follow the custom to represent this as

$$F(a, b, c; x) = \sum \frac{(a, n)(b, n)}{(c, n)(1, n)} x^n.$$

Thus I use the parameters a, b and c.

(2) When I regard the hypergeometric integrals of type $(k, n) = (2, n)$ as periods of the curves defined by $s^d = \prod_{j=1}^{n} (t - x_j)^{d\mu_j}$, I would like to write the integrals as

$$\int \prod_{j=1}^{n} (t - x_j)^{-\mu_j} dt.$$

Thus I use the parameters μ_j satisfying

$$\mu_1 + \cdots + \mu_n = 2 \quad (= k),$$

as is done in [DM1, 2]. The relation between the μ_j and (a, b, c) is given in §8 (see also §2 of Chapter VI).

(3) When I present, in §§1 and 5 of Chapter VIII, the system of differential equations satisfied by the hypergeometric integrals of type (k, n), I would like to write the integral as

$$\int \prod_{j=1}^{n} L_j^{\alpha_j - 1} dt.$$

Thus I use the parameters α_j satisfying

$$\alpha_1 + \cdots + \alpha_n = n - k,$$

as is done in [Gel]. Note that

$$\mu_j = 1 - \alpha_j.$$

(4) When we are concerned only about the monodromy behavior of the multi-valued integrands in §§4 – 7, I write the integrands as

$$\prod (t - x_j)^{\alpha_j},$$

because there would be no point to write $\alpha_j - 1$ in place of α_j. In this case, the α_j has no restraint as in the previous case and $-\alpha_1 - \cdots - \alpha_n$ is considered to be the exponent at infinity.

(5) When I would like to transform the hypergeometric integral into a $GL(k) \times H_n$-invariant form, it is very important to introduce the parameters β_j satisfying

$$\beta_1 + \cdots + \beta_n = 0,$$

as we did in §1.

12. Toward Generalizations

If you are a Gauss or an Euler, you should do mathematics as they did. Unfortunately, most mathematicians are not so gifted. So I will assume that you are an ordinary mathematician. What we do is to make *evolutions*, or, perhaps I should say analogies and generalizations (although I hate these words, because they sound so shabby).

I will assume also that you are impressed by the story of modular interpretations. Let us think of its evolutions. It is quite natural (it may appear to be too easy) to consider $n (\geq 5)$ points $x_1, \ldots, x_n \in \mathbb{P}^1$, instead of four points, and the family of curves

$$S_x : s^d = \prod_{j=1}^{n} (t - x_j)^{d\mu_j}, \qquad (\textstyle\sum \mu_j = 2),$$

and think of what happens with regard to the period map defined on the configuration space $X(n)$. The first attempt on this problem was made by E. Picard for the case

$$n = 5, \quad d = 3, \quad 3\mu = (2,1,1,1,1).$$

More general cases were studied by many authors ([Trd], [DM], [Shi], etc). Some of their results will be given in Chapter VI. These give isomorphisms of the configuration spaces $X(n)$ and the quotients of the n-ball

$$\mathbb{B}_n = \{ z_0 : \cdots : z_n \in \mathbb{P}^n \mid |z_0|^2 - |z_1|^2 - \cdots - |z_n|^2 > 0 \}$$

under discontinuous groups.

Are there any stories about the configuration space $X(k,n)$ for k greater than 2? Well, yes. In fact, the main purpose of this book is to tell the story of $X(3,6)$. Why $X(3,6)$? Because this is the only such story I am able to tell (cf. [SYY]).

Comment: I believe that developments of mathematics are made by generalizations followed by specializations. You should jump and fly like an eagle and then fly down toward a game. To establish a story of modular interpretation of $X(3,6)$ we must jump at least as a grasshopper.

Part 2

The Story of the Configuration Space $X(2,n)$ of n Points on the Projective Line

CHAPTER V

The Configuration Space $X(2,5)$

The configuration space $X(4) = X(2,4)$ was studied in Chapter I; it is a simple space. Indeed it is the unique configuration space which is 1-dimensional. In the next chapter we study modular interpretations of the configuration spaces $X(2,n)(n \geq 5)$. Recall that the configuration space of n distinct points on the line is given as

$$X(2,n) = GL(2)\backslash M^*(2,n)/H_n,$$

where $M^*(2,n)$ is the space of $2 \times n$ complex matrices such that any 2×2-minor does not vanish. Since every point $x \in X(2,n)$ can be represented by

$$\begin{pmatrix} 1 & 0 & 1 & 1 & \cdots & 1 \\ 0 & 1 & 1 & x_1 & \cdots & x_{n-3} \end{pmatrix},$$

the space can be seen as an open set in $\mathbb{C}^{n-3}$:

$$X(2,n) \cong \{(x_1,\ldots,x_{n-3}) \in \mathbb{C}^{n-3} \mid x_j \neq 0,\ 1,\ x_k(j \neq k)\}.$$

You may think that this space is simple enough, just the complement of hyperplanes in $\mathbb{C}^{n-3}$. However, you shall know how difficult it is to truly understand this space.

In this chapter we study the least complicated space not yet considered, $X(2,5)$, a 2-dimensional space. I would like to present several nice geometries on $X(2,5)$. To understand the rest of this book, the present chapter is not quite necessary, at least logically. But these geometries will give you a feeling of the configuration spaces. We will later study the space $X(3,6)$. This is our main playground. One cannot expect too much to visualize it, however, since it is a complex 4-dimensional space.

1. Juzu Sequences

Suppose we have n beads, $1,\ldots,n$ each painted in a different color. A *juzu* is a closed string of such beads (Figure 1.1).

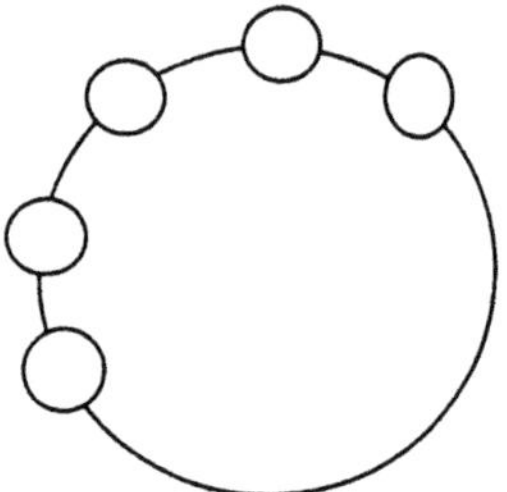

FIGURE 1.1. A juzu

There are several ways to make a juzu. You first arrange the beads on a line and glue the two ends (Figure 1.2). The following arrangements of

FIGURE 1.2. n beads on a line

beads, for instance, give the same juzu:

$$12\cdots n, \quad 23\cdots n1, \quad \cdots, \quad n1\cdots(n-1),$$
$$n\cdots 21, \quad 1n\cdots 32, \quad \cdots, \quad (n-1)\cdots 1n.$$

A *juzu sequence* of $1,\ldots,n$ (or simply an n-juzu) is the class of sequences of the n beads which form the same juzu; there are

$$\frac{n!}{2n} = \frac{(n-1)!}{2}$$

such classes. For example, there is only one 3-juzu

$$123 = 231 = 312 = 321 = 132 = 213,$$

there are three 4-juzus

$$1234, \quad 1423, \quad 1243$$

and there are twelve 5-juzus.

Note. In Buddhism, a *juzu* consisting of 108 beads, each representing a human desire, is used as a rosary. In this chapter, we have only five desires, say to

$$1 = \text{sleep}, \quad 2 = \text{eat}, \quad 3 = \text{drink}, \quad 4 = \text{make love}, \quad 5 = \text{do mathematics}.$$

Let us restrict ourselves to consider 5-juzus. There is an involution, called *duality*, on the set of 5-juzus defined by

$$\text{abcde} \quad \longleftrightarrow \quad \text{acebd}.$$

This can be visualized as in Figure 1.3. Note that on the set of n-juzus ($n \neq 5$) we cannot define such an involution; in this sense, 5 is a magic number. Two 5-juzus are said to be *even* with respect to each other if

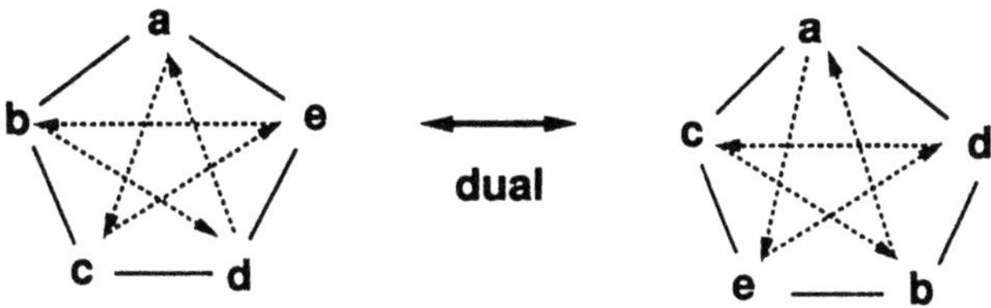

FIGURE 1.3. The duality of 5-juzus

one is an even permutation of the other. Otherwise they are said to be *odd* with respect to each other. With respect to any given 5-juzu, there are six even ones and six odd ones. Check that dual juzus are odd with respect to each other. Two juzus are said to be *adjacent* if one is obtained from the other by switching two neighboring beads. A juzu is adjacent to five juzus with respect to which it is odd; the remaining juzu with respect to which it is odd is its dual.

It is natural to arrange the twelve juzus nicely so that all adjacent juzus are situated in adjacent positions. Try to do this yourself; an answer will be given in §3.

The real projective plane. The projective plane $\mathbb{P}^2_{\mathbb{R}}$ is defined in Chapter I. It is a space of ratios $\{\lambda_1 : \lambda_2 : \lambda_3\}$, something studied in grammar school mathematics. But you are now asked to visualize its geometric image. Here are some hints:

1) It is compact. A line, which is by definition the zero locus of a linear form in the λ_i, is isomorphic to $\mathbb{P}^1_{\mathbb{R}}$, a circle.

2) Two lines, if they do not coincide, meet at a point.

3) If you remove any line, you are left with the affine plane $\mathbb{R}^2$.

4) If you walk straight along a line with your dog on your right, you eventually come back to the starting point, but with your dog on your left; i.e., a tubular neighborhood of a line is a *Möbius strip*.

4') If you remove a disc, then you are left with a Möbius strip.

∞) Without permitting self-intersections, you cannot make a model in our space $\mathbb{R}^3$. If you permit self-intersections, you can do it. I advice that you do not try to check this; it is highly non-trivial. (See [Apé].)

2. Blowing Up and Down

Consider two real planes S and U coordinatized by (s,t) and (u,v), respectively, and a map $p : S \to U$ defined by

$$p : (s,t) \mapsto (u,v) = (s, st).$$

Any line $\{t = \text{constant}\}$ in S is mapped into a line in U passing through the origin. Any point on the line $\{s = 0\}$ is mapped to the origin of U

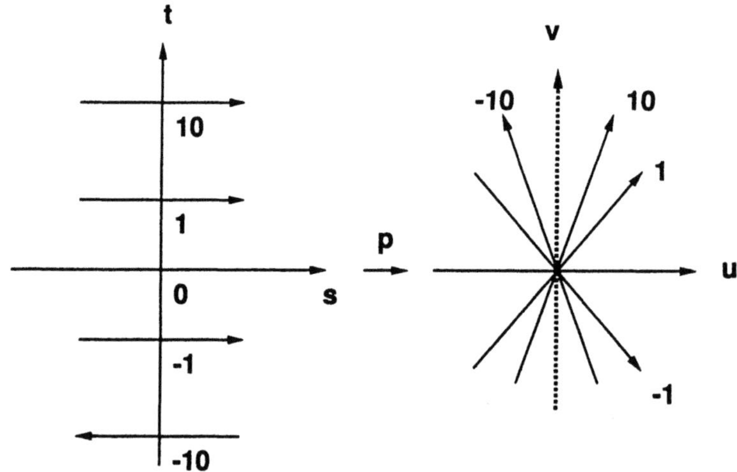

FIGURE 2.1. Blowing down p

(see Figure 2.1).

In order to see what happens when t tends to $\pm\infty$, we consider the map (see Figure 2.2)

$$p' : S' \ni (s',t') \mapsto (u,v) = (s't', s') \in U.$$

Glue S and S' together by identifying

$$(s,t) \in S - \{t = 0\} \quad \text{with} \quad (s',t') \in S' - \{t' = 0\}$$

if

$$tt' = 1, \quad s' = st \quad (\text{or equivalently } s = s't')$$

to obtain $\tilde{U}$. One can readily check that $p(s,t) = p'(s',t')$ on $S \cap S'$; let $\tilde{p} : \tilde{U} \to U$ be the glued map. Note that

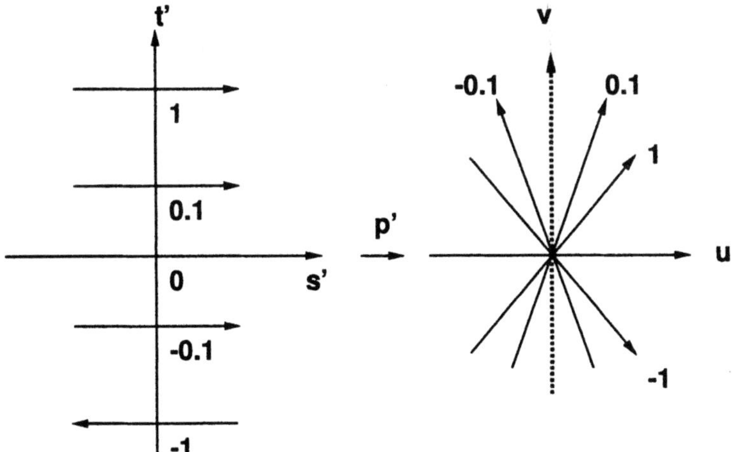

FIGURE 2.2. Blowing down p'

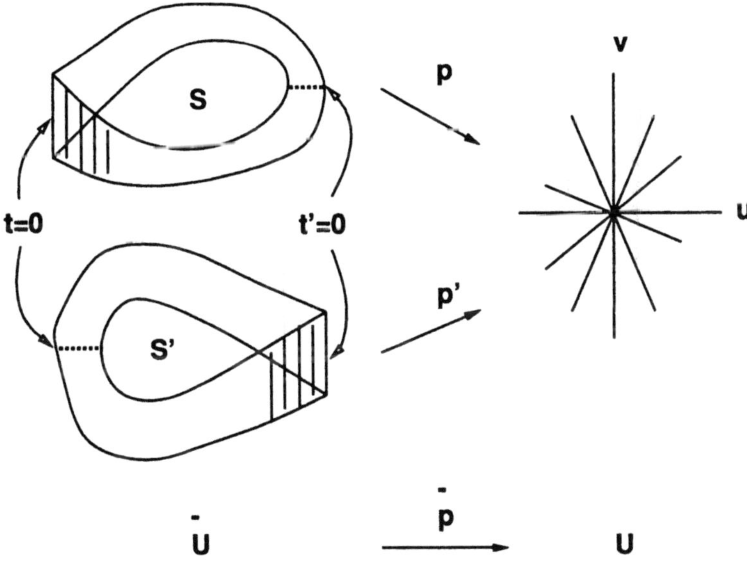

FIGURE 2.3. Blowing down $\tilde{p}$

$$\tilde{p}^{-1}(0,0) = \{(s,t) \in S \mid s = 0\} \cup \{(s',t') \in S' \mid s' = 0\}$$

form a circle (Figure 2.3), and that we have the isomorphism

$$\tilde{p} : \tilde{U} - \tilde{p}^{-1}(0,0) \xrightarrow{\sim} U - \{(0,0)\}.$$

We call $\tilde{U}$ the *blowing up* of U at the origin, and $\tilde{p}$ the *blowing down* of $\tilde{U}$ along the circle. At the beginning, we assumed that U is a whole plane, but now we know that blowing up is a local operation. You blow up a disc at a point in the disc and get a Möbius strip; the point is replaced by a projective line, a circle.

We shall blow up a point when three lines intersect. The six domains bounded by the three lines are transformed by the blowing up as shown in Figure 2.4.

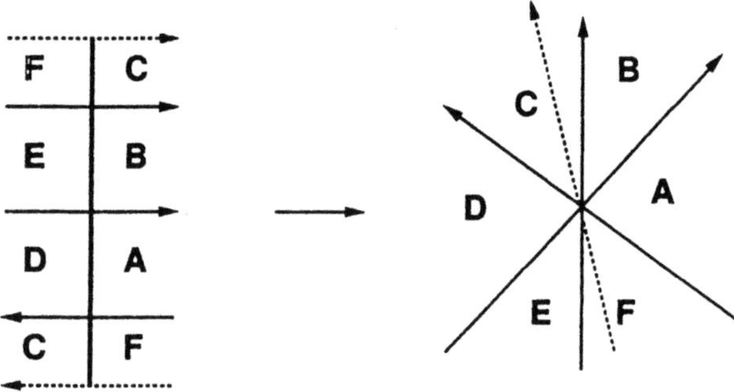

FIGURE 2.4. A, B, C, D, E, F

3. The Democratic Compactification $\overline{X}_{\mathbb{R}}$

Recall that the configuration space of five distinct points on the "real" projective line is defined by

$$X_{\mathbb{R}} = PGL(2, \mathbb{R})\backslash\{(\mathbb{P}^1_{\mathbb{R}})^5 - \Delta_{\mathbb{R}}\},$$

where the action is diagonal, and

$$\Delta_{\mathbb{R}} = \{(x_1, \ldots, x_5) \in (\mathbb{P}^1_{\mathbb{R}})^5 \mid x_i = x_j \text{ for some } i \neq j\}.$$

In order to see $X_{\mathbb{R}}$ situated nicely in a compact surface, the standard procedure is to make a democratic projective embedding. We can make such an embedding, and it is useful and interesting in its own right (cf. [AVY1]), but since our situation is fairly simple, a democratic compactification can be given directly by adding suitable degenerate configurations as follows. Consider

$$\overline{X}_{\mathbb{R}} = PGL(2, \mathbb{R})\backslash\{(\mathbb{P}^1_{\mathbb{R}})^5 - \Delta'_{\mathbb{R}}\},$$

where

$$\Delta'_{\mathbb{R}} = \{(x_1, \ldots, x_5) \in \Delta_{\mathbb{R}} \mid \text{at least three points coincide}\}.$$

Let $L_{\mathbb{R}}(ij) = L_{\mathbb{R}}(ji)$ be a subset of $\overline{X}_{\mathbb{R}}$ defined by

$$x_i = x_j, \quad 1 \leq i \neq j \leq 5.$$

$L_{\mathbb{R}}(ij)$ is isomorphic to the compactification of $X_{\mathbb{R}}(2,4)$; it is a circle. We have

$$\overline{X}_{\mathbb{R}} - \cup_{i,j} L_{\mathbb{R}}(ij) = X_{\mathbb{R}}.$$

Let $U_{\mathbb{R}}(i)$ be the subset of $\overline{X}_{\mathbb{R}}$ consisting of $x = (x_1, \ldots, x_5)$ such that

$$x_j \neq x_i, \quad 1 \leq j \leq 5, \quad j \neq i.$$

Notice that

$$\overline{X}_{\mathbb{R}} - U_{\mathbb{R}}(i) = \cup_{j \neq i} L_{\mathbb{R}}(ij) \quad \text{(disjoint sum)}.$$

Normalizing x by setting $x_i = \infty$, one can regard $U_{\mathbb{R}}(i)$ as part of $\mathbb{P}^2_{\mathbb{R}}$, as seen in Figure 3.1. One finds four holes which correspond to the four

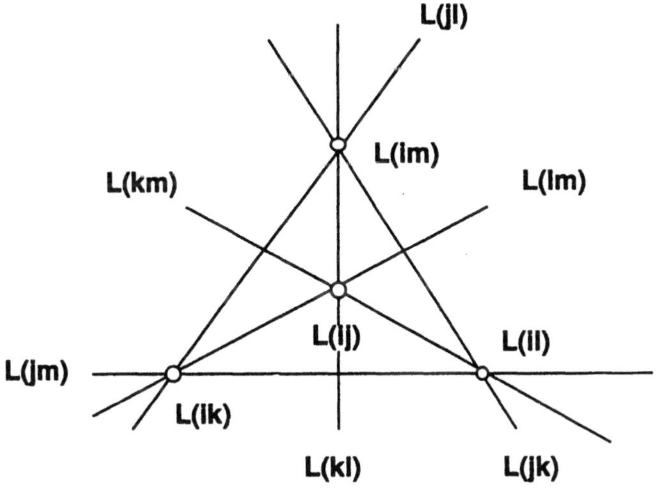

FIGURE 3.1. $U_{\mathbb{R}}(i) \subset \mathbb{P}^2_{\mathbb{R}}$

curves $\{L_{\mathbb{R}}(ij) \mid j \neq i\}$. The surface $\overline{X}_{\mathbb{R}}$ is now obtained by glueing the five $U_{\mathbb{R}}(i)$, $i = 1, \ldots, 5$.

Note that the projective plane in which $U_{\mathbb{R}}(i)$ lives is the blowing down of $\overline{X}_{\mathbb{R}}$ along the four curves $\{L_{\mathbb{R}}(ij) \mid j \neq i\}$, that is, the space $\overline{X}_{\mathbb{R}}$ is the blowing up of the projective plane at four points *in general position*, with no three points lying on a line.

The configuration space

$$X_{\mathbb{R}} = \overline{X}_{\mathbb{R}} - \cup_{i,j} L_{\mathbb{R}}(ij)$$
$$= U_{\mathbb{R}}(i) - \cup_{k,j\neq i} L_{\mathbb{R}}(kj)$$

is now divided into twelve disjoint connected components. Let us call these *chambers*. Though each chamber looks like a triangle in $U_{\mathbb{R}}(i)$, it is bounded by five edges in $\overline{X}_{\mathbb{R}}$. Though I want to draw a picture of $\overline{X}_{\mathbb{R}}$, I cannot make a neat one since it contains many Möbius strips. (I can visualize the orientable double of $\overline{X}_{\mathbb{R}}$, see [AY].) Each chamber can be labeled by a 5-juzu sequence according to the arrangement of the five points $x_1, \ldots, x_5$ on $\mathbb{P}^1_{\mathbb{R}}$. For a juzu J, denote the corresponding chamber by $D(J)$. In Figure 3.2, the chamber $D(12345)$ is shown with its dual chamber. For example, when $J = (12345)$, the chamber $D(J)$ is

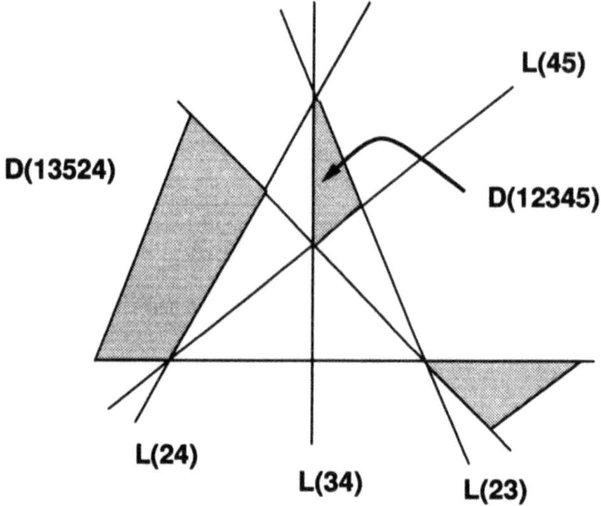

FIGURE 3.2. Chambers $D(J)$ in $U_{\mathbb{R}}(1) \subset \mathbb{P}^2_{\mathbb{R}}$

a pentagon bounded by the five curves

$$L_{\mathbb{R}}(12), \ L_{\mathbb{R}}(34), \ L_{\mathbb{R}}(15), \ L_{\mathbb{R}}(23), \text{ and } \ L_{\mathbb{R}}(45).$$

Note the order of the five curves (see Figure 3.3).

4. The Democratic Compactification $\overline{X}$

We now consider the configuration space $X = X(2,5)$ of five points on the *complex* projective line. To describe this space, we need only remove

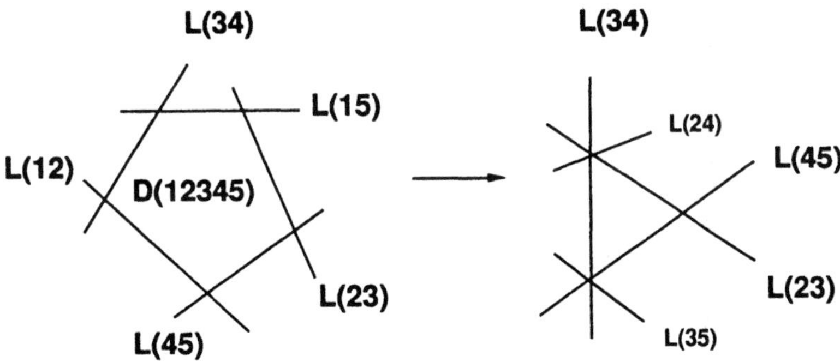

FIGURE 3.3. $D(12345)$

the suffix $\mathbb{R}$ from the statements in the previous two sections. Let us reintroduce some of these statements.

The configuration space X of five distinct colored points on the complex projective line $\mathbb{P}^1$ is defined by

$$X = PGL(2, \mathbb{C})\backslash\{(\mathbb{P}^1)^5 - \Delta\},$$

where the action is diagonal, and

$$\Delta = \{(x_1, \ldots, x_5) \in (\mathbb{P}^1)^5 \mid x_i = x_j \text{ for some } i \neq j\}.$$

The symmetric group S_5 acts cn X as permutations of five points. A smooth compactification of X is given by

$$\overline{X} = PGL(2, \mathbb{C})\backslash\{(\mathbb{P}^1)^5 - \Delta'\},$$

where

$$\Delta' = \{(x_1, \ldots, x_5) \in \Delta \mid \text{at least three points coincide}\}.$$

$\overline{X}$ is isomorphic to the surface obtained by blowing up four points in general position (i.e., no three points lying on a line) on the complex projective plane. Thus there are ten rational curves in $\overline{X}$ with self-intersection number -1 (if you do not know the term 'self-intersection number', don't worry). Let us name these curves as follows:

$$L(ij) = L(ji) := \text{ the orbit of } \{x_i = x_j\}, \quad 1 \leq i, j \leq 5, \quad i \neq j.$$

Two such distinct curves do not meet or meet normally at a single point:

$$L(ij) \cap L(pq) = \text{ a point } \iff \{i, j\} \cap \{p, q\} = \emptyset.$$

Note that

$$X = \overline{X} - \cup L(ij).$$

The smooth action of S_5 on $\overline{X}$ induces a transitive action on the set $\{L(ij)\}$.

You might think that the configuration space X and its compactification $\overline{X}$ are simple objects presented clearly enough. If so, please answer the question

"How do the real manifolds $X_{\mathbb{R}}$ and $\overline{X}_{\mathbb{R}}$ live in X and $\overline{X}$?"

The configuration space $\overline{X}_{\mathbb{R}}(2,4) \cong \mathbb{P}^1_{\mathbb{R}}$ divides $\overline{X}(2,4) \cong \mathbb{P}^1$ into two pieces, of course, but our space $\overline{X}_{\mathbb{R}}(2,5)$ does not. Let us try to see what is happening.

The manifolds $X_{\mathbb{R}}$, $\overline{X}_{\mathbb{R}}$ and $L_{\mathbb{R}}(ij)$ are the sets of fixed points in X, $\overline{X}$ and $L(ij)$, respectively, of the involution c induced by the complex conjugation

$$c : (x_1, \ldots, x_5) \mapsto (\overline{x}_1, \ldots, \overline{x}_5).$$

Let $cr_j(x)$ be a cross-ratio of four points $\{x_i \mid 1 \le i \le 5,\ i \ne j\}$. For four points on the projective line, if no three points coincide, one can define six cross-ratios.

N. Takayama, a friend of mine, found that

$$\overline{X} - \cup_{j=1}^{5}\{x \in \overline{X} \mid \Im cr_j(x) = 0\},$$

where $\Im \bullet$ is the imaginary part of $\bullet$, is the disjoint union of twenty simply connected real 4-dimensional open subsets (let us call them chambres) of $\overline{X}$. He also found that the symmetric group S_5 acts transitively on the twenty chambres. (Look out! We use the French spelling, *chambre*, so please pronounce it as the French.) Note that the set $\{x \in \overline{X} \mid \Im cr_j(x) = 0\}$ does not depend on the choice of the cross-ratio.

I must confess that neither I nor Takayama know any intrinsic explanation behind these facts, which can be shown by brute force computation. In this chapter, I will not give proofs but will instead explain in enough detail that the reader can easily believe. The propositions in this chapter will be called "facts."

FACT 4.1. *For each curve $L(ij)$, there are exactly two chambres that do not touch the curve, i.e., their closures do not intersect $L(ij)$. These two chambres are permuted by the involution c. Let us call them $C(ij)^+ = C(ji)^+$, $C(ij)^- = C(ji)^-$, and put*

$$\mathcal{C} = \{C(ij)^+, C(ij)^- \mid 1 \le i < j \le 5\}.$$

5. The Orbifold $\overline{X}/\langle c\rangle$

In this section we study the quotient space $\overline{X}/\langle c\rangle$. Since $\overline{X}_{\mathbb{R}}$ is the set of fixed points of c, we regard $\overline{X}_{\mathbb{R}}$ as part of $\overline{X}/\langle c\rangle$. Recall that the configuration space

$$X_{\mathbb{R}} = \overline{X}_{\mathbb{R}} - \cup L_{\mathbb{R}}(ij)$$

of five distinct points on the real projective line consists of twelve disjoint chambers (look out! English spelling) $D(J)$, each corresponding to a juzu sequence J of five letters.

Let π be the projection $\overline{X} \to \overline{X}/\langle c\rangle$ and

$$C(ij) := \pi C(ij)^{+} = \pi C(ij)^{-}, \quad \underline{\mathcal{C}} := \{C(ij)\}.$$

We are going to construct a graph $\underline{G} = \underline{G}(\underline{\mathcal{C}})$ whose vertices are the ten elements of $\underline{\mathcal{C}}$ such that vertices are joined with an edge when the intersection of their closures (in $\overline{X}/\langle c\rangle$) is real 3-dimensional.

FACT 5.1. *$C(ij)$ is connected with exactly three vertices, $C(pq), C(qr)$ and $C(rp)$, where $\{i,j,p,q,r\} = \{1,\ldots,5\}$.*

This fact is fairly surprising. An n-simplex is bounded by $n + 1$ walls: a triangle by three lines and a tetrahedron by four planes. This chambre $C(ij)$ however, is bounded by three walls only. This implies that the walls are fairly sharply curved! To give you an idea of how this can be, I normalize as follows:

$$x_1 = \infty, \quad x_2 = 0, \quad x_3 = 1, \quad x_4 = x, \quad x_5 = y,$$

and put

$$\varphi_1 = \Im cr_1(x) = \Im\frac{y(x-1)}{x-y}, \quad \varphi_2 = \Im cr_2(x) = \Im\frac{x-1}{x-y},$$

$$\varphi_3 = \Im cr_3(x) = \Im\frac{y}{x}, \quad \varphi_4 = \Im cr_4(x) = \Im y, \quad \varphi_5 = \Im cr_5(x) = \Im x.$$

The inequalities

$$\varphi_1 < 0, \quad \varphi_2 > 0, \quad \varphi_3 > 0, \quad \varphi_4 < 0, \quad \varphi_5 > 0$$

define a chambre. You can check that the first two are redundant. Thus the chambre is defined by only the latter three:

$$\Im\frac{y}{x} > 0, \quad \Im y < 0, \quad \Im x > 0;$$

indeed we have

$$\Im y(x-1)(\overline{x}-\overline{y}) = |x|^2\Im y - \Im y\overline{x} - |y|^2\Im x,$$

$$\Im(x-1)(\overline{x}-\overline{y}) = \Im x + \Im\frac{y}{x} - \Im y.$$

Put

$$x = se^{i\alpha}, \quad y = te^{i\beta}, \quad s,t > 0, \quad 0 \le \alpha, \beta < 2\pi.$$

Then the chambre is the product of the s-space ($\cong \mathbb{R}_{>0}$), the t-space ($\cong \mathbb{R}_{>0}$), and the triangle

$$\{(\alpha,\beta) \mid \pi < \beta < 2\pi,\ 0 < \alpha < \pi,\ 0 < \beta - \alpha < \pi\}.$$

Let us visualize $\underline{G}$ by putting the $C(ij)$ on the vertices of the regular dodecahedron, same ones on two antipodal vertices of the dodecahedron. The edges of $\underline{G}$ are the edges of the dodecahedron. Then identify two antipodal points; the graph is now drawn on the dodecahedral (real) projective plane. There are two ways, up to the extended dodecahedral group ($\cong S_5$), to do this, and either will do (Figure 5.1). The difference

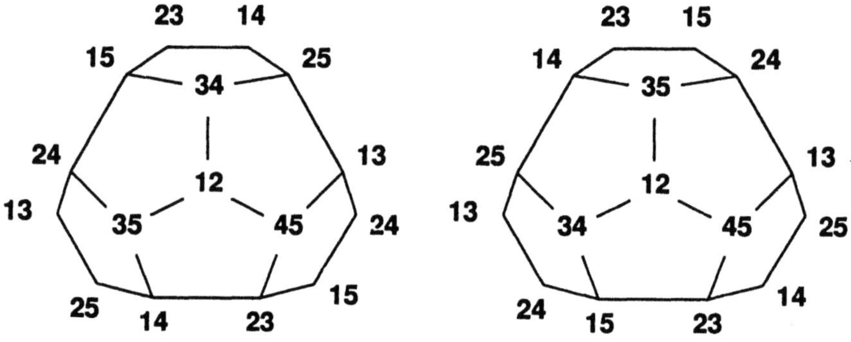

FIGURE 5.1. Graph $\underline{G}$ on two dodecahedra

between these realizations is that a pentagon realized as the boundary of a pentagonal face in one realization is an equator in the other realization.

Note that in this graph, one can find twelve pentagons. It is easy to see the boundaries of six faces of the dodecahedron and six equators such as

$$C(14) - C(35) - C(12) - C(45) - C(23) - C(14).$$

These pentagons correspond to the twelve juzu sequences, for example,

$$(12345) \Longleftrightarrow C(12) - C(34) - C(15) - C(23) - C(45) - C(12).$$

FACT 5.2. *Let $\overline{C}(ij)$ and $\overline{D}(J)$ denote the closure of $C(ij)$ and $D(J)$ in $\overline{X}/\langle c\rangle$. We have*

$$\overline{C}(ij) \cap \overline{X}_{\mathbb{R}} = \cup_J \overline{D}(J),$$

where J runs through the six juzu sequences for which i and j are not adjacent. In Figure 5.2 the chambers $\overline{D}(J)$ are shown for the case $(ij) = (12)$; note that those form a Möbius strip.

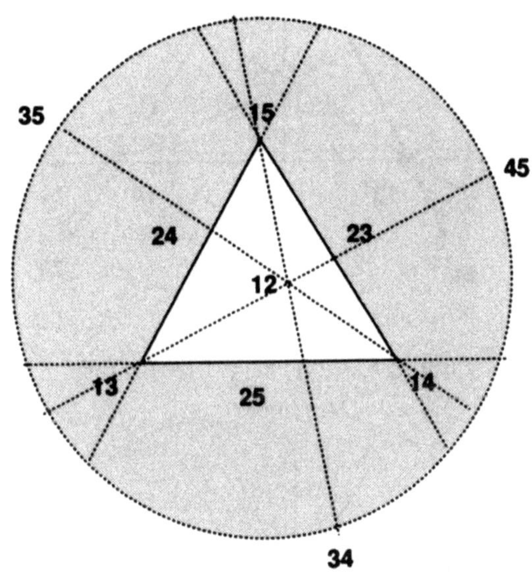

FIGURE 5.2. $\overline{C}(12) \cap \overline{X}_{\mathbb{R}} = \cup_J \overline{D}(J)$

FACT 5.3. *For $\{i,j\} \cap \{p,q\} = \emptyset$, the intersection $\overline{C}(ij) \cap \overline{C}(pq)$ is a simply connected real 3-dimensional object bounded by four real surfaces*

$$L(nm)/\langle c\rangle, \quad n \in \{i,j\}, \; m \in \{p,q\}$$

and four $\overline{D}(J)$, where neither i and j nor p and q are adjacent in J. The union of the four $\overline{D}(J)$ is simply connected. Figure 5.3 illustrates this union for the case $(ij) = (12), (pq) = (34)$.

FACT 5.4. *For $\{i,j,p,q,r\} = \{1,\ldots,5\}$,*

$$\overline{C}(qr) \cap \overline{C}(rp) = \overline{C}(qr) \cap \overline{C}(rp) \cap \overline{C}(ij) = \overline{D}(irjpq) \cup \overline{D}(irjqp).$$

The set $\overline{C}(12) \cap \overline{C}(14) = \overline{D}(31524) \cup \overline{D}(31542)$ is shown in Figure 5.4.

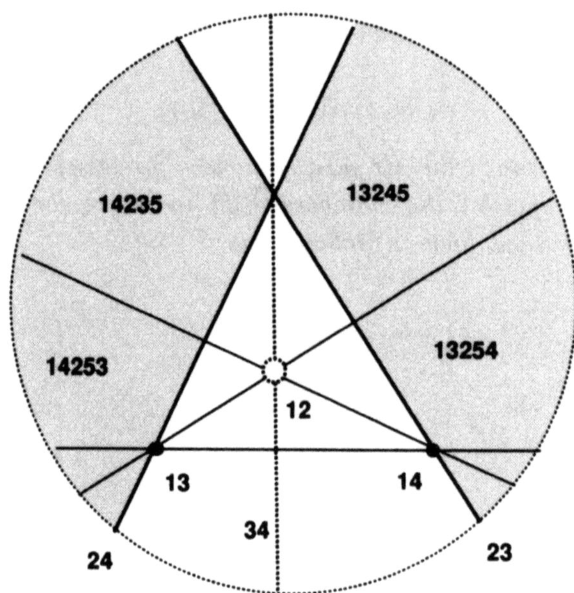

FIGURE 5.3. $(ij) = (12), (pq) = (34)$

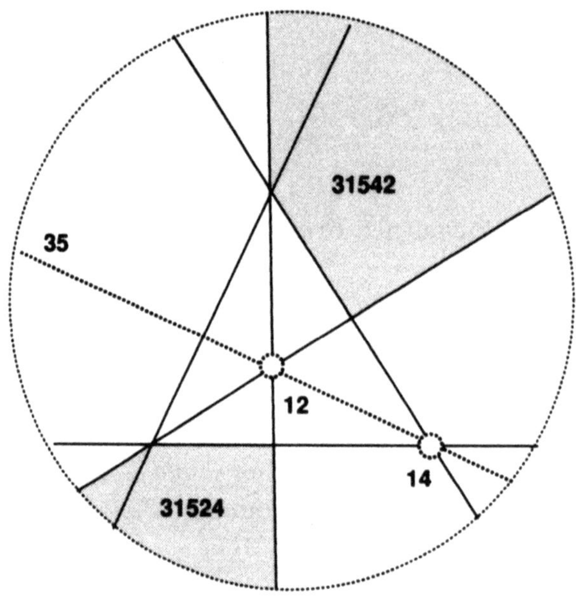

FIGURE 5.4. $\overline{C}(12) \cap \overline{C}(14) = \overline{D}(31524) \cup \overline{D}(31542)$

Recall that (Fact 5.1)

$$\partial \overline{C}(ij) = \{\overline{C}(pq) \cap \overline{C}(ij)\} \cup \{\overline{C}(qr) \cap \overline{C}(ij)\} \cup \{\overline{C}(rp) \cap \overline{C}(ij)\}.$$

Consider a juzu sequence say, $J = (12345)$. Note that there is a chain of chambers $C(12) - C(34) - C(15) - C(23) - C(45) - C(12)$ in the graph $\underline{G}$ and that the pentagon $D(J) \subset \overline{X}_{\mathbb{R}}$ is bounded by

$$L_{\mathbb{R}}(12), L_{\mathbb{R}}(34), L_{\mathbb{R}}(15), L_{\mathbb{R}}(23), L_{\mathbb{R}}(45).$$

FACT 5.5. *We have*

$$\overline{C}(12) \cap \overline{C}(34) \cap \overline{C}(15) \cap \overline{C}(23) \cap \overline{C}(45) = \overline{D}(J'),$$

where J' is the juzu sequence dual to J, $J' = (13524)$ (see Figure 5.5).

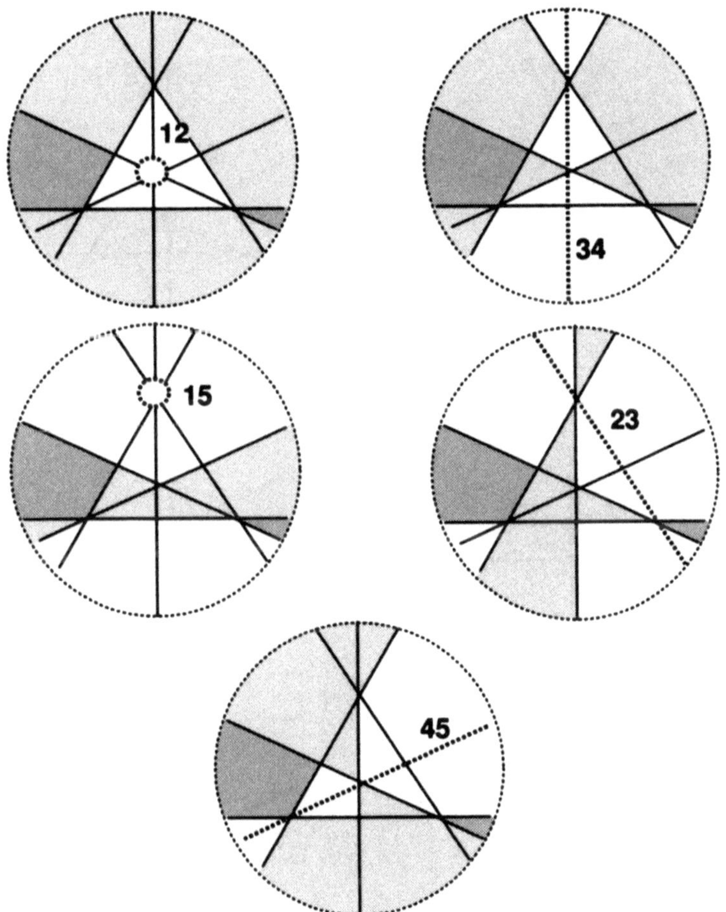

FIGURE 5.5. $\overline{C}(12) \cap \cdots \cap \overline{C}(45) = \overline{D}(J')$

In the graph $\underline{G}$, for each $C(ij)$ there are six vertices $C(ab)$, $a \in \{i,j\}$, $b \in \{p,q,r\}$, which are not adjacent to $C(ij)$. These form a hexagon (see Figure 5.1):

$$C(ip) - C(jq) - C(ir) - C(jp) - C(iq) - C(jr) - C(ip).$$

FACT 5.6. *We have*

$$\cap\{\overline{C}(ab) \mid a \in \{i,j\},\ b \in \{p,q,r\}\} = L(ij)/\langle c\rangle.$$

Figure 5.6 illustrates, for the case $(ij) = (12)$, the cycle

$$C(13) - C(24) - C(15) - C(23) - C(14) - C(25) - C(13).$$

This describes how these six chambers are arranged around the line $L(ij)$.

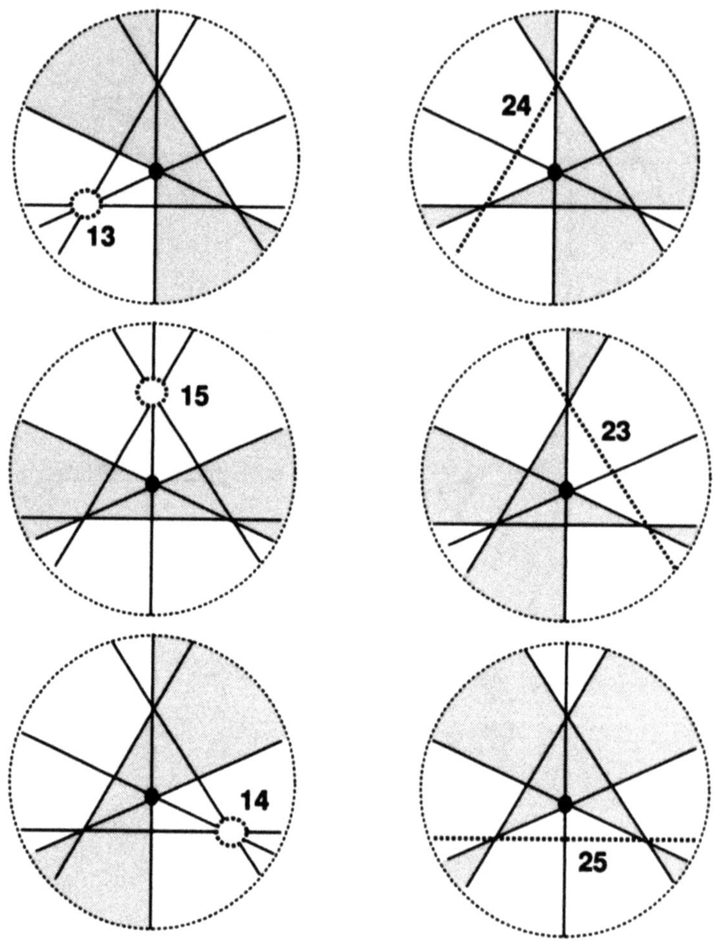

FIGURE 5.6. $C(13) \cap C(24) \cap C(15) \cap C(23) \cap C(14) \cap C(25)$

6. The Graph G

We study the graph $G = G(\mathcal{C})$. The vertices of this graph are the twenty elements of $\mathcal{C}$, and two of these vertices are joined with an edge when the intersection of their closures (in $\overline{X}$) is 3-dimensional.

We are going to construct a double cover of the projective plane branching at six points. Since six is an even number, such a cover exists, and as a topological manifold it is unique. As a covering, however, it is not uniquely determined. (Between two such covers, there does not necessarily exist a homeomorphism which is equivariant with the projections.) Compare this fact with the unique existence of the double cover of the complex projective line branching at an even number of points, a hyper-elliptic curve. At any rate, in order to specify a double cover of the dodecahedral projective plane branching at the barycenter of the six pentagonal faces, one must give extra information: slits cutting the six pentagons. The double cover is obtained by gluing two copies of the slitted projective plane along the slits using a standard method that can be found in any elementary book on Riemann surfaces.

FACT 6.1. *The graph G is the double cover of $\underline{G}$ drawn on the dodecahedral projective plane branching at the barycenters of the six pentagons with the five slits cutting the five edges, as shown in Figure 6.1.*

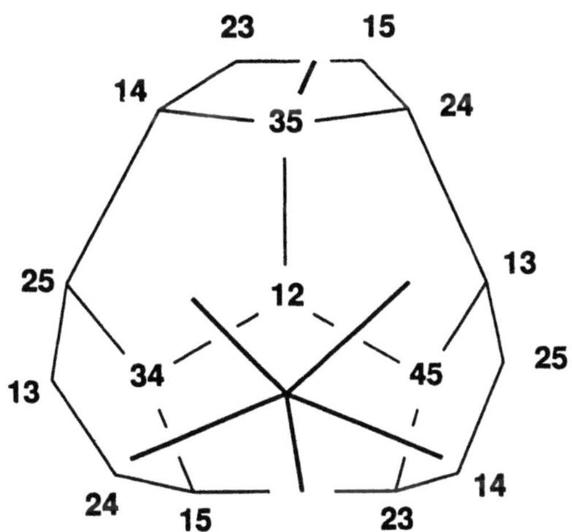

FIGURE 6.1. Dodecahedron with cuts

Slits are not uniquely determined. In fact, the three slits cutting the three edges 14-35, 13-45 and 15-34 work as well, while the three slits cutting the three edges 15-23, 14-25 and 15-23, for example, give a different covering (see Figure 6.2). In fact, the inverse image of the pentagon 14-35-12-45-13-14 consists of two copies of the pentagon on this covering, while it is a 10-gon in G. The graph G is given in Figure 6.3. (In the fig-

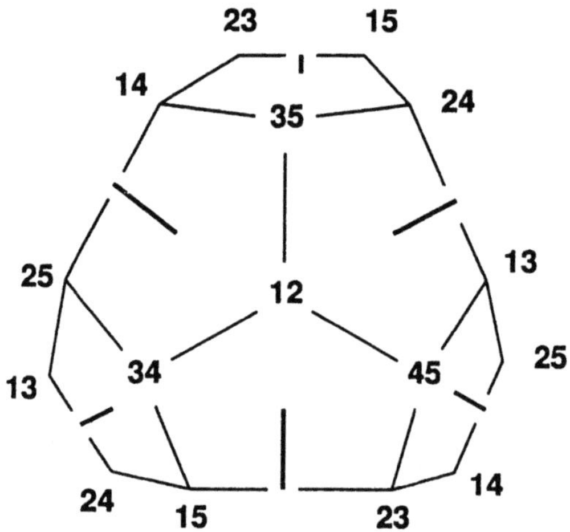

FIGURE 6.2. Dodecahedron with false cuts

ure, the three vertices denoted by 12 at the top, left, and right represent the same vertex, situated at infinity.)

7. A Presentation of the Fundamental Group of X

Each loop in X is expressed by a cycle in G.

FACT 7.1. *The twelve pentagons in $\underline{G}$ coded by juzu sequences, e.g.,*

$$J = (12345): \quad C(12) - C(34) - C(15) - C(23) - C(45) - C(12),$$

are lifted to 10-gons in G. For each 10-gon, the closures of the ten chambres have $D(J')$ in common, where J' is the juzu sequence dual to $J = (12345)$. The union of the closures in X of these ten chambres is contractible in X.

FACT 7.2. *The ten hexagons in $\underline{G}$ coded by $\{i,j\}$, e.g.,*

$$\{1,2\}: \quad C(13) - C(24) - C(15) - C(23) - C(14) - C(25) - C(13),$$

are lifted to two hexagons in G, corresponding to two loops around $L(ij)$. One of these loops is homotopic to the loop obtained by reversing the orientation of the other.

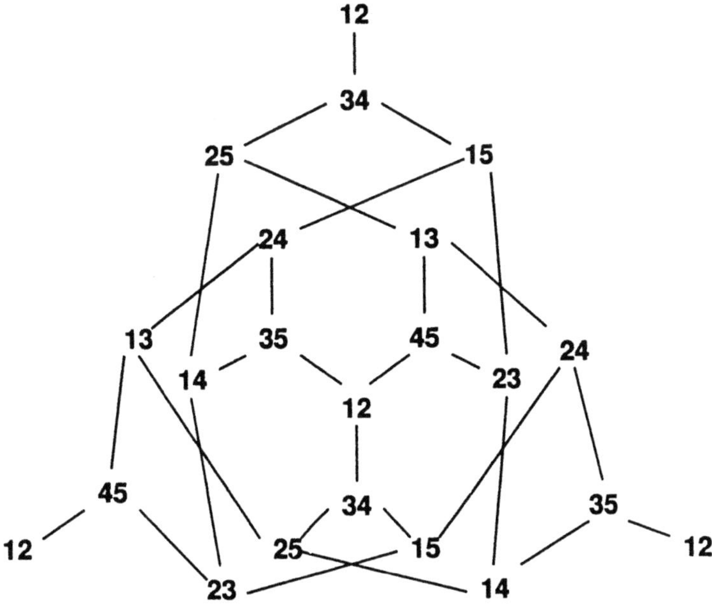

FIGURE 6.3. Graph G

In fact Figure 7.1 shows that the two hexagons coded by $\{1,2\}$ are homotopic (after reversing the orientation of one of the two) by using two 10-gons coded by (13524) and (15324). By permuting $\{3,4,5\}$, one can take pairs of 10-gons coded by (13425) and (14325), as well as those coded by (14523) and (15423).

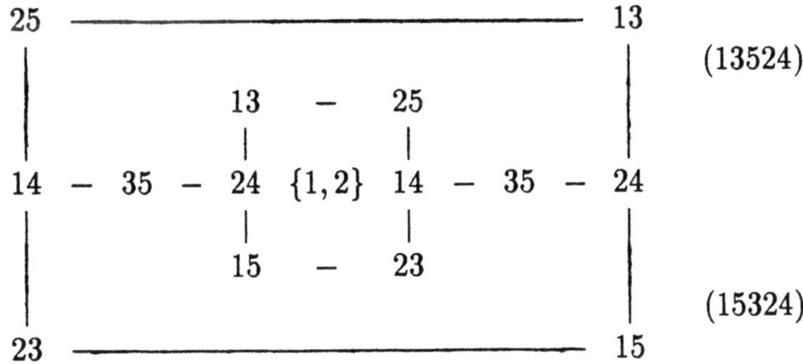

FIGURE 7.1. The two loops represented by 13-24-15-23-14-25

To this point our representation has in all cases been symmetric under S_5. That is, all objects we have considered remain unchanged under the action of any element of this group. Now we are going to destroy the symmetry, but maintain symmetry under cyclic permutations of $1, 2, 3, 4, 5$.

Around the 10-gon coded by the dual of (12345), construct a maximal tree B in G using the ten edges issuing from the vertices of the 10-gon, as is shown in Figure 7.2. The vertices of the 10-gon are printed in darker

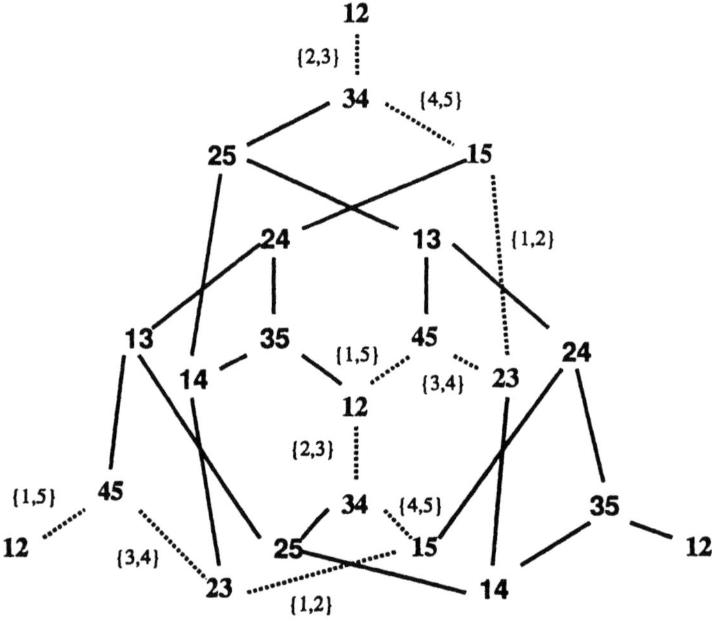

FIGURE 7.2. Tree B in Graph G

font, while those of the complementary 10-gon are printed in lighter font. The edges of the tree are represented by heavy lines and the sides of the complementary 10-gon are represented by dotted lines. We shall regard this tree, representing a simply connected domain in X, as a base, and study the fundamental group $\pi_1(X, B)$. We find ten hexagons, each of which is composed by five heavy edges and one dotted edge. They are two-by-two coded by

$$\{1,2\}: \quad h(12) = 15 - 24 - 13 - 25 - 14 - 23 \cdots 15,$$
$$\{4,5\}: \quad h(45) = 34 - 25 - 13 - 24 - 35 - 12 \cdots 34,$$
$$\{2,3\}: \quad h(23) = 12 - 35 - 24 - 13 - 25 - 34 \cdots 12,$$
$$\{1,5\}: \quad h(15) = 45 - 13 - 25 - 14 - 35 - 12 \cdots 45,$$
$$\{3,4\}: \quad h(34) = 45 - 13 - 24 - 35 - 14 - 23 \cdots 45.$$

Notice that the hexagon $h(i, i+1)$ is obtained from $h(i-1, i)$ by applying the cyclic permutation $1 \to 2 \to \cdots \to 5 \to 1$.

In the graph G, one finds the subgraph given in Figure 7.1, which is shown again in Figure 7.3 with additional information. A close look at the

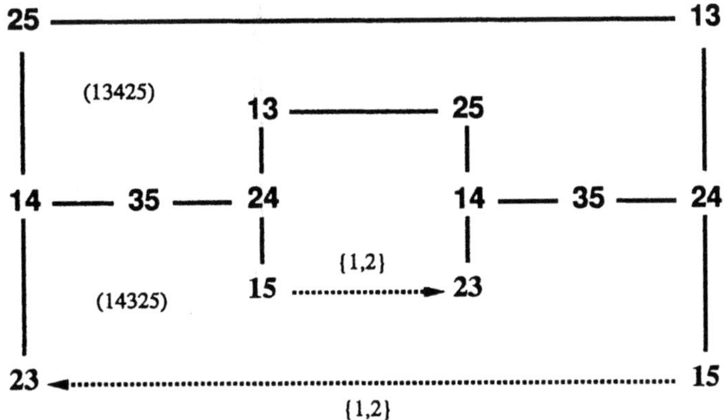

FIGURE 7.3. $\{1, 2\}$

figure (note that the 10-gon coded by (14325) is contractible) reveals that the two oriented edges $15 \to 23$, representing a loop around $L(12)$, are inverses of each other as elements of $\pi_1(X, B)$; let us call these loops $\gamma(12)$ and $\gamma(12)^{-1}$. By cyclic permutations of $1, 2, 3, 4, 5$, we obtain the corresponding facts for the other four cases, yielding $\gamma(45), \gamma(23), \gamma(15), \gamma(34)$. We take these as a set of generators of $\pi_1(X, B)$. Fact 7.1 implies that after suitably fixing the orientations of the $\gamma(ij)$, we have the pentagonal relation

$$\gamma(12)\gamma(45)^{-1}\gamma(23)\gamma(15)^{-1}\gamma(34)\gamma(12)^{-1}\gamma(45)\gamma(23)^{-1}\gamma(15)\gamma(34)^{-1} = 1.$$

It is easy to check the commutation relations

$$\gamma(ij)\gamma(pq) = \gamma(pq)\gamma(ij) \quad \text{if} \quad \{i, j\} \cap \{p, q\} = \emptyset.$$

For instance the relation $\gamma(12)\gamma(45) = \gamma(45)\gamma(12)$ can be derived using the fact that the 10-gon coded by (15234) is contractible.

The commutation relations and the pentagonal relation generate the relations of the five generators of $\pi_1(X, B)$.

CHAPTER VI

Modular Interpretation of the Configuration Space $X(2, n)$

For $n + 3(\geq 5)$ distinct points $x_1, \ldots, x_{n+3} \in \mathbb{P}^1$ and $n + 3$ rational numbers $\mu_1, \ldots, \mu_{n+3}$ satisfying

$$0 < \mu_j < 1, \qquad \sum_{j=1}^{n+3} \mu_j = 2,$$

consider the family of curves

$$S_x : s^d = \prod_{j=1}^{n+3} (t - x_j)^{d\mu_j}$$

and $n + 1$ periods defined by the following hypergeometric integrals:

$$u_k(x) = \int_{x_k}^{x_{k+1}} \prod_{j=1}^{n+3} (t - x_j)^{-\mu_j} dt \quad (k = 1, \ldots, n + 1).$$

The ratio of the $n + 1$ periods defines the period map

$$\varphi(\mu) : X(2, n + 3) \longrightarrow \mathbb{P}^n.$$

Let $\Gamma(\mu)$ be its monodromy group. For some μ, the closure of the image of the period map is projectively equivalent to the n-ball

$$\mathbb{B}_n = \{z_0 : \cdots : z_n \in \mathbb{P}^n \mid |z_0|^2 - |z_1|^2 - \cdots - |z_n|^2 > 0\},$$

and the period map gives an isomorphism of $X(2, n + 3)$ into $\mathbb{B}_n / \Gamma(\mu)$.

In this chapter we collect such μ and state, without proof, some of the known facts about the period maps (see [Trd], [DM] and [CH] for further details).

1. Admissible Sequences

Let $\mu_1, \ldots, \mu_{n+3}$ be $n+3$ rational numbers satisfying

$$0 < \mu_j < 1, \qquad \sum_{j=1}^{n+3} \mu_j = 2.$$

A sequence of rational numbers $\mu_1, \ldots, \mu_{n+3}$ is said to be *admissible* if there is an integer n_{ij} greater than or equal to 2 (including ∞) such that

$$|\mu_i + \mu_j - 1| = \frac{1}{n_{ij}}$$

for any distinct i and j. If $n \geq 2$, there are only finitely many such sequences. We shall tabulate all of them. For such a sequence, $\mu_1, \ldots, \mu_{n+3}$, if there are indices i and j $(i < j)$ such that $\mu_i + \mu_j < 1$, then the sequence

$$\mu_i + \mu_j, \mu_1, \ldots, \widehat{\mu_i}, \ldots \widehat{\mu_j}, \ldots, \mu_{n+3}$$

(μ_i and μ_j are removed) is also admissible. This sequence will be called the *restriction* with respect to i and j. In the table below, the sequence of rational numbers

$$\mu_1 = m_1/d, \quad \ldots, \quad \mu_{n+3} = m_{n+3}/d$$

is expressed by

$$d, \qquad m_1 + \cdots + m_{n+3},$$

where $d, m_1, \ldots, m_{n+3}$ are natural numbers, and d is a common multiple of the denominators of the μ_j. The arrows '$\rightarrow$' point from admissible sequences to their restrictions. The following is a list of all admissible sequences for $n \geq 2$ together with their restrictions. For $n = 1$, we also give the characteristic (n_{12}, n_{23}, n_{31}) of the corresponding Schwarz triangle group.

Table 1: The admissible sequences obtained by repeatedly restricting the unique admissible sequence with $n = 5$.

$$d = 4, \qquad \overbrace{1 + \cdots + 1}^{8}$$
$$\downarrow$$
$$d = 4, \qquad 2 + \overbrace{1 + \cdots + 1}^{6}$$

$$d = 4, \qquad 2 + 2 + \overbrace{1 + \cdots + 1}^{4} \qquad\qquad\qquad 3 + \overbrace{1 + \cdots + 1}^{5}$$
$$\downarrow \qquad\qquad\qquad\qquad\qquad\qquad \downarrow$$
$$d = 4, \qquad 2 + 2 + 2 + 1 + 1 \qquad\qquad\qquad\qquad 3 + 2 + 1 + 1 + 1$$
$$\downarrow \qquad\qquad\qquad\qquad\qquad\qquad \downarrow$$
$$d = 4, \qquad 2 + 2 + 2 + 2 \qquad 3 + 2 + 2 + 1 \qquad 3 + 3 + 1 + 1$$
$$(\infty, \infty, \infty) \qquad (4, 4, \infty) \qquad (2, \infty, \infty)$$

Table 2: The restrictions of the admissible sequences with $n = 3$ excluding the ones appearing in Table 1.

$$d = 3, \qquad 1 + 1 + 1 + 1 + 1 + 1 \rightarrow 2 + 1 + 1 + 1 + 1$$

$$d = 6, \qquad 3 + 2 + 2 + 2 + 2 + 1 \rightarrow \begin{cases} 3 + 3 + 2 + 2 + 2 \\ 4 + 2 + 2 + 2 + 2 \\ 4 + 3 + 2 + 2 + 1 \\ 5 + 2 + 2 + 2 + 1 \end{cases}$$

$$d = 8, \qquad 5 + 5 + 5 + 5 + 5 + 1 \rightarrow \begin{cases} 4 + 3 + 3 + 3 + 3 \\ 6 + 3 + 3 + 3 + 1 \end{cases}$$

$$d = 12, \qquad 5 + 5 + 5 + 3 + 3 + 3 \rightarrow \begin{cases} 6 + 5 + 5 + 5 + 3 \\ 8 + 5 + 5 + 3 + 3 \\ 10 + 5 + 3 + 3 + 3 \end{cases}$$

$$d = 12, \qquad 7 + 5 + 3 + 3 + 3 + 3 \rightarrow \begin{cases} 7 + 6 + 5 + 3 + 3 \\ 8 + 7 + 3 + 3 + 3 \\ 10 + 5 + 3 + 3 + 3 \end{cases}$$

Table 3: The restrictions of the admissible sequences with $n = 2$ excluding the ones appearing in Table 1.

$$
\begin{array}{lll}
d = 3, & 2+1+1+1+1 \rightarrow & 2+2+1+1 \quad (3,\infty,\infty) \\
d = 5, & 2+2+2+2+2 \rightarrow & 4+2+2+2 \quad (5,5,5) \\[1ex]
d = 6, & 3+3+2+2+2 \rightarrow & \begin{cases} 4+3+3+2 & (6,6,\infty) \\ 5+3+2+2 & (3,6,6) \end{cases} \\[2ex]
d = 6, & 3+3+3+2+1 \rightarrow & \begin{cases} 3+3+3+3 & (\infty,\infty,\infty) \\ 4+3+3+2 & (6,6,\infty) \\ 5+3+3+1 & (3,3,\infty) \end{cases} \\[3ex]
d = 6, & 4+3+2+2+1 \rightarrow & \begin{cases} 4+3+3+2 & (6,6,\infty) \\ 4+4+2+2 & (3,\infty,\infty) \\ 4+4+3+1 & (3,6,6) \\ 5+3+2+2 & (3,6,6) \\ 5+4+2+1 & (2,6,\infty) \end{cases} \\[4ex]
d = 6, & 5+2+2+2+1 \rightarrow & \begin{cases} 5+3+2+2 & (3,6,6) \\ 5+4+2+1 & (2,6,\infty) \end{cases} \\[2ex]
d = 8, & 4+3+3+3+3 \rightarrow & \begin{cases} 6+4+3+3 & (4,8,8) \\ 7+3+3+3 & (4,4,4) \end{cases} \\[2ex]
d = 8, & 5+5+2+2+2 \rightarrow & \begin{cases} 5+5+4+2 & (4,8,8) \\ 7+5+2+2 & (2,8,8) \end{cases} \\[2ex]
d = 8, & 6+3+3+3+1 \rightarrow & \begin{cases} 6+4+3+3 & (4,8,8) \\ 6+6+3+1 & (2,8,8) \\ 7+3+3+3 & (4,4,4) \end{cases} \\[3ex]
d = 9, & 4+4+4+4+2 \rightarrow & \begin{cases} 6+4+4+4 & (9,9,9) \\ 8+4+4+2 & (3,3,9) \end{cases}
\end{array}
$$

$$d = 10, \qquad 7+4+4+4+1 \rightarrow \begin{cases} 7+5+4+4 & (5,10,10) \\ 8+4+4+4 & (5,5,5) \\ 8+7+4+1 & (2,5,10) \end{cases}$$

$$d = 12, \qquad 5+5+5+5+4 \rightarrow \begin{cases} 9+5+5+5 & (6,6,6) \\ 10+5+4+4 & (4,4,6) \end{cases}$$

$$d = 12, \qquad 6+5+5+4+4 \rightarrow \begin{cases} 8+6+5+5 & (6,12,12) \\ 9+6+5+4 & (4,6,12)* \\ 10+5+5+4 & (4,4,6) \\ 10+6+4+4 & (3,6,6) \\ 11+5+4+4 & (3,4,4) \end{cases}$$

$$d = 12, \qquad 6+5+5+5+3 \rightarrow \begin{cases} 8+6+5+5 & (6,12,12) \\ 9+5+5+5 & (6,6,6) \\ 10+6+5+3 & (3,4,12) \\ 11+5+5+3 & (3,3,6) \end{cases}$$

$$d = 12, \qquad 7+5+4+4+4 \rightarrow \begin{cases} 8+7+5+4 & (4,12,\infty)* \\ 9+7+4+4 & (3,12,12) \\ 11+5+4+4 & (3,4,4) \end{cases}$$

$$d = 12, \qquad 7+6+5+3+3 \rightarrow \begin{cases} 7+6+6+5 & (12,12,\infty)* \\ 8+7+6+3 & (4,6,12)* \\ 9+7+5+3 & (3,6,\infty)* \\ 10+6+5+3 & (3,4,12) \\ 11+7+3+3 & (2,6,6) \end{cases}$$

$$d = 12, \qquad 7+7+4+4+2 \rightarrow \begin{cases} 7+7+6+4 & (6,12,12) \\ 8+7+7+2 & (4,4,6) \\ 9+7+4+4 & (3,12,12) \\ 11+7+4+2 & (2,4,12) \end{cases}$$

$$d = 12, \qquad 8+5+5+3+3 \rightarrow \begin{cases} 8+6+5+5 & (6,12,12) \\ 8+8+5+3 & (3,12,12) \\ 10+8+3+3 & (2,12,12) \\ 11+5+5+3 & (3,3,6) \end{cases}$$

$$d = 12, \quad 8 + 5 + 5 + 5 + 1 \rightarrow \begin{cases} 8 + 6 + 5 + 5 & (6, 12, 12) \\ 9 + 5 + 5 + 5 & (6, 6, 6) \\ 10 + 8 + 5 + 1 & (2, 4, 12) \end{cases}$$

$$d = 12, \quad 8 + 7 + 3 + 3 + 3 \rightarrow \begin{cases} 8 + 7 + 6 + 3 & (4, 6, 12)* \\ 10 + 8 + 3 + 3 & (2, 12, 12) \\ 11 + 7 + 3 + 3 & (2, 6, 6) \end{cases}$$

$$d = 12, \quad 10 + 5 + 3 + 3 + 3 \rightarrow \begin{cases} 10 + 6 + 5 + 3 & (3, 4, 12) \\ 10 + 8 + 3 + 3 & (2, 12, 12) \end{cases}$$

$$d = 15, \quad 8 + 6 + 6 + 6 + 4 \rightarrow \begin{cases} 10 + 8 + 6 + 6 & (5, 15, 15)* \\ 12 + 6 + 6 + 6 & (5, 5, 5) \\ 12 + 8 + 6 + 4 & (3, 5, 15)* \\ 14 + 6 + 6 + 4 & (3, 3, 5) \end{cases}$$

$$d = 18, \quad 11 + 8 + 8 + 8 + 1 \rightarrow \begin{cases} 11 + 9 + 8 + 8 & (9, 18, 18) \\ 12 + 8 + 8 + 8 & (9, 9, 9) \\ 16 + 11 + 8 + 1 & (2, 3, 18) \end{cases}$$

$$d = 20, \quad 14 + 11 + 5 + 5 + 5 \rightarrow \begin{cases} 14 + 11 + 10 + 5 & (4, 5, 20)* \\ 16 + 14 + 5 + 5 & (2, 20, 20)* \\ 19 + 11 + 5 + 5 & (2, 5, 5) \end{cases}$$

$$d = 24, \quad 14 + 9 + 9 + 9 + 7 \rightarrow \begin{cases} 16 + 14 + 9 + 9 & (4, 24, 24)* \\ 18 + 14 + 9 + 7 & (3, 8, 24) \\ 21 + 9 + 9 + 9 & (4, 4, 4) \\ 23 + 9 + 9 + 7 & (3, 3, 4)* \end{cases}$$

In the tables, the non-arithmetic triangle groups are marked by asterisks. (Cf. [Take] and [DM]. If you are not interested in non-arithmeticity, you need not concern yourself with this point.) Note that if $n \geq 6$ there is no admissible sequence, for $n = 5$ there is one, for $n = 4$ there is one, which is the restriction of the former, for $n = 3$ there are seven, and for $n = 2$ there are 27.

2. Families of Curves and Their Periods

Let $x_1, \ldots, x_{n+3}$ be $n + 3$ $(n \geq 2)$ distinct points on $\mathbb{P}^1$, and consider for each admissible sequence $\mu = \{\mu_1, \ldots, \mu_{n+3}\}$ the family of curves

$$S_x : \quad s^d = \prod_{j=1}^{n+3} (t - x_j)^{d\mu_j},$$

where d is the smallest common denominator of the μ_j, and the following $n + 1$ periods of S_x:

$$u_k(x) = \int_{x_k}^{x_{k+1}} \prod_{j=1}^{n+3} (t - x_j)^{-\mu_j} dt, \quad (k = 1, \ldots, n + 1).$$

Let $\Gamma(\mu)$ be the projectivization of the monodromy group of the (multivalued) vector function $(u_1, \ldots, u_{n+1})$. It turns out that this group is a discrete subgroup of $Aut(\mathbb{B}_n)$. Let $\mathbb{B}_n^0(\mu)$ be the open dense subset of $\mathbb{B}_n$ on which the group $\Gamma(\mu)$ acts freely. The complement of $\mathbb{B}_n^0(\mu)$ in $\mathbb{B}_n$ is the union of countably many hyperplanes (locally finite) passing through the ball $\mathbb{B}_n$. Let X_n be the configuration space $X(2, n + 3)$ of $n + 3$ distinct points on the projective line:

$$X_n = \{(x_1, \ldots, x_{n+3}) \in (\mathbb{P}^1)^{n+3} | x_i \neq x_j (i \neq j)\}/PGL(2).$$

Then the period map

$$x \mapsto u_1(x) : \cdots : u_{n+1}(x) \in \mathbb{P}^n$$

gives the isomorphism

$$X_n \to \mathbb{B}_n^0(\mu)/\Gamma(\mu).$$

If μ possesses a restriction with respect to i and j, i.e., if the sequence

$$\mu_1, \ldots, \mu_{i-1}, \mu_i + \mu_j, \mu_{i+1}, \ldots, \widehat{\mu_j}, \ldots, \mu_{n+3}$$

is admissible, then attach to X_n the manifold

$$\{(x_1, \ldots, x_{n+3}) \in (\mathbb{P}^1)^{n+3} \mid x_i = x_j, \text{ all others distinct}\}/PGL(2),$$

which is isomorphic to X_{n-1} in the obvious way. If μ possesses a further restriction, i.e., if the sequence

$$\mu_1, \ldots, \mu_i + \mu_j, \ldots, \mu_k + \mu_l, \ldots, \widehat{\mu_j}, \ldots, \widehat{\mu_l}, \ldots, \mu_{n+3}$$

or

$$\mu_1, \ldots, \mu_i + \mu_j + \mu_k, \ldots, \widehat{\mu_j}, \ldots, \widehat{\mu_k}, \ldots, \mu_{n+3}$$

is admissible, then attach to X_n

$$\{(x_1, \ldots, x_{n+3}) \in (\mathbb{P}^1)^{n+3} \mid x_i = x_j \text{ and } x_k = x_l, \text{ all others distinct}\}/PGL(2)$$

or

$$\{(x_1, \ldots, x_{n+3}) \in (\mathbb{P}^1)^{n+3} \mid x_i = x_j = x_k, \text{ all others distinct}\}/PGL(2),$$

respectively, each of which is isomorphic to X_{n-2}. Continue this procedure: for every iterated restriction of μ of length $r+3$ ($r = 0, 1, \ldots, n-1$)

$$\lambda = (\lambda_1, \ldots, \lambda_{r+3}), \quad \lambda_i = \mu_{k_i(1)} + \cdots + \mu_{k_i(r_i)},$$

attach the manifold

$$X_\lambda := \{(x_j)_j \in (\mathbb{P}^1)^{n+3} \mid x_{k_i(1)} = \cdots = x_{k_i(r_i)}, \text{ all others distinct}$$
$$(i = 1, \ldots, r)\}/PGL(2),$$

and you end up with a quasi-projective variety $X'(\mu)$. The period map extends to

$$X'_n(\mu) \longrightarrow \mathbb{B}_n/\Gamma(\mu).$$

The variety $X'(\mu)$ can be compactified to a projective variety $\overline{X}_n(\mu)$ by adding a finite number of points; the points to be added correspond to the ways of dividing the set $\{1, \ldots, n + 3\}$ into two subsets I_1 and I_2 satisfying

$$\sum_{i \in I_1} \mu_i = \sum_{i \in I_2} \mu_i = 1.$$

The smallest compactification $\overline{\mathbb{B}_n/\Gamma(\mu)}$ of the quotient space $\mathbb{B}_n/\Gamma(\mu)$ is obtained by adding the cusps, a finite number of points. The period map extends to an isomorphism between the two projective varieties:

$$\overline{X}_n(\mu) \longrightarrow \overline{\mathbb{B}_n/\Gamma(\mu)}.$$

If we normalize x_j ($j = 1, \ldots, n + 3$) by setting $x_{n+1} = 0, x_{n+2} = 1, x_{n+3} = \infty$, then the periods $u_1(x), \ldots, u_{n+1}(x)$, as functions in $x = (x_1, \ldots, x_n)$, are linearly independent solutions of *Appell-Lauricella's hypergeometric system* $E_D^n(a, b_1, \ldots, b_n, c)$ of differential equations, where

$$a = \alpha_{n+2}, \quad b_j = 1 - \alpha_j \ (1 \le j \le n), \quad c = \alpha_{n+1} + \alpha_{n+2},$$

$$\alpha_k = 1 - \mu_k.$$

This hypergeometric system $E_D^n(a, b_1, \ldots, b_n, c)$ is a system of linear differential equations in n variables $x_1, \ldots, x_n$ of *rank* (dimension of the space of solutions at a generic point) $n+1$ which has singularities along the hyperplanes

$$x_j = 0, \; x_j = 1, \; x_j = x_k \; (j \neq k),$$

and the hyperplane at infinity. This system can be written as

$$\{(a + \sum_{i=1}^{n} D_i)(b_j + D_j) - (c + \sum_{i=1}^{n} D_i)(1 + D_j)\frac{1}{x_j}\}u = 0, \quad j = 1,\ldots,n,$$

where $D_i = x_i \partial/\partial x_i$. It is a generalization of the hypergeometric equation; indeed when $n = 1$ it is identically the hypergeometric equation. If you still remember the technique presented in §2 of Chapter III, you can readily show that it admits the power-series solution

$$F_D(a, b_1,\ldots, b_n, c; x) = \sum_{m_1=0}^{\infty} \cdots \sum_{m_n=0}^{\infty} \frac{(a, \sum_{j=1}^{n} m_j) \prod_{j=1}^{n}(b_j, m_j)}{(c, \sum_{j=1}^{n} m_j) \prod_{j=1}^{n}(1, m_j)} \prod_{j=1}^{n} x_j^{m_j},$$

which in the case $n = 1$ is just the hypergeometric series $F(a, b, c; x)$.

If an admissible sequence $\mu_1,\ldots,\mu_{n+3}$ has an iterated restriction λ of length $r + 3$, then the restriction of a linear combination of the periods $u_1(x),\ldots, u_{n+1}(x)$ which is holomorphic along $X_\lambda (\cong X_r)$ satisfies the hypergeometric system E_D^r with the corresponding parameters. The system can be thought of as the restriction of the system E_D^n along X_λ in the sense that any solution of the restricted system is the restriction of a solution of the system E_D^n which is holomorphic along X_λ.

REMARK 2.1. The inverse map of the period map should be written in terms of automorphic forms on $\mathbb{B}_n$; explicit formulae are known only for the following three cases:

$$(\frac{1}{3}, \frac{1}{3}, \frac{1}{3}, \frac{1}{3}, \frac{1}{3}, \frac{1}{3}), \quad \longrightarrow \quad (\frac{2}{3}, \frac{1}{3}, \frac{1}{3}, \frac{1}{3}, \frac{1}{3}),$$
$$(\frac{2}{4}, \frac{2}{4}, \frac{2}{4}, \frac{1}{4}, \frac{1}{4}).$$

(See [Gon], [Shi], [Mat1], respectively.)

3. Typical Examples Modeled after $\mathbb{B}_n$

We consider the cases when all the μ_j are equal:

$$n = 1, \; \mu_j = \frac{1}{2}; \qquad n = 2, \; \mu_j = \frac{2}{5};$$
$$n = 3, \; \mu_j = \frac{1}{3}; \qquad n = 5, \; \mu_j = \frac{1}{4}.$$

Case 1 ($n = 1$). The monodromy group is the principal congruence subgroup $\Gamma(2)$ of level 2 of the elliptic modular group $SL(2,\mathbb{Z})$; its projectivization is the triangle group with characteristic (∞, ∞, ∞), which has $\binom{4}{2}/2 = 3$ cusps. This is the case we treated in §3 of Chapter I in detail. Note that the quotient group $SL(2,\mathbb{Z})/\Gamma(2)$ is isomorphic to the symmetric group of degree 3 acting naturally as the full permutation group of the 3 cusps.

In what follows, for a hermitian matrix A of signature $(n+, 1-)$, $\mathbb{B}_n^A$ represents the complex n-ball

$$\{z = (z_0, \ldots, z_n) \in \mathbb{C}^{n+1} \mid zA^t\bar{z} < 0\}/\mathbb{C}^\times.$$

Case 2 ($n = 2$). Put

$$\zeta_5 = \exp(\frac{2\pi i}{5}), \quad A = \operatorname{diag}(1, 1, u), \quad u = \frac{1 - \sqrt{5}}{2}.$$

The monodromy group is the principal congruence subgroup

$$\Gamma(1 - \zeta_5) = \{g \in \Gamma \mid g \equiv I \mod (1 - \zeta_5)\}$$

with respect to the ideal $(1 - \zeta_5) \subset \mathbb{Z}[\zeta_5]$ of the group

$$\Gamma = \{g \in GL(3, \mathbb{Z}[\zeta_5]) \mid g^*Ag = A\}.$$

The quotient space $\mathbb{B}_2^A/\Gamma(1 - \zeta_5)$ is compact and is isomorphic to the space $\overline{X}_2 = \overline{X}(2,5) = \overline{X}$, which we studied in detail in Chapter V. The quotient group $\Gamma/\Gamma(1 - \zeta_5)$ is isomorphic to the symmetric group of degree 5 acting transitively on the set of the ten curves $L(ij)$ on $\overline{X}$ (cf. [YY]).

Case 3 ($n = 3$). Put

$$\zeta_3 = \exp(\frac{2\pi i}{3}), \quad A = \operatorname{diag}(1, 1, 1, -1).$$

The monodromy group is the principal congruence subgroup

$$\Gamma(1 - \zeta_3) = \{g \in \Gamma \mid g \equiv I \mod (1 - \zeta_3)\}$$

with respect to the ideal $(1 - \zeta_3) \subset \mathbb{Z}[\zeta_3]$ of the group

$$\Gamma = \{g \in GL(4, \mathbb{Z}[\zeta_3]) \mid g^*Ag = A\}.$$

The quotient space $\mathbb{B}_3^A/\Gamma(1 - \zeta_3)$ has $\binom{6}{3}/2 = 10$ cusps. The quotient group $\Gamma/\Gamma(1 - \zeta_3)$ is isomorphic to the symmetric group of degree 6 acting transitively on the set of 10 cusps. For more detail, see [Gon] and [Hun1].

Case 4 ($n = 5$). Put

$$A = \begin{pmatrix} 0 & 0 & 0 & 1-i \\ 0 & H & 0 & 0 \\ 0 & 0 & H & 0 \\ 1+i & 0 & 0 & 0 \end{pmatrix} \quad \text{where} \quad H = \begin{pmatrix} 2 & 1-i \\ 1+i & 2 \end{pmatrix}.$$

The monodromy group is the principal congruence subgroup

$$\Gamma(1-i) = \{g \in \Gamma \mid g \equiv I \mod (1-i)\}$$

with respect to the ideal $(1-i) \subset \mathbb{Z}[i]$ of the group

$$\Gamma = \{g \in GL(6, \mathbb{Z}[i]) \mid g^* A g = A\}.$$

The quotient space $\mathbb{B}_5^A/\Gamma(1-i)$ has $\binom{8}{4}/2 = 35$ cusps. The quotient group $\Gamma/\Gamma(1-i)$ is isomorphic to the symmetric group of degree 8 acting transitively on the set of 35 cusps. For more detail, see [MY].

REMARK 3.1. We can consider the 'real part' of the story of Case 2. Let $\mathbb{B}_{\mathbb{R}}^A$ be the real locus of the 2-ball $\mathbb{B}_2^A$:

$$\mathbb{B}_{\mathbb{R}}^A = \{x = (x_0, x_1, x_2) \in \mathbb{R}^3 \mid x \Lambda {}^t x < 0\}/\mathbb{R}^\times.$$

Let us 'restrict' our groups Γ and $\Gamma(1 - \zeta_5)$ to this real 2-ball $\mathbb{B}_{\mathbb{R}}^A$; we put

$$\Lambda := \Gamma \cap GL(3, \mathbb{R}), \quad \Lambda(\sqrt{5}) := \Gamma(1 - \zeta_5) \cap GL(3, \mathbb{R}).$$

These groups act (discontinuously, of course) on $\mathbb{B}_{\mathbb{R}}^A$. Since we have

$$\mathbb{Z}[\zeta_5] \cap \mathbb{R} = \mathbb{Z}[u], \quad (1 - \zeta_5)\mathbb{Z}[\zeta_5] \cap \mathbb{R} = \sqrt{5}\mathbb{Z}[u],$$

these groups have the following presentations:

$$\Lambda = \{g \in GL(3, \mathbb{Z}[u]) \mid {}^t g A g = A\},$$
$$\Lambda(\sqrt{5}) = \{g \in \Lambda \mid g \equiv I \mod \sqrt{5}\}.$$

Thanks to the presentations above, we can show the isomorphism

$$\Lambda/\Lambda(\sqrt{5}) \cong \Gamma/\Gamma(1 - \zeta_5) \cong S_5$$

and the commutative diagram

$$\begin{array}{ccc} \mathbb{B}_{\mathbb{R}}^A/\Lambda(\sqrt{5}) & \hookrightarrow & \mathbb{B}_2^A/\Gamma(1 - \zeta_5) \\ \cong \downarrow & & \downarrow \cong \\ \overline{X}_{\mathbb{R}}(2,5) & \hookrightarrow & \overline{X}(2,5). \end{array}$$

That is, the quotient of the real 2-ball $\mathbb{B}_{\mathbb{R}}^A$ under the group $\Lambda(\sqrt{5})$ is isomorphic to the compactification of the real configuration space $X_{\mathbb{R}}(2,5)$, on which the symmetric group S_5 acts. You see, this is just the 'real part' of the story stated in Case 2. For more detail, see [AY].

Part 3

The Story of the Configuration Space $X(3,6)$ of Six Lines on the Projective Plane

CHAPTER VII

The Configuration Space $X(3,6)$

We study the configuration space of six points on the complex projective plane. Let $M(3,6)$ be the space of 3×6-complex matrices, $D_x(ijk)$ the (i,j,k)-minor of $x \in M(3,6)$, and

$$M^*(3,6) = \{x \in M(3,6) \mid D_x(ijk) \neq 0 \text{ for } 1 \leq i < j < k \leq 6\}.$$

The configuration space can be expressed by

$$X := X(3,6) = GL(3,\mathbb{C})\backslash M^*(3,6)/H_6, \quad H_6 = (\mathbb{C}^\times)^6.$$

We have referred to this as the configuration space of six points, regarding each column $x_j = {}^t(x_{1j}, x_{2j}, x_{3j})$ of $x \in M^*(3,6)$ as the homogeneous coordinate of a "point" of $\mathbb{P}^2$. If you consider the column x_j to represent the linear equation

$$x_{1j}t_1 + x_{2j}t_2 + x_{3j}t_3 = 0, \quad t_1 : t_2 : t_3 \in \mathbb{P}^2,$$

then the space $X(3,6)$ can be considered as the configuration space of six "lines" on the projective plane. $D_x(ijk) = 0$ implies that the three points x_i, x_j and x_k lie on a line or the three lines x_i, x_j and x_k meet at a point. We shall regard X as either the configuration space of six points or that of six lines according to convenience.

This space can be seen as an open subset of the affine space $\mathbb{C}^4$, since for any point $z \in M^*(3,6)$, one can find $g \in GL(3)$ and $h \in H_6$ such that

$$gxh = \begin{pmatrix} x^1 & x^2 & 1 & 1 & 0 & 0 \\ x^3 & x^4 & 1 & 0 & 1 & 0 \\ 1 & 1 & 1 & 0 & 0 & 1 \end{pmatrix}.$$

This implies in particular that X is 4-dimensional and that $\mathbb{C}^4 - X$ is the union of not only hyperplanes but also a quadric hypersurface.

In this chapter, we study the space X in terms of classical algebraic geometry. Several studies have been made of this space (see [DO], [Sek], [ST], [MSY1]). We give a democratic projective embedding of X, obtain a compactification $\overline{X}$ of X, and study the structure of the divisor $\overline{X} - X$. We find that $\overline{X}$ is the double cover of $\mathbb{P}^4$ branching along a quartic hypersurface.

Since X is a complex 4-dimensional space (real 8-dimensional), we cannot expect to understand it intuitively, as we were able to do for $X(2,5)$ in Chapter V. However, thanks to the fact that X (and $\overline{X}$) is defined over the real numbers, many features of X can be understood on its real locus $X_{\mathbb{R}}$, as we saw in §2.3 and §2.4 of Chapter V for $X(2,5)$. In §8 we study $X_{\mathbb{R}}$ from an intuitive geometric viewpoint so that we can really see it.

1. One Attempt to Make a Democratic Projective Embedding

Recall that when we made a democratic projective embedding of $X(2,4)$, we considered the left $GL(2)$ and the right H_4 invariant map

$$M^*(2,4) \ni x \longmapsto \, : D_x(ij)D_x(kl) : \, \in \mathbb{P}^{3-1},$$

where

$$i < j, \ k < l, \ \{i,j,k,l\} = \{1,\ldots,4\}, \quad \text{and} \quad 3 = \binom{4}{2}/2.$$

This map induces an embedding of $X(2,4)$ into the line

$$D(12)D(34) - D(13)D(24) + D(14)D(23) = 0.$$

This is the Plücker relation.

Let us consider the map

$$p_r : M^*(3,6) \ni z \longmapsto \, : D_z(ijk)D_z(lmn) : \, \in \mathbb{P}^{10-1},$$

where

$$i < j < k, \ l < m < n, \ \{i,j,k,l,m,n\} = \{1,\ldots,6\}, \quad \text{and} \quad 10 = \binom{6}{3}/2,$$

hoping that it induces an embedding of $X(3,6)$; at least it is democratic! This map is invariant under the left $GL(3)$ and the right H_6 actions since we have

$$D_{gxh}(ijk)D_{gxh}(lmn) = (\det g)^2 D_x(ijk)D_x(lmn)h_1 \cdots h_6,$$

where

$$g \in GL(3), \quad h = diag(h_1, \cdots, h_6).$$

The *Plücker relations* read

$$Plk(ij) : D(ijk)D(lmn) - D(ijl)D(mnk)$$
$$+ D(ijm)D(nkl) - D(ijn)D(klm) = 0.$$

(This can be checked directly using a normal form.) This system $\{Plk(ij)\}$ is of rank 5 (showing this is a simple exercise in linear algebra). Thus the image of $M^*(3,6)$ under p_r lies in a 4-dimensional linear subspace Z of $\mathbb{P}^{10-1}$. Moreover, p_r is an open map to Z. These facts can be checked by the help of another expression for p_r given in the next section. So far, so good. Then does p_r give an embedding of $X = X(3,6)$ into $Z \cong \mathbb{P}^4$? Unfortunately, the answer is "no."

$$\textit{Life is not so easy.}$$

Our map $p_r : X \to Z$ is not one-to-one but two-to-one, as we see in the next section.

2. A Non-democratic Embedding of X

We shall make a democratic embedding of X in §4. Its codimension will be very high. On the other hand, the normalization appearing in the beginning of this chapter gives an embedding $X \subset \mathbb{C}^4$. This is an embedding of codimension 0, but it destroys the symmetry to too great a degree. The following embedding is of codimension 1. It is not democratic, but is symmetric with respect to $\{1,2,3\}$ and $\{4,5,6\}$. With the help of this embedding, we see that our map $p_r : X \to Z$ is two-to-one.

Let us normalize matrices in $M^*(3,6)$ into

$$z = (x, I_3) = \begin{pmatrix} 11 & 12 & 13 & 1 & 0 & 0 \\ 21 & 22 & 23 & 0 & 1 & 0 \\ 31 & 32 & 33 & 0 & 0 & 1 \end{pmatrix}, \quad \text{where} \quad x = (x_{ij}), \quad ij = x_{ij},$$

and consider the following embedding (*not democratic!*) of X into the 5-dimensional projective space coordinatized by $(p; n) = (p_1, p_2, p_3; n_1, n_2, n_3)$:

$$(x, I_3) \mapsto (p; n) = (11 \cdot 22 \cdot 33, \ 13 \cdot 21 \cdot 32, \ 12 \cdot 23 \cdot 31;$$
$$13 \cdot 22 \cdot 31, \ 12 \cdot 21 \cdot 33, \ 11 \cdot 23 \cdot 32).$$

The closure of the image, let us call it W, is defined by a single equation:

$$W : p_1 p_2 p_3 = n_1 n_2 n_3.$$

Note: The reader who is familiar with the theory of torus embeddings can play with W (cf. [CMY]).

If one normalizes x as $31 = 32 = 33 = 13 = 23 = 1$, we have

$$p_2 = 21, \quad p_3 = 12, \quad n_1 = 22, \quad n_3 = 11.$$

Therefore it is an embedding.

For convenience's sake, we tabulate the values of $D(ijk) = D_z(ijk)$. For

$$z = (x, I_3) \in M^*(3,6), \quad x = (x_{ij}), \quad ij := x_{ij},$$

we have

$$D(123) = \det x, \quad D(456) = 1,$$

$$D(k56) = 1k, \qquad D(k64) = 2k, \qquad D(k45) = 3k,$$

$$D(ij4) = \begin{vmatrix} 2i & 2j \\ 3i & 3j \end{vmatrix}, \quad D(ij5) = \begin{vmatrix} 3i & 3j \\ 1i & 1j \end{vmatrix}, \quad D(ij6) = \begin{vmatrix} 1i & 1j \\ 2i & 2j \end{vmatrix},$$

for $k = 1, 2, 3$ and $1 \le i \ne j \le 3$. Explicitly, these yield

$$D(156) = 11, \qquad D(256) = 12, \qquad D(356) = 13,$$
$$D(164) = 21, \qquad D(264) = 22, \qquad D(364) = 23,$$
$$D(145) = 31, \qquad D(245) = 32, \qquad D(345) = 33,$$

$$D(234) = \begin{vmatrix} 22 & 23 \\ 32 & 33 \end{vmatrix}, \quad D(314) = \begin{vmatrix} 23 & 21 \\ 33 & 31 \end{vmatrix}, \quad D(124) = \begin{vmatrix} 21 & 22 \\ 31 & 32 \end{vmatrix},$$

$$D(235) = \begin{vmatrix} 32 & 33 \\ 12 & 13 \end{vmatrix}, \quad D(315) = \begin{vmatrix} 33 & 31 \\ 13 & 11 \end{vmatrix}, \quad D(125) = \begin{vmatrix} 31 & 32 \\ 11 & 12 \end{vmatrix},$$

$$D(236) = \begin{vmatrix} 12 & 13 \\ 22 & 23 \end{vmatrix}, \quad D(316) = \begin{vmatrix} 13 & 11 \\ 23 & 21 \end{vmatrix}, \quad D(126) = \begin{vmatrix} 11 & 12 \\ 21 & 22 \end{vmatrix}.$$

Our democratic map p_r is now defined on W as follows:

$$W \ni (p; n) \mapsto D(p; n) : p_1 - n_1 : p_1 - n_2 : \cdots : p_3 - n_3 \in Z \subset \mathbb{P}^{10-1},$$

where

$$D(p; n) := p_1 + p_2 + p_3 - (n_1 + n_2 + n_3).$$

If $(n; p) \in W$ and $(n'; p') \in W$ have the same image, then we have

$$D(p', n') = aD(p, n), \quad p'_i - n'_j = a(p_i - n_j), \quad i, j = 1, 2, 3$$

for some $a \neq 0$. Since we have

$$p'_i - ap_i = n'_j - an_j, \quad i, j = 1, 2, 3,$$

there is a constant b such that

$$p'_i = ap_i + b, \quad n'_j = an_j + b, \quad i, j = 1, 2, 3.$$

Putting these expressions into the equation $p'_1 p'_2 p'_3 = n'_1 n'_2 n'_3$, we have

$$(ap_1 + b)(ap_2 + b)(ap_3 + b) = (an_1 + b)(an_2 + b)(an_3 + b).$$

The equation $p_1 p_2 p_3 = n_1 n_2 n_3$ leads to

$$b\{aq(p; n) + bD(p; n)\} = 0,$$

where (note that $D \neq 0$)

$$q(p; n) := p_1 p_2 + p_2 p_3 + p_3 p_1 - (n_1 n_2 + n_2 n_3 + n_3 n_1).$$

Therefore we conclude that if $q(p; n) = 0$ (and $b \neq 0$), then

$$b = 0, \quad \text{i.e.,} \quad (p'; n') = (p; n),$$

and if $q(p, n) \neq 0$, then

$$p'_i = D(p; n)p_i - q(p; n), \quad n'_i = D(p; n)n_i - q(p; n), \quad i, j = 1, 2, 3.$$

In other words, our map $p_r : X \to Z$ is two-to-one outside the variety

$$Q : q(p; n) = 0$$

and is one-to-one on this variety. That is, p_r gives a double cover branching along the variety Q.

If you wish to write $q(p; n)$ in terms of the x_{ij}, then you have

$$q(z) = q(p; n) = 123 \cdot 312 + 312 \cdot 231 + 231 \cdot 123$$
$$-321 \cdot 132 - 132 \cdot 213 - 213 \cdot 321,$$

where

$$z = (x, I_3) \in M^*(3, 6), \quad x = (x_{ij}), \quad ijk := x_{1i} x_{2j} x_{3k}.$$

We have found an *involution* (automorphism of order 2) on X defined, in terms of $(p; n)$, by

$$(p_i, n_j) \mapsto (Dp_i - q, Dn_j - q), \quad D = D(p; n), \quad q = q(p; n).$$

In the next section, we give its geometric meaning.

Since $Q : q = 0$ is the branching locus of the double cover $p_r : X \to Z$, the expression q^2 must be a quartic form on Z. Of course one can write this expression directly, for instance, as

$$
\begin{aligned}
\{ &- (p_1 - n_3)(p_3 - n_1) - (p_3 - n_1)(p_3 - n_2) + (p_3 - n_2)(p_3 - n_3) \\
&+ (p_3 - n_3)(p_2 - n_1) - (p_2 - n_1)(p_1 - n_3)\}^2 \\
&- 4(p_2 - n_1)(p_1 - n_2)(p_1 - n_3)(p_3 - n_1) \\
= \{ &p_1 p_2 + p_2 p_3 + p_3 p_1 - (n_1 n_2 + n_2 n_3 + n_3 n_1)\}^2 = q^2,
\end{aligned}
$$

modulo $p_1 p_2 p_3 - n_1 n_2 n_3$. However, there is no symmetry in this formula; a nicer expression will be given at the end of §4.

3. The Involution $*$

Let us recall the duality on the projective plane: We arbitrarily fix a non-singular conic C on $\mathbb{P}^2$. The dual of a point $p \in \mathbb{P}^2$, with respect to C, is the line joining the two tangent points of the two tangents of C passing through p (if p is on the conic C, the dual is the tangent of C at p).

The *involution* $*$ on X is defined as follows: For a given system l of six lines $l_1, \ldots, l_6$ in general position on the plane, let us name six intersection points (out of fifteen) as

$$
\begin{aligned}
p_1 &:= l_2 \cap l_3, & p_2 &:= l_3 \cap l_1, & p_3 &:= l_1 \cap l_2, \\
p_4 &:= l_5 \cap l_6, & p_5 &:= l_6 \cap l_4, & p_6 &:= l_4 \cap l_5,
\end{aligned}
$$

(see Figure 3.1). The system $*l$ is the system of six lines which are dual,

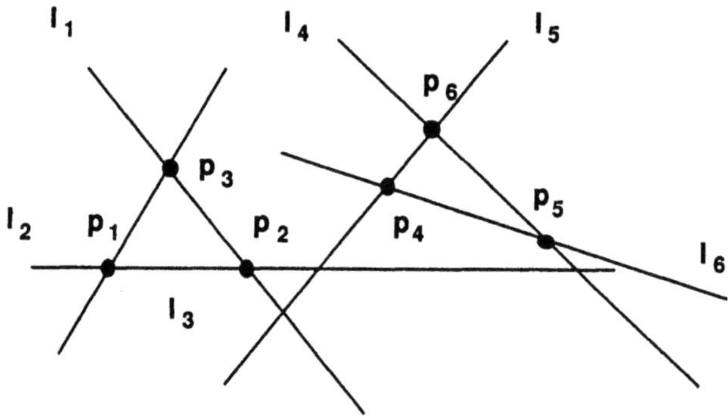

FIGURE 3.1. The involution $*$: Getting $\{p_j\}$ from $\{l_j\}$

with respect to a conic, to $p_1, \ldots, p_6$. Since two non-singular conics are equivalent under projective transformations, the involution $*$ on X is independent of the choice of the non-singular conic.

If we fix homogeneous coordinates on the projective plane and define the conic C by the quadratic form I_3, we can easily trace the above procedure in terms of coordinates and obtain

PROPOSITION 3.1. *If* $(x, y) \in M^*(3,6)$ *represents a system* $l \in X$, *then* $(\tilde{x}, \tilde{y})$ *represents* $*l$, *where* x *and* y *are* 3×3 *matrices, and* $\tilde{x}$ *and* $\tilde{y}$ *are cofactor matrices of* x *and* y, *respectively:*

$$\tilde{x} = \begin{pmatrix} \begin{vmatrix} 22 & 23 \\ 32 & 33 \end{vmatrix} & \begin{vmatrix} 23 & 21 \\ 33 & 31 \end{vmatrix} & \begin{vmatrix} 21 & 22 \\ 31 & 32 \end{vmatrix} \\[2mm] \begin{vmatrix} 32 & 33 \\ 12 & 13 \end{vmatrix} & \begin{vmatrix} 33 & 31 \\ 13 & 11 \end{vmatrix} & \begin{vmatrix} 31 & 32 \\ 11 & 12 \end{vmatrix} \\[2mm] \begin{vmatrix} 12 & 13 \\ 22 & 23 \end{vmatrix} & \begin{vmatrix} 13 & 11 \\ 23 & 21 \end{vmatrix} & \begin{vmatrix} 11 & 12 \\ 21 & 22 \end{vmatrix} \end{pmatrix}, \quad x = \begin{pmatrix} 11 & 12 & 13 \\ 21 & 22 & 23 \\ 31 & 32 & 33 \end{pmatrix}, \quad ij = x_{ij}.$$

PROPOSITION 3.2. *This involution* $*$ *coincides with the Grassmann isomorphism defined in* §13 *of Chapter I.*

PROOF. Under the right action of H_6 on $M^*(3,6)$, $(\tilde{x}, \tilde{y})$ is equivalent to $((\det x)^{-1}\tilde{x}, -(\det y)^{-1}\tilde{y})$, which is equal to $({}^t x^{-1}, -{}^t y^{-1})$. We have

$$(x, y)^t({}^t x^{-1}, -{}^t y^{-1}) = I_3 - I_3 = 0.$$

□

Though the definition of the involution $*$ given in this section apparently depends on the partition $\{1, 2, 3\} \cup \{4, 5, 6\}$, the definition of the Grassmann isomorphism tells that our $*$ is intrinsic. More precisely, we have

PROPOSITION 3.3. *The action of* $*$ *commutes with that of* S_6.

You can also check this proposition by a straightforward computation. This will become clearer in the next section.

Note that this involution $*$ on X is the very involution on W which appears at the end of the previous section. This can be checked for

$z = (x, I_3)$ by computing the (p,n)-coordinates of $*z = (\tilde{x}, I_3)$. For instance, $p_1 = 11 \cdot 22 \cdot 33$ is transformed into

$$*p_1 = \begin{vmatrix} 22 & 23 \\ 32 & 33 \end{vmatrix} \begin{vmatrix} 33 & 31 \\ 13 & 11 \end{vmatrix} \begin{vmatrix} 11 & 12 \\ 21 & 22 \end{vmatrix},$$

and we have

$$*p_1 = 11 \cdot 22 \cdot 33 \cdot D - q,$$

where

$$D = \det x = p_1 + \cdots - (n_1 + \cdots), \quad q = p_1 p_2 + \cdots - (n_1 n_2 + \cdots).$$

The fixed points of $*$ have a clear geometric characterization:

PROPOSITION 3.4. $l = (l_1, \ldots, l_6) \in X$ is a fixed point of the involution $*$ if and only if there is a conic tangent to l_j $(1 \le j \le 6)$.

PROOF. Let (x, I_3) represent a system l. Since a conic is by definition a linear form in quadratic monomials,

$$y_1^2, \; y_2^2, \; y_3^2, \; y_1 y_2, \; y_2 y_3, \; y_3 y_1, \quad \text{where} \quad y_1 : y_2 : y_3 \in \mathbb{P}^2,$$

there is a conic tangent to l_j $(1 \le j \le 6)$ if and only if

$$qq := \begin{vmatrix} 11^2 & 12^2 & 13^2 & 1 & 0 & 0 \\ 21^2 & 22^2 & 23^2 & 0 & 1 & 0 \\ 31^2 & 32^2 & 33^2 & 0 & 0 & 1 \\ 11 \cdot 21 & 12 \cdot 22 & 13 \cdot 23 & 0 & 0 & 0 \\ 21 \cdot 31 & 22 \cdot 32 & 23 \cdot 33 & 0 & 0 & 0 \\ 31 \cdot 11 & 32 \cdot 12 & 33 \cdot 13 & 0 & 0 & 0 \end{vmatrix} = 0, \quad ij = x_{ij}, \quad x = (x_{ij}).$$

Put $ijk := 1i \cdot 2j \cdot 3k$. Then

$$\begin{aligned} qq &= 11 \cdot 21 \cdot 22 \cdot 32 \cdot 33 \cdot 13 + \cdots \\ &\quad - 13 \cdot 23 \cdot 22 \cdot 32 \cdot 31 \cdot 11 - \cdots \\ &= 123 \cdot 312 + 312 \cdot 231 + 231 \cdot 123 \\ &\quad - 321 \cdot 132 - 132 \cdot 213 - 213 \cdot 321. \end{aligned}$$

This is precisely the q which appeared in the previous section. $\square$

Let Q denote the set of fixed points of $*$ in X. Since there is only one non-singular conic up to projective transformations, and since six lines tangent to a conic are determined by six points on the conic, we know that

$$Q \cong X(2,6).$$

Note that Q is non-singular in X.

4. A Democratic Embedding

In this section, the previously announced projective embedding of X is given. This embedding is S_6-symmetric at the cost of its high-dimensional target (29-dimensional) and unavoidable complicated notation. The reader who is ready to believe the results can skip the proofs, which will not be used later.

Consider a regular tetrahedron in euclidean 3-space and label the six edges by $1,\ldots,6$ (see Figure 4.1). We identify two *labeled tetrahedra* if

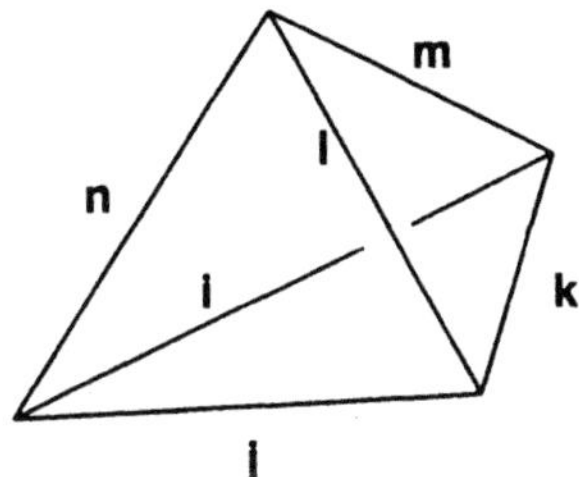

FIGURE 4.1. A labeled tetrahedron

one can be transformed into the other by an element of the extended tetrahedral group, which is isomorphic to the symmetric group S_4 on four letters. In this way we have $|S_6/S_4| = 30$ labeled tetrahedra.

Since the Poincaré dual of a tetrahedron is again a tetrahedron, associating the same number to the corresponding edges of two dual tetrahedra, we define the involution $*$ acting on the set of labeled tetrahedra (see Figure 4.2). Note that the dual of a labeled tetrahedron is obtained

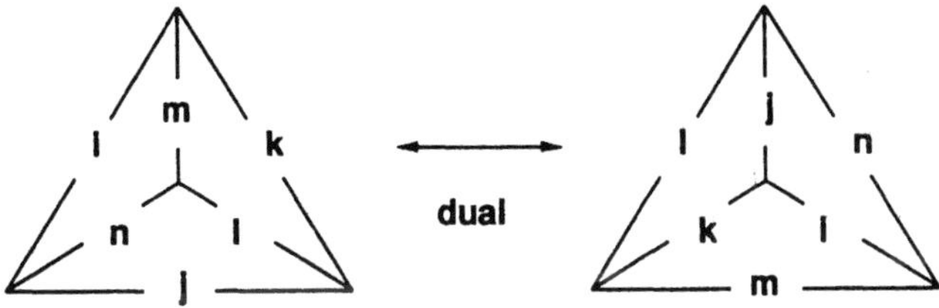

FIGURE 4.2. Duality on labeled tetrahedra

by exchanging the labels of a pair of antipodal edges. (Antipodal edges are defined to be edges which do not meet.) The symmetric group S_6 on six numerals $\{1,\ldots,6\}$ acts on the set of labeled tetrahedra. Note that the action of $*$ commutes with that of S_6.

Let $\mathbb{P}^{30-1}$ be the projective space with the 30 labeled tetrahedra as homogeneous coordinates. For printing convenience, we denote the labeled tetrahedron shown in Figure 4.1 (also the one on the left in Figure 4.2) by

$$
\begin{array}{cccc}
 & m & & \\
i & & k & \\
n & j & l &
\end{array}\;.
$$

We define a map $f : X \to \mathbb{P}^{30-1}$ as follows. Let $z = (z_{ij}) \in M^*(3,6)$ be a representative of a six-point system $P \in X$. For each labeled tetrahedron we associate the product of four 3×3 minors of the matrix z corresponding to the four triangular faces which are oriented according to an orientation of the euclidean space where the tetrahedron lives, i.e.,

$$
f : P \mapsto \left(\begin{array}{cccc} & m & & \\ i & & k & \\ n & & j & l \end{array} = D_z(min)D_z(njl)D_z(lkm)D_z(kji)\right)_{(ijklmn)\in S_6/S_4}\;.
$$

LEMMA 4.1. *The map f is well defined and continuous.*

PROOF. Continuity is obvious by the definition of quotient topology. The following assertions can be easily checked. (1) The map f is independent of the choice of the representative z. In other words it is invariant under the action on $M^*(3,6)$ of $GL(3)$ and H_6. (2) It is independent of the choice of orientation of the euclidean space. $\square$

LEMMA 4.2. *The map f is compatible with the action of S_6 and $*$.*

PROOF. The assertion for the S_6-action is obvious from the definition. The involution $*$ on X is defined by $*(x, I_3) = (\tilde{x}, I_3)$, where $\tilde{x}$ is the cofactor matrix of x. The involution $*$ on $\mathbb{P}^{30-1}$ is defined by the Poincaré duality:

$$
*\left(\begin{array}{cccc} & m & & \\ i & & k & \\ n & j & l & \end{array}\right) = \begin{array}{cccc} & j & & \\ l & & n & \\ k & m & i & \end{array} = D(jlk)D(kmi)D(inj)D(nml).
$$

Then we have only to compare the expression of the $D(ijk)$ given in §2 and the explicit form of the cofactor matrix $\tilde{x}$ given in Proposition 3.1. $\square$

By the explicit expression of the $D_z(ijk)$ for $z = (x, I_3)$ in §2, we have

LEMMA 4.3. *For a labeled tetrahedron T,*

$$T(x, I_3) - *T(x, I_3) = \pm q,$$

where q is defined in §2.

We will see that f is injective, i.e., that it is an embedding of X. Let $Y \subset \mathbb{P}^{30-1}$ be the closure of the image of X under f. We study the relation between $p_r : X \to Z$ and $f : X \to Y$. We wish to define a map $h : Y \to Z$ such that the following diagram commutes:

$$
\begin{array}{ccc}
& f & \\
X & \longrightarrow & Y \subset \mathbb{P}^{30-1} \\
p_r \downarrow & \swarrow\ h & \\
\mathbb{P}^{10-1} \supset Z & &
\end{array}
$$

Since p_r is defined by quadratic forms of the $D(ijk)$ and f by quartic forms, a standard way to define such a map is to make a Veronese embedding of Z and to construct a map from Y to the embedded variety. Let us carry out this program.

Let $\mathbb{P}^{45-1}$ be the projective space with homogeneous coordinates $\{ij; kl\}$, where $\{i, j, k, l\}$ is a subset of $\{1, \ldots, 6\}$ and

$$
\begin{aligned}
\{ij; kl\} &= \{ji; kl\} \\
&= \{ij; lk\} = \{kl; ij\}.
\end{aligned}
$$

There are $\binom{6}{2}\binom{4}{2}/2 = 45$ such coordinates. Let us define a map

$$g : \mathbb{P}^{10-1} \longrightarrow \mathbb{P}^{45-1}$$

by

$$\{ij; mn\} = h(ijk; lmn)h(ijl; kmn),$$

where

$$h(ijk; lmn) = h(lmn; ijk) = h(kij; lmn) = -h(jik, lmn)$$

are homogeneous coordinates on $\mathbb{P}^{10-1}$. The 4-dimensional linear subvariety Z of $\mathbb{P}^{10-1}$ is defined by the Plücker relation

$$Plk(ij) : h(ijk; lmn) - h(ijl; mnk) + h(ijm; nkl) - h(ijn; klm) = 0.$$

LEMMA 4.4. *The restriction of g on $Z \subset \mathbb{P}^{10-1}$ gives the isomorphism $Z \to g(Z)$.*

PROOF. By associating, to the coordinates $h(ijk; lmn)$, all their quadratic monomials $h(ijk; lmn)h(ijl; kmn)$ and $h(ijk; lmn)^2$, we have an embedding (the Veronese embedding) of $\mathbb{P}^{10-1}$ into $(10 \cdot 11/2 - 1)$-dimensional projective space. On the other hand, multiplying the Plücker relation above by $h(ijk; lmn)$, we have

$$h(ijk; lmn)^2 = h(ijl; kmn)h(ijk; lmn)$$
$$+ h(ijm; kln)h(ijk; mln) + h(ijn; klm)h(ijk; nlm).$$

This shows that the squares $h(ijk; lmn)^2$ are redundant on Z. $\square$

Let

$$\pi : \mathbb{P}^{30-1} \supset Y \longrightarrow \mathbb{P}^{45-1}$$

be defined by

$$\{ij, mn\} = \begin{matrix} & & l \\ & k & & j \\ m & & i & & n \end{matrix} + \begin{matrix} & & l \\ & k & & i \\ m & & j & & n \end{matrix}.$$

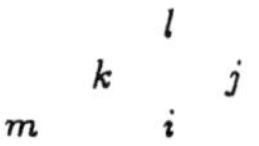

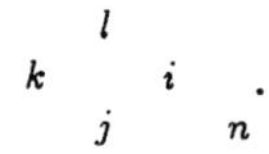

PROPOSITION 4.5. *The following diagram is commutative.*

$$
\begin{array}{ccc}
X & \xrightarrow{\ f\ } & Y \\
p_r \downarrow & & \downarrow \pi \\
Z & \xrightarrow[g]{\ \sim\ } & g(Z) = \pi(Y) \subset \mathbb{P}^{45-1}.
\end{array}
$$

PROOF. Putting $j = k$ in the Plücker relation $Plk(ij)$ in §1, we find

$$-D(ikl)D(mnk) + D(ikm)D(nkl) - D(ikn)D(klm) = 0.$$

Multiplying $D(nji)D(mjl)$, we have

$$0 = -D(ikl)D(mnk)D(nji)D(mjl) + D(ikm)D(nkl)D(nji)D(mjl)$$
$$- D(ikn)D(klm)D(nji)D(mjl)$$

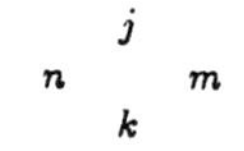

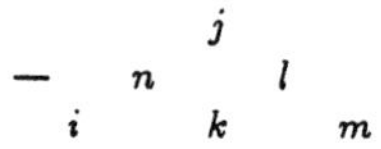

$$= - \begin{matrix} & j \\ n & & m \\ i & k & l \end{matrix} - \begin{matrix} & j \\ n & & l \\ i & k & m \end{matrix} - D(lmk)D(jin)D(lmj)D(kni)$$

$$= -\{lm; in\} + D(lmk)D(jin)D(lmj)D(kin),$$

which gives an expression for $\pi \circ f$. On the other hand, $g \circ p$ is expressed by

$$\{ij; mn\} = h(ijk; lmn)h(ijl; kmn)$$
$$= D(ijk)D(lmn)D(ijl)D(kmn).$$

These two expressions agree. $\square$

COROLLARY 4.6. *The map π on Y is $*$-invariant. It is two-to-one on $Y - Q$, while it is one-to-one on $Q \subset Y$.*

LEMMA 4.7. *On $Y \subset \mathbb{P}^{30-1}$, we have*

$$
\begin{pmatrix} & l & \\ k & & i \\ m & j & n \end{pmatrix} + * \begin{pmatrix} & l & \\ k & & i \\ m & j & n \end{pmatrix} = \{im; jl\} - \{jn; ik\} + \{kl; jm\}
$$
$$
- \{nl; ij\} + \{nm; jk\},
$$
$$
\begin{pmatrix} & l & \\ k & & i \\ m & j & n \end{pmatrix} \cdot * \begin{pmatrix} & l & \\ k & & i \\ m & j & n \end{pmatrix} = \{ij; ml\}\{jl; mi\},
$$

where $\{ij; kl\}$ are the linear forms on $\mathbb{P}^{30-1}$ defined just above Proposition 4.5.

PROOF. The result follows from a straightforward computation. $\square$

COROLLARY 4.8. $f : X \longrightarrow Y$ *is an injection.*

PROPOSITION 4.9. *The defining equation of $Q \subset Z$ is a quartic in $h(ijk; lmn)$ given by*

$$
(T - {*}T)^2 = (\{im; jl\} - \{jn; ik\} + \{kl; jm\} - \{nl; ij\} + \{nm; jk\})^2
$$
$$
- 4\{ij; ml\}\{jl; mi\}.
$$

where $\{ij; mn\} = h(ijk; lmn)h(ijl; kmn)$.

PROOF. Since, for each labeled tetrahedron T, we have

$$
(T - {*}T)^2 = (T + {*}T)^2 - 4T \cdot {*}T,
$$

Lemma 4.7 leads to the conclusion. $\square$

The expression for q^2 given in §2 is obtained by taking $(i, j, k, l, m, n) = (1, 2, 4, 3, 5, 6)$ in the above expression:

$$
\{D(154)D(623)D(156)D(423) - D(263)D(514)D(265)D(314)
$$
$$
+ D(431)D(625)D(436)D(125) - D(634)D(512)D(635)D(412)
$$
$$
+ D(651)D(324)D(653)D(124)\}^2
$$
$$
- 4D(124)D(653)D(126)D(453)D(234)D(651)D(236)D(451)
$$
$$
= \left\{ -31 \begin{vmatrix} 12 & 13 \\ 22 & 23 \end{vmatrix} 11 \begin{vmatrix} 22 & 23 \\ 32 & 33 \end{vmatrix} - \cdots \right\}^2
$$
$$
- 4 \cdot 13 \begin{vmatrix} 21 & 22 \\ 31 & 32 \end{vmatrix} 33 \begin{vmatrix} 11 & 12 \\ 21 & 22 \end{vmatrix} 11 \begin{vmatrix} 22 & 23 \\ 32 & 33 \end{vmatrix} 31 \begin{vmatrix} 12 & 13 \\ 22 & 23 \end{vmatrix}.
$$

5. Degenerate Arrangements

Since we obtained an embedding f of X into $Y \subset \mathbb{P}^{30-1}$, we would like to add degenerate arrangements to

$$X = X(3,6) = G\backslash M^*/H, \quad M^* = M^*(3,6), \ G = GL(3,\mathbb{C}), \ H = H_6$$

in order to make an isomorphism onto Y. We do not add all the degenerate arrangements in $M(3,6) - M^*(3,6)$. This is similar to the situation in Chapter V where, when we made $\overline{X}(2,5)$ out of $X(2,5)$, we did not add arrangements for which three points coincide. Do you remember? We will add the following seven kinds of arrangements:

1) $X_3^{ijk} = G\backslash\{x \in M^* \mid D_x(ijk) = 0, \text{ all other } D \neq 0\}/H, \quad D = D_x,$

where $\{i,j,k\} \subset \{1,\ldots,6\}$. This is the set of arrangements where three lines labeled i, j and k meet at a point, but nothing further special occurs.

2) $X_{2\alpha}^{ijk;lmn} = G\backslash\{x \in M^* \mid D(ijk) = D(lmn) = 0, \text{ all other } D \neq 0\}/H,$

where $\{i,j,k,l,m,n\} = \{1,\ldots,6\}$. This is the set of arrangements where three lines labeled i, j and k meet at a point and three lines labeled l, m and n meet at another point, but nothing further special occurs.

3) $X_{2\beta}^{ijk;imn} = G\backslash\{x \in M^* \mid D(ijk) = D(imn) = 0, \text{ all other } D \neq 0\}/H,$

where $\{i,j,k,m,n\} \subset \{1,\ldots,6\}$. This is the set of arrangements where three lines labeled i, j and k meet at a point and three lines labeled i, m and n meet at another point, but nothing further special occurs.

4) $X_{1\alpha}^{ij} = G\backslash\{x \in M^* \mid D(ijk) = D(lmn) = 0, \text{ for all } k,l,m,n \notin \{i,j\},$
$$\text{all other } D \neq 0\}/H,$$

where $\{i,j\} \subset \{1,\ldots,6\}$. This is the set of arrangements where two lines labeled i and j coincide and four other lines meet at a point, but nothing further special occurs.

5) $X_{1\beta}^{ijk;klm;mni} = G\backslash\{x \in M^* \mid D(ijk) = D(klm) = D(mni) = 0,$
$$\text{all other } D \neq 0\}/H,$$

where $\{i,j,k,l,m,n\} = \{1,\ldots,6\}$. This is the set of arrangements where three lines labeled i,j,k meet at a point, three lines labeled k,l,m meet at a point, and three lines labeled m,n,i meet at a point, but nothing further special occurs.

$$6) \quad X_{0\alpha}^{ij;kl;mn} = G\backslash\{x \in M^* \mid D(pqr) \neq 0,\ \text{for } p \in \{i,j\},\ q \in \{k,l\},$$
$$r \in \{m,n\},\ \text{all other } D = 0\}/H,$$

where $\{i,j,k,l,m,n\} = \{1,\ldots,6\}$. This is the set of arrangements where two lines labeled i and j coincide, two lines labeled k and l coincide, and two lines labeled m and n coincide, but nothing further special occurs.

$$7) \quad X_{0\beta}^{ijk;klm;mni;jln} = G\backslash\{x \in M^* \mid D(ijk) = D(klm) = D(mni)$$
$$= D(jln) = 0,\ \text{all other } D \neq 0\}/H,$$

where $\{i,j,k,l,m,n\} = \{1,\ldots,6\}$. This is the set of arrangements where three lines labeled i,j,k meet at a point, three lines labeled k,l,m meet at a point, three lines labeled m,n,i meet at a point, and three lines labeled j,l,n meet at a point, but nothing further special occurs.

6. A Democratic Compactification $\overline{X}(3,6)$

Put
$$X_3 = \cup X_3^{ijk} \quad [20 \text{ summands}]$$

$$X_{2\alpha} = \cup X_{2\alpha}^{ijk;lmn} \quad [10] \qquad X_{2\beta} = \cup X_{2\beta}^{ijk;imn} \quad [90]$$
$$X_{1\alpha} = \cup X_{1\alpha}^{ij} \quad [15] \qquad X_{1\beta} = \cup X_{1\beta}^{ijk;klm;mni} \quad [120]$$
$$X_{0\alpha} = \cup X_{0\alpha}^{ij;kl;mn} \quad [15] \qquad X_{0\beta} = \cup X_{0\beta}^{ijk;klm;mni;jln} \quad [30]$$

and define $\overline{X}'$ and $\overline{X}$ to be the disjoint unions

$$\overline{X}' = X \cup X_3 \cup X_{2\alpha} \cup X_{2\beta} \cup X_{1\beta} \cup X_{0\beta},$$
$$\overline{X} = X \cup X_3 \cup X_{2\alpha} \cup X_{2\beta} \cup X_{1\alpha} \cup X_{1\beta} \cup X_{0\alpha} \cup X_{0\beta}$$

equipped with the quotient topology as subspaces of $G\backslash M/H$.

Let $\overline{p_r} : \overline{X} \to Z$ and $\overline{f} : \overline{X} \to Y$ be the extensions of p_r and f and let them be defined by the formulae defining p_r and f given in §1 and §4, respectively. By the definition of quotient topology, $\overline{p_r}$ and $\overline{f}$ are continuous.

The following proposition is not difficult to prove, but it requires a case by case study for each stratum, so I ask you to simply believe it.

PROPOSITION 6.1. *1) The maps $\overline{p_r}$ and $\overline{f}$ are well-defined, i.e., at least one coordinate does not vanish.*
2) $\overline{f}$ is a homeomorphism onto Y. Through $\overline{f}$, the space $\overline{X}$ naturally has the structure of a projective variety.
3) The space $\overline{X}'$ has the structure of a non-singular algebraic variety.

The following proposition is easy to show.

PROPOSITION 6.2. *1) The involution $*$ can be extended to a holomorphic involution on $\overline{X}$; this will also be denoted by $*$. This involution sends X_3^{ijk} to X_3^{lmn}, and its fixed locus is*

$$\overline{Q} = \text{ closure of } Q \text{ in } \overline{X}$$
$$= Q \cup X_{2\alpha} \cup X_{1\alpha} \cup X_{0\alpha}.$$

2) $l = (l_1, \ldots, l_6) \in \overline{X}$ is a fixed point of the involution $*$ if and only if there is a (not necessarily non-singular) conic tangent to l_j $(1 \leq j \leq 6)$.
3) $\overline{p_r}$ is a double covering branching along $\overline{Q}$.
4) The map $\overline{p_r} : \overline{X} \to Z$ induces the isomorphism

$$\overline{X}/\langle * \rangle \cong Z.$$

Indeed, for example, since

$$*(X, I_3) \sim ({}^t X^{-1}, I_3) \sim (I_3, {}^t X),$$

X_3^{123} is sent to X_3^{456}.

The variety $\overline{X}$ admits the stratifications given above. These are illustrated in Figure 6.1.

Regarding the strata $X_3^{ijk}, \ldots$ as subvarieties of $\overline{X}$, their boundaries are given as follows:

$$\partial X_3^{ijk} = \overline{X}_{2\alpha}^{ijk;lmn} \cup \bigcup_{\#\{i,j,k\} \cap \{p,q,r\}=1} \overline{X}_{2\beta}^{ijk;pqr},$$

$$\partial X_{2\alpha}^{ijk;lmn} = \bigcup_{\{p,q\} \subset \{i,j,k\}} \overline{X}_{1\alpha}^{pq} \cup \bigcup_{\{p,q\} \subset \{l,m,n\}} \overline{X}_{1\alpha}^{pq},$$

$$\partial X_{2\beta}^{ijk;imn} = \overline{X}_{1\alpha}^{jk} \cup \overline{X}_{1\alpha}^{mn} \cup \bigcup_{p=j,k;q=m,n} \overline{X}_{1\beta}^{ijk;imn;lpq},$$

$$\partial X_{1\alpha}^{ij} = X_{0\alpha}^{ij;kl;mn} \cup X_{0\alpha}^{ij;km;ln} \cup X_{0\alpha}^{ij;kn;lm},$$

$$\partial X_{1\beta}^{ijk;klm;mni} = X_{0\alpha}^{ij;kl;mn} \cup X_{0\alpha}^{jk;lm;ni} \cup X_{0\beta}^{ijk;klm;mni;jln},$$

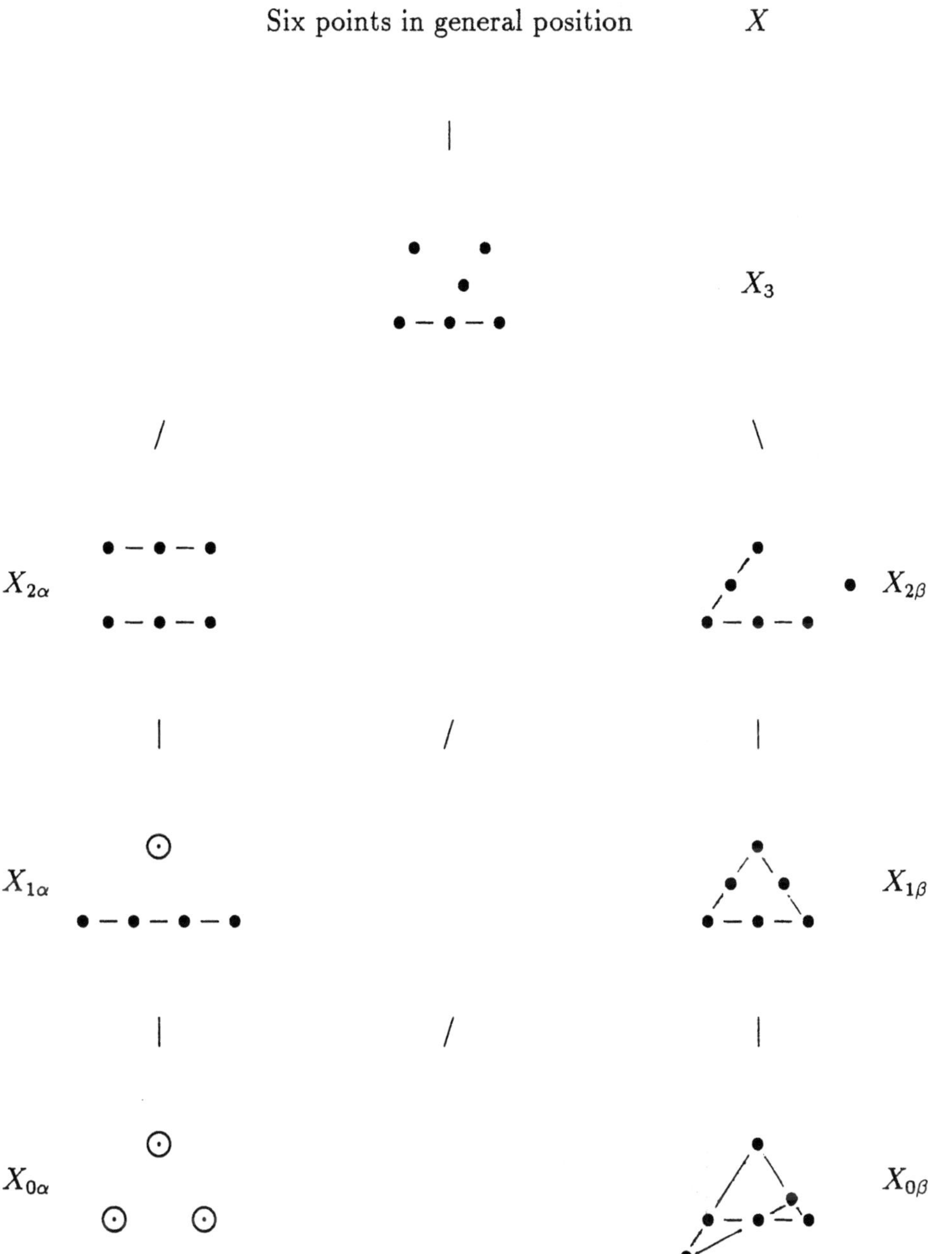

FIGURE 6.1. The stratification: A circle • stands for a line, • — • — •
for three lines with a point in common, and ⊙ for a double line (two
lines coincide). Where two figures are joined by a segment, the one
below is a limit of the one above.

where ∂ and the bar (over X) represent respectively the boundary and closure in $\overline{X}$.

7. Intersection Pattern of the Divisors $\overline{X}_3^{abc}$ and $\overline{Q}$ in $\overline{X}$

The indices a, b, c, p, q, r are assumed to be given as

$$\{a, b, c, p, q, r\} = \{1, 2, 3, 4, 5, 6\}.$$

The divisors $X^{abc} := \overline{X}_3^{abc}$ are isomorphic to $\mathbb{P}^3$. (This fact can be seen by considering their images under the map $\overline{p_r} : \overline{X} \to Z$.)

The intersection

$$X^{abc} \cap X^{pqr} = X^{abc} \cap \overline{Q} \cong \mathbb{P}^1 \times \mathbb{P}^1,$$

which is the closure of $X_{2\alpha}^{abc;pqc}$, is a non-singular quadric in X^{abc}. Note that the six lines

$$X^{ab} := \overline{X}_{1\alpha}^{ab}, X^{bc}, X^{ca}, \quad X^{pq}, X^{qr}, X^{rp} \cong \mathbb{P}^1$$

lie on this quadric (see Figure 7.1).

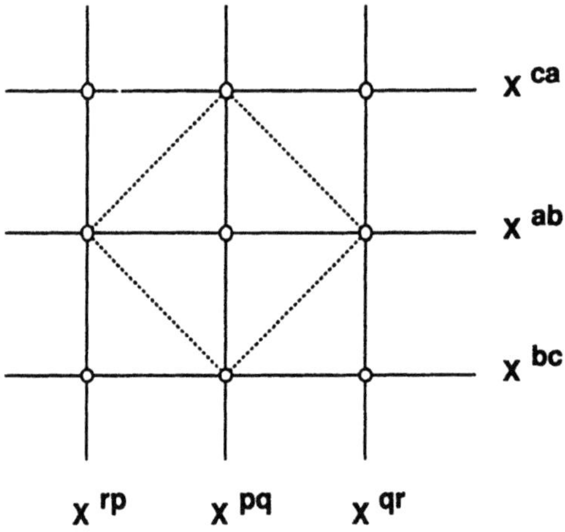

FIGURE 7.1. $X^{abc} \cap X^{pqr} = \mathbb{P}^1 \times \mathbb{P}^1, \circ \in X_{0\alpha}$

The intersection

$$X^{abc} \cap X^{abr}$$

is the union of two projective lines X^{ab} and X^{pq} intersecting at the point $X_{0\alpha}^{ab;pq;cr}$.

The intersection

$$X^{abc} \cap X^{pqc} \cong \mathbb{P}^2,$$

which is the closure of $X_{2\beta}^{abc;pqc}$, is a projective plane spanned by the two lines X^{ab} and X^{pq} in $X^{abc} \cong \mathbb{P}^3$. On the plane $X^{abc} \cap X^{pqc}$, the intersections with other divisors are arranged as shown in Figure 7.2 (also see Figure 7.3).

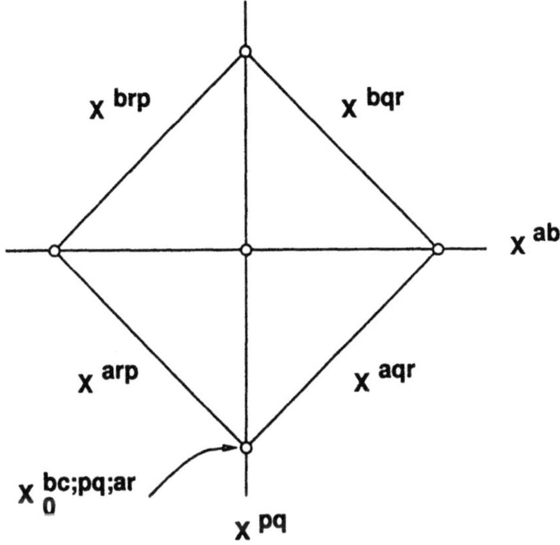

FIGURE 7.2. $X^{abc} \cap X^{pqc} = \mathbb{P}^2$, $\circ \in X_{0\alpha}$

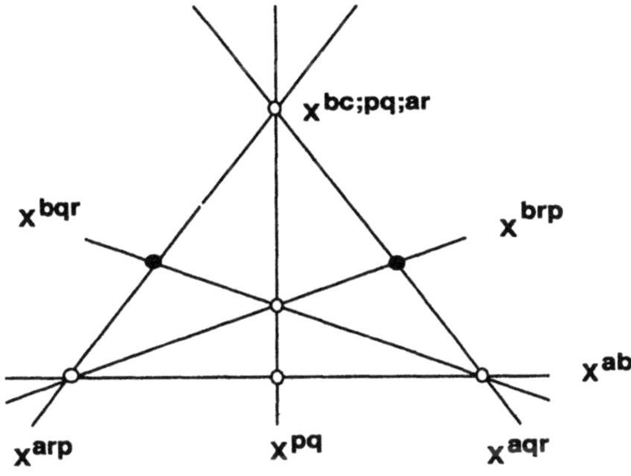

FIGURE 7.3. $X^{abc} \cap X^{pqc} = \mathbb{P}^2$, $\circ \in X_{0\alpha}$, $\bullet \in X_{0\beta}$

Recall that $\overline{p_r} : \overline{X} \to Z \cong \mathbb{P}^4$ is the double cover which ramifies along the quartic $\overline{p_r}(\overline{Q})$; since $\overline{p_r}(\overline{Q})$ is isomorphic to $\overline{Q}$, for notational simplicity we call it $\overline{Q}$. Define

$$Z^{abc} := \overline{p_r}(X^{abc}).$$

With the convention of indices introduced above, we have $Z^{abc} = Z^{pqr}$. Each divisor Z^{abc}, which is linearly embedded in Z (indeed, such a divisor is defined by $D(abc)D(pqr) = 0$), is tangent to the divisor $\overline{Q} \subset Z$ along a quadric. On the quadric there are six lines

$$Z^{ab} := \overline{p_r}(X^{ab}),\ldots,Z^{rp},$$

as shown in Figure 7.1. For a divisor Z^{ijk} distinct from Z^{abc}, the intersection

$$Z^{abc} \cup Z^{ijk}$$

is a plane spanned by two of the six lines above. Indeed, since

$$\{i,j,k\} \neq \{a,b,c\},\{p,q,r\},$$

either $|\{a,b,c\} \cap \{i,j,k\}| = 2$ or $|\{p,q,r\} \cap \{i,j,k\}| = 2$. Let us assume, for example, $i = a, j = b, k = r$. Then $Z^{abc} \cup Z^{ijk}$ is spanned by the two lines Z^{ab} and Z^{pq} (see Figure 7.4).

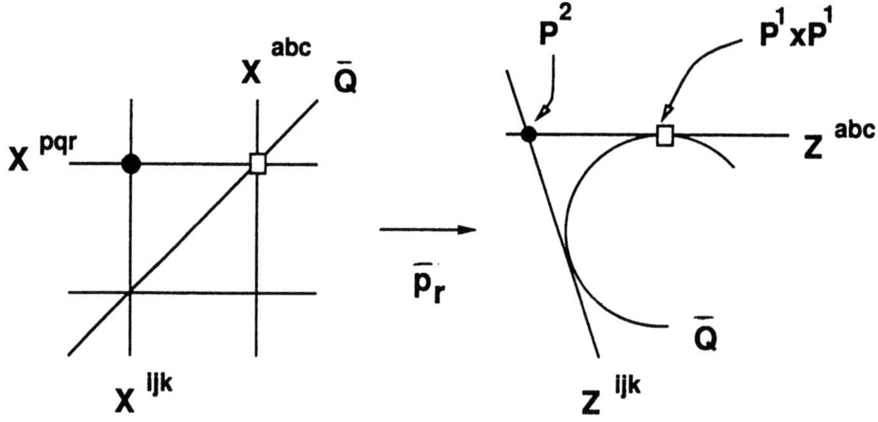

FIGURE 7.4. Intersection pattern of X^{abc}, Z^{abc} and $\overline{Q}$

8. The Structure of $X_{\mathbb{R}}(3,6) \subset \overline{X}_{\mathbb{R}}(3,6)$

In this section, we study the real locus $X_{\mathbb{R}}$ of X. The space $X_{\mathbb{R}}$ is the configuration space of six lines in general position on the real projective plane $\mathbb{P}^2_{\mathbb{R}}$. The 4-dimensional space $X_{\mathbb{R}}$ is the disjoint union of 372 contractible chambers. The symmetric group S_6 acts on the set of these chambers; there are four orbits. (Compare this fact with that for the space $X_{\mathbb{R}}(2,5)$: It is the disjoint union of twelve pentagons, and the action of S_5 on this set of pentagons is transitive.) We describe the shape of a chamber in each orbit. Since the compactification $\overline{X}$ and its strata $X_3^{ijk}, \ldots$ are also defined over the real numbers, we can see these strata on the boundaries of the chambers in $\overline{X}_{\mathbb{R}}$. Furthermore, we can find where the real locus $Q_{\mathbb{R}}$ of Q lives. In this section, we state facts without proof. The proofs can be found in [SekY].

8.1. Four Types of Arrangements. Given an arrangement $l \in X_{\mathbb{R}}$ of six lines $\{l_1, \ldots, l_6\}$ on $\mathbb{P}^2_{\mathbb{R}}$, the totality of arrangements in $X_{\mathbb{R}}$ which can be obtained by continuous deformations of the l_j forms a connected component (referred to as a *chamber*) of $X_{\mathbb{R}}$. Each component is a 4-dimensional cell. The symmetric group S_6 acts naturally on the set of chambers. There are four orbits, referred to as O, I, II and III. Each orbit can be characterized by the polygons cut out by any of its arrangements:

type	O	I	II	III
hexagon	1	0	0	0
pentagons	0	2	3	6
rectangles	9	8	6	0
triangles	6	6	7	10

An arrangement (and also a chamber) is said to be of *type* $T \in \{O, I, II, III\}$ if it belongs to the orbit T. Figures 8.1, 8.2, 8.3 and 8.4 show arrangements of the four types.

Note that an arrangement of type III (Figure 8.4) is the projectivization (identification of antipodal points) of the icosidodecahedron shown in Figure 8.5.

Here we tabulate the cardinality of the chambers of each type:

type	O	I	II	III	total
cardinality	60	180	120	12	372

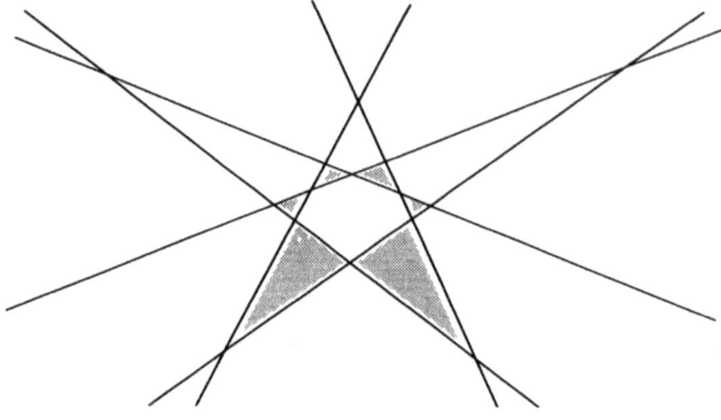

FIGURE 8.1. An arrangement of type O

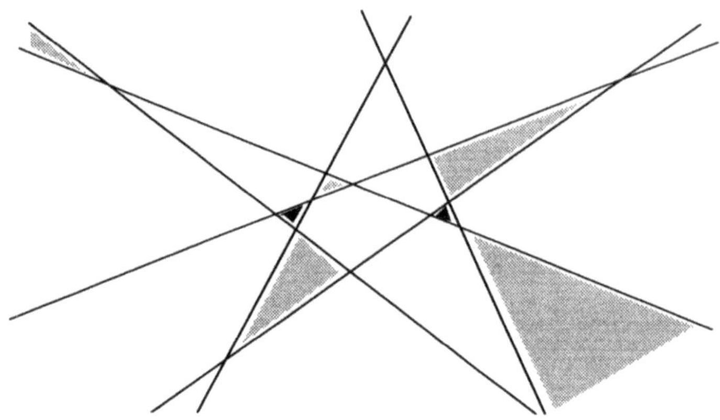

FIGURE 8.2. An arrangement of type I

The involution $*$ acts on $X_{\mathbb{R}}$ and the set of chambers. It preserves the four types: It acts freely on the orbits I, II and III, while trivially on the orbit O. The set $Q_{\mathbb{R}}$ of fixed points in $X_{\mathbb{R}}$ lies only in chambers of type O, and each chamber of type O is divided into two parts by $Q_{\mathbb{R}}$.

8.2. Adjacency of Chambers. The intersection of the closures (in $\overline{X}_{\mathbb{R}}$) of two chambers C_1 and C_2 is 3-dimensional if and only if a(ny) arrangement in C_1 is transformed by a *switch* of a triangle into a(ny) arrangement in C_2, where a switch of a triangle $l_i l_j l_k$ is the exchange of lines depicted in Figure 8.6. Two such chambers are said to be *adjacent* to each other.

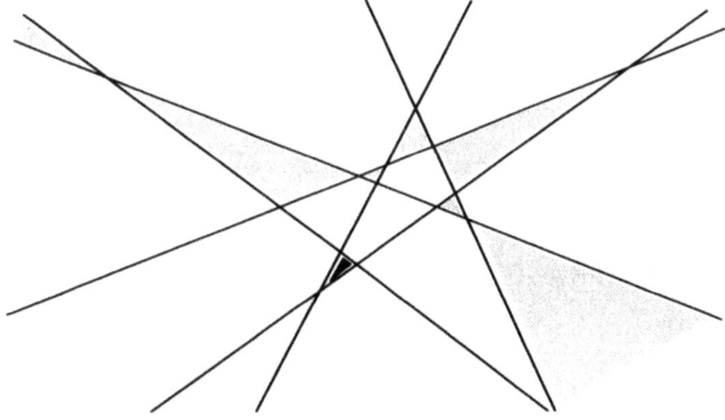

FIGURE 8.3. An arrangement of type II

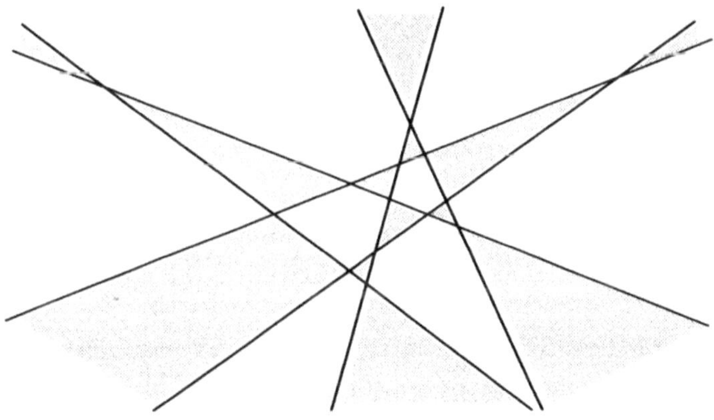

FIGURE 8.4. An arrangement of type III

Since there are six triangles on an arrangement of type O, there are six ways to switch. These switches always yield an arrangement of type I. For an arrangement of type I, there are two ways to switch into those of type O and four ways to switch into those of type II. For type II, there is a unique way to switch it into type III and six ways to switch it into type I. Type III is switched into type II in ten ways. These facts can be

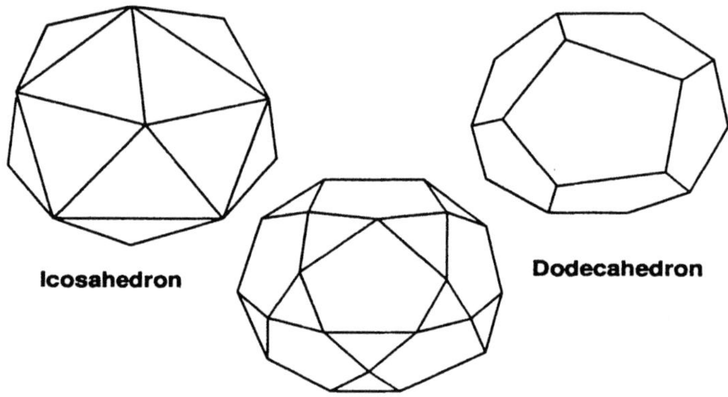

FIGURE 8.5. An icosidodecahedron

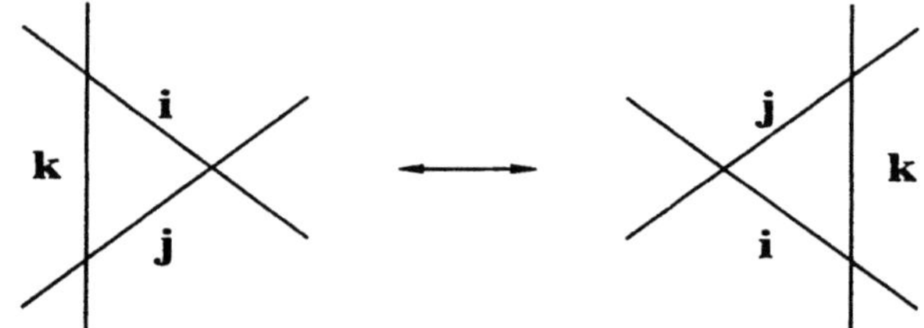

FIGURE 8.6. A switch

visualized as follows:

```
         I                            O
   I     |     I            II        |        II
     \       /                 \           /
         O                            I
     /       \                 /           \
   I     |     I            II        |        II
         I                            O
```

```
         I                            II
   I     |     I            II        |        II
     \       /            II  \           /  II
 III ─── II                      III
     /       \            II  /           \  II
   I     |     I            II        |        II
         I                            II
```

By coding these 372 chambers, I can make the adjacency graph, which can be found in [StiY]. The following consideration hopefully gives you an idea of the adjacency of the chambers. Regarding the space $X(3,6)$ as the configuration of six *points* in general position in the (projective) plane, we consider a map

$$p : X(3,6) \longrightarrow X(3,5)$$

defined by forgetting the sixth point. This map defines a fibration of $X(3,6)$ with base $X(3,5)$. Thanks to the Grassmann isomorphism (see §13 of Chapter I), the base space $X(3,5)$ is isomorphic to the space $X(2,5)$, which is well-studied in Chapter V. Let us see the fiber of p on a point of $X(3,5)$. Let $x_1,\ldots,x_5$ be five (ordered) points in the plane with no three points collinear, representing a point x of $X(3,5)$. Note that there is a unique nonsingular conic passing through these five points. The fiber $p^{-1}(x)$ consists of points x_6 in the plane such that no three of $\{x_1,\ldots,x_6\}$ are collinear, that is, it is the complement (in the plane) of the union of all the lines joining x_i and x_j $(1 \le i < j \le 5)$. This is visualized in Figure 8.7; you will find 31 polygons cut out by the ten lines. I will tell you, without giving a proof, the types of chambers which include these polygons. In the center of the figure you find a unique pentagon which is a (2-dimensional) section of a chamber of type III. For simplicity we will describe this by saying 'the pentagon is of type III.' Around the pentagon there are five triangles of type II. Adjacent to these, there are five triangles of type I. The conic passing through the five points lies in the five triangles of type O. Adjacent to these, there are ten triangles of type I. There are five rectangles of type II surrounded by these ten triangles. As a whole we find 31 polygons. Since the cardinality of the connected components of $X(2,5)$ is twelve, we have $31 \times 12 = 372$ chambers, as expected.

8.3. Intersections of the Closures of Adjacent Chambers. In this subsection, we describe the intersection of the closures of two adjacent chambers, and the intersection of a chamber of type O and $Q_{\mathbb{R}}$. In the figures in this and the following subsections, a white vertex denotes a point in $X_{0\alpha}$, a black vertex denotes a point in $X_{0\beta}$, a thick edge denotes part of $X_{1\alpha}$, and a thin edge denotes part of $X_{1\beta}$. These objects will be referred to by names such as 'α-vertex' or 'vertex of the α-type', 'β-edge' or 'edge of the β-type', and so on.

When two adjacent chambers are of types III and II, the intersection of their closures is the *double tetrahedron* shown in Figure 8.8. The six

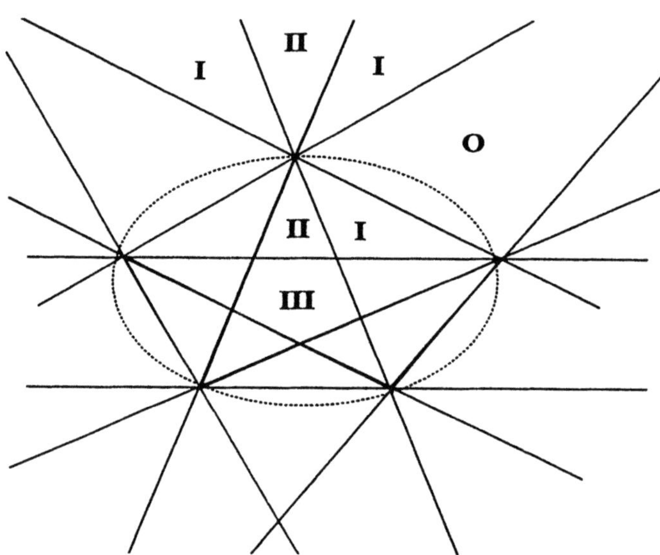

FIGURE 8.7. A fiber of $p : X(3,6) \to X(3,5)$ divided into 31 components

triangular faces and the nine edges are of the β-type.

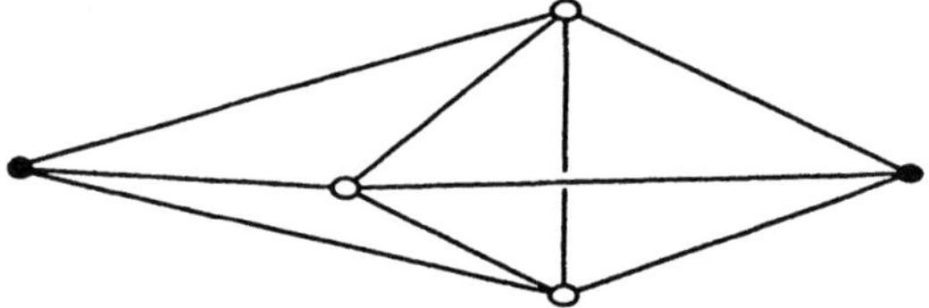

FIGURE 8.8. A double tetrahedron

When two adjacent chambers are of types II and I, the intersection of their closures is the tetrahedron shown in Figure 8.9. The four triangular faces are of the β-type.

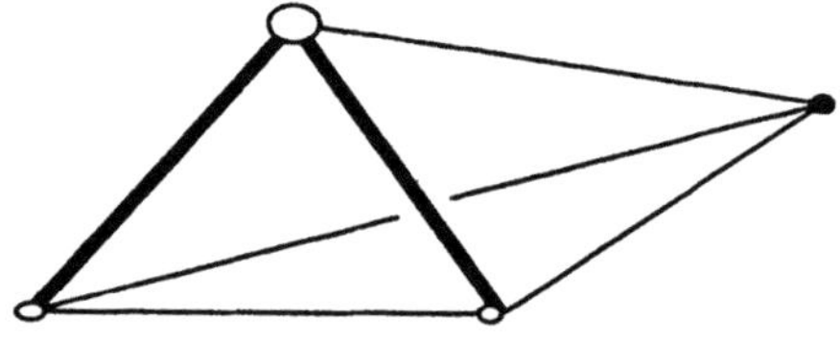

FIGURE 8.9. A tetrahedron

When two adjacent chambers are of types I and O, the intersection of their closures is the 3-face body, referred to as a *piece-of-orange*, shown in

Figure 8.10. The rectangular face is of the α-type, and the two triangular faces are of the β-type; there is only one β-edge. In this figure, the sketch on the right is the stereographic image of the piece-of-orange with center at the midpoint of the β-edge.

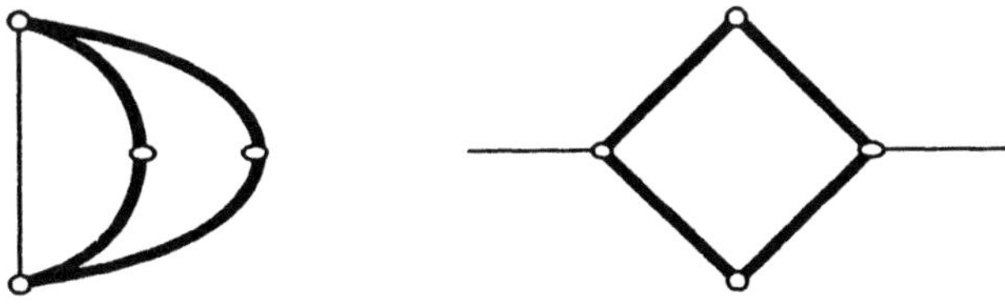

FIGURE 8.10. A piece-of-orange

The intersection of a chamber of type O and $Q_{\mathbb{R}}$ is the *3-rectangular-face-body* shown in Figure 8.11. In this figure, the sketch on the right is the stereographic image of the 3-rectangular-face-body with center at a vertex on the equator. $Q_{\mathbb{R}}$ is the union of 60 copies of the 3-rectangular-face-body.

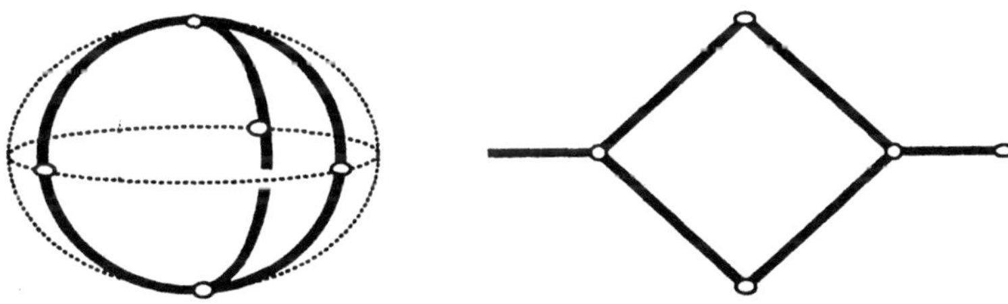

FIGURE 8.11. A 3-rectangular-face-body

8.4. The Shapes of Chambers. In this subsection, we describe the shapes of chambers by giving the cellular decompositions of their boundaries in $\overline{X}_{\mathbb{R}}$, which are homeomorphic to the 3-dimensional sphere S^3. With stereographic projections, we draw the decompositions on $\mathbb{R}^3$.

A chamber of type III is adjacent to ten chambers of type II. Hence its boundary is the union of ten double tetrahedra, filling the sphere S^3. These ten double tetrahedra are situated as follows (see Figure 8.12): Place the five α-vertices on the vertices of a tetrahedron T and its barycenter. One can then visualize ten triangles, the four triangular faces of T and the six triangles formed by the barycenter and the six edges of T. These ten triangles will be referred to as (triangular) *films*. The union of the ten films cuts the sphere S^3 into five tetrahedra (the four small tetrahedra inside T and the infinitely large one outside T).

Next, place five β-vertices at the barycenters of these five tetrahedra (the point at infinity is considered to be the barycenter of the infinitely large tetrahedron). For each triangular film, there are two β-vertices which are barycenters of the two tetrahedra adjacent through the film. Form two cones (small tetrahedra) with the film and the two β-vertices, and glue these cones along the film. Then one obtains a double tetrahedron bounded by six triangular faces with three α-vertices and two β-vertices. In this way, the sphere S^3 is divided into ten double tetrahedra.

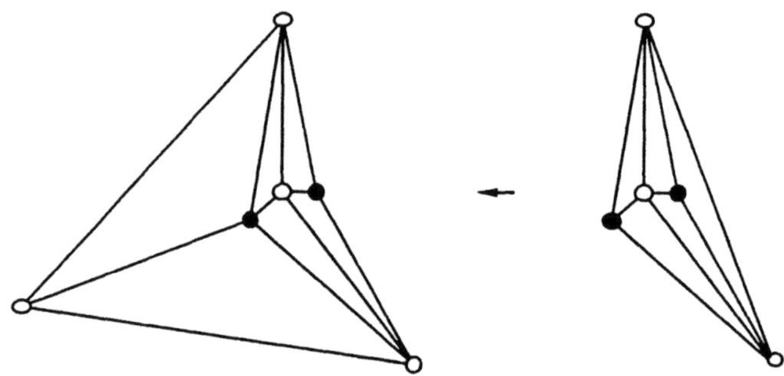

FIGURE 8.12. The boundary of a chamber of type III: Six double tetrahedra are shown. A β-vertex is at ∞.

A chamber of type II is adjacent to a chamber of type III and six chambers of type I. Thus its boundary is the union of a double tetrahedron and six tetrahedra, filling the sphere S^3. The double tetrahedron and the six tetrahedra are situated as follows (see Figure 8.13): Place the double tetrahedron in $\mathbb{R}^3$ and an α-vertex at the point at infinity. Make six cones from the point ∞ to the six triangular faces of the double tetrahedron. In Figure 8.13, the point ∞ is at the center.

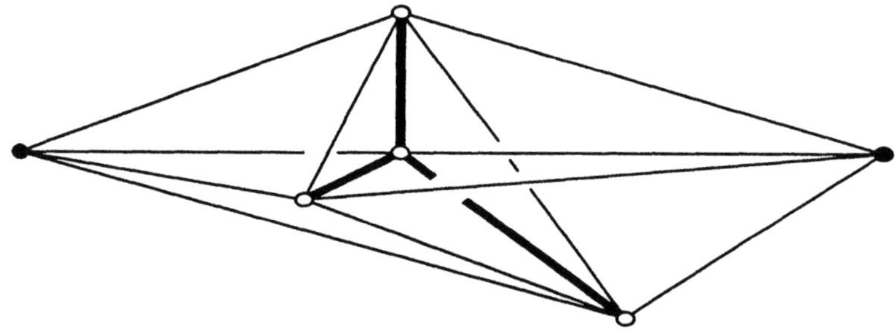

FIGURE 8.13. The boundary of a chamber of type II: The six tetrahedra are glued to form the complement of the double tetrahedron

A chamber of type I is adjacent to four chambers of type II and two chambers of type O. Thus its boundary is the union of four tetrahedra and two pieces-of-orange situated as follows (see Figure 8.14): The β-vertex is common to all four tetrahedra. These four small tetrahedra are glued to produce one large tetrahedron. Put the large tetrahedron in $\mathbb{R}^3$. The four α-edges form a rectangle. Make an (infinitely large) wall with this rectangle as the boundary. The outside of this large tetrahedron is divided by this wall into two pieces-of-orange.

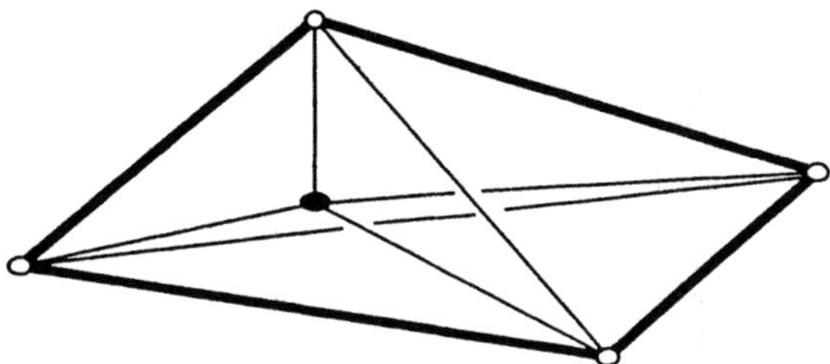

FIGURE 8.14. The boundary of a chamber of type I: Four tetrahedra form a large tetrahedron. Outside are two pieces-of-orange.

A chamber of type O is adjacent to six chambers of type I. Therefore its boundary is the union of six pieces-of-orange situated as follows (see Figure 8.15): Three pieces-of-orange are glued along a common β-edge to form an orange (ball). The remaining three pieces-of-orange are glued along another common β-edge to form another orange. Gluing the two oranges along the three rectangular faces yields S^3.

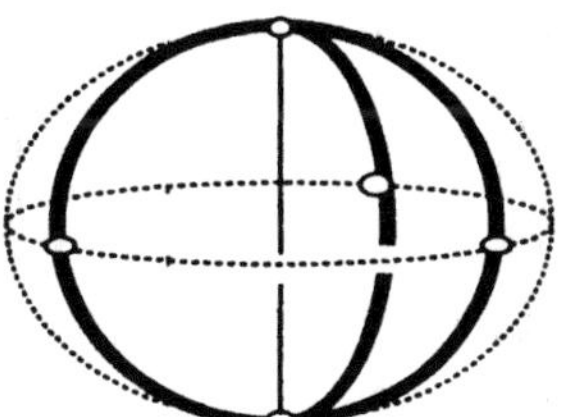

FIGURE 8.15. The boundary of a chamber of type O: An orange (the union of the three pieces-of-orange) is shown.

A chamber of type O is divided into two pieces by $Q_\mathbb{R}$. These pieces are mapped through the involution $*$ to each other. Therefore the boundary of each piece (referred to as a chamber of type $\frac{1}{2}$O) is the union of three pieces-of-orange and a 3-rectangular-face-body situated as follows: Glue the three pieces-of-orange to form an orange, as done above. Then glue the orange and the 3-rectangular-face-body along the three rectangular faces.

8.5. The Action of the Weyl Group $W(E_6)$. Let us introduce the bi-rational involution s on $\overline{X}$ defined by

$$(I_3, (x_{ij})) \longmapsto (I_3, (\frac{1}{x_{ij}})).$$

This involution is bi-regular on $X - Q = \overline{X} - \overline{Q}$. It can be easily checked that s interchanges Q and X_3^{456}. The group G generated by S_6 and s is isomorphic to the Weyl group $W(E_6)$ of type E_6 (note that $* \in G$).

$$(12) \quad \text{---} \quad (23) \quad \text{---} \quad (34) \quad \text{---} \quad (45) \quad \text{---} \quad (56)$$
$$|$$
$$s$$
$$|$$
$$*$$

There is a minimal smooth compactification $\mathcal{C}$ of $X - Q$ such that G acts bi-regularly. There is also a bi-rational morphism π:

$$\pi : \mathcal{C} \to \overline{X}.$$

The variety $\mathcal{C}$ can be obtained from $\overline{X}$ through blowing-ups along $X_{2\alpha}, X_{1\alpha}$ and at $X_{0\alpha}$.

The variety $\mathcal{C}$ is defined over the real numbers; let $\mathcal{C}_{\mathbb{R}}$ be its real locus. The group G acts transitively on the set of 432 chambers, 120 chambers of types $\frac{1}{2}O$, 180 chambers of type I, 120 chambers of type II, and 12 chambers of type III. This implies, in particular, that every chamber has the same shape in $\mathcal{C}_{\mathbb{R}}$.

Let P represent the closure of a (any) chamber in $\mathcal{C}_{\mathbb{R}}$ and let us describe P by making use of the boundary of a chamber of type III. For each double tetrahedron, we truncate the three α-vertices. The consequent 3-body, referred to as a *terada3*, has six pentagonal faces and three square faces (see Figure 8.16). Now the boundary ∂P is the union of five cubes and ten terada3s. In short, it is obtained from the boundary of a chamber of type III by simply replacing the five α-vertices with cubes.

If you start from another chamber, you can eventually obtain the same P by performing the following procedure:

(1) Replace (rectangular) α-faces with cubes.

(2) Replace α-edges with terada3s.

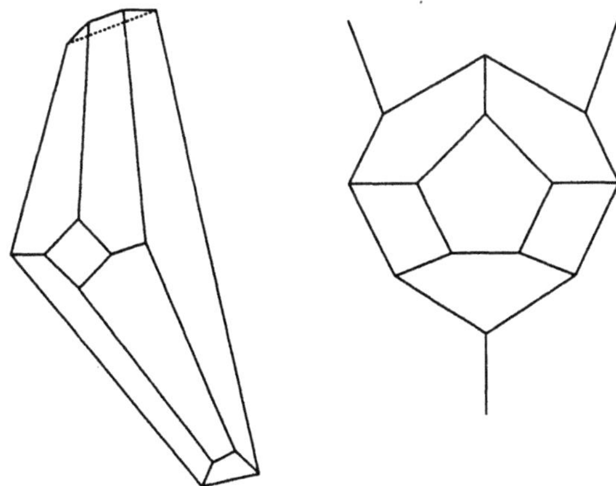

FIGURE 8.16. Terada3, the right-hand side is its stereographic image

(3) Replace α-vertices with cubes.

By making paper-models, you can realize the procedure above (if you can do it in your head, you are great!). Hints can be found in [SekY].

The above truncation procedure describes how the variety $\overline{X}$ is blown up to become $\mathcal{C}$.

Comment: Indeed, 'terada3' is a funny name for a 3-body. This is a nickname of the '3-dimensional Terada model'. If you pronounce 3 in Japanese, terada3 means Mr. Terada, who is my old friend and is a pioneer of the theory of the modular interpretations of $X(2,n)$ (cf. [Trd]). The 3-dimensional space $X(2,6)_\mathbb{R}$ is the union of $(6-1)!/2 = 60$ copies of terada3s.

CHAPTER VIII

Hypergeometric Functions of Type $(3,6)$

When we studied the hypergeometric function (of type $(2,4)$) in Chapters III and IV, we started from the hypergeometric series $F(a,b,c;x)$, then derived the differential equation annihilating the series, and found the Euler integral representation, the hypergeometric integral of type $(2,4)$.

In this chapter, we start from the hypergeometric integral of type $(3,6)$ and proceed the other way round, i.e., we shall find a system of differential equations annihilating these integrals.

The description of loaded cycles given in Chapter IV will be generalized to higher dimensional (mainly 2-dimensional) loaded cycles.

1. Hypergeometric Integrals of Type $(3,6)$

For a matrix $x = (x_{ij}) \in M^*(3,6)$, consider the six linear forms

$$L_j = x_{1j}t^1 + x_{2j}t^2 + x_{3j}t^3, \quad (1 \le j \le 6)$$

on the projective plane $\mathbb{P}^2$ with homogeneous coordinates $t^1 : t^2 : t^3$ and the formal integral

$$\int_\Delta \prod_{j=1}^6 (L_j)^{\alpha_j - 1} dt,$$

where

$$dt = t^1 dt^2 \wedge dt^3 + t^3 dt^1 \wedge dt^2 + t^2 dt^3 \wedge dt^1,$$

and the α_j are parameters satisfying

$$\alpha_1 + \cdots + \alpha_6 = 6 - 3 = 3.$$

This condition guarantees that the integrand is invariant under the multiplication of the t_j by constants.

In general, integrals

$$\int_\Delta \prod_{j=1}^{n} (L_j)^{\alpha_j - 1} dt,$$

$$dt = \sum (-)^p t^p t^1 \wedge \cdots \wedge dt^{p-1} \wedge dt^{p+1} \cdots dt^k,$$

$$\alpha_1 + \cdots + \alpha_n = n - k$$

of the product of complex powers of n linear forms

$$L_j = \sum_{i=1}^{k} x_{ij} t^i$$

are called *hypergeometric integrals* of type (k, n). You might think these are simple objects. Yes, they are not particularly difficult things, but as you shall see, they are sufficiently complicated. In any case, the integral given at the beginning of this section is a hypergeometric integral of type $(3, 6)$, the original one studied in Chapter IV is of type $(2, 4)$, and the beta function that appeared in §2.1 of Chapter IV is of type $(2, 3)$.

Let us see what happens if the x in a hypergeometric integral of type (k, n) is replaced by gx and xh, where $g \in GL(k)$ and $h \in H_n \cong (\mathbb{C}^\times)^n$, both very close to the identity. The form

$$\eta(x) := \prod_{j=1}^{n} (L_j)^{\alpha_j - 1} dt$$

satisfies the following transformation formulae:

$$\eta(gx) = (\det g)\eta(x),$$

$$\eta(xh) = \eta(x) \prod_{j=1}^{n} (h_j)^{\alpha_j - 1}, \quad h = (h_1, \ldots, h_n).$$

Thus this integral is not quite invariant under the left $GL(k)$-action and right H_n-action on $x \in M(k, n)$. We can modify the integral to be invariant, as we did in §1 of Chapter IV. However, since we are going to consider integrals with different domains of integration Δ and their *ratio* in the case $(k, n) = (3, 6)$, we can think of this ratio as a map of the configuration space $X(3, 6)$ into a projective space.

The hypergeometric integrals appearing above satisfy a system of linear differential equations. You may think that once you know a system of linearly independent solutions to such a system of differential equation, no further knowledge can be obtained by knowing the differential equations themselves. Well, yes, if the differential equations were too complicated and unmanageable. Fortunately however, the system of differential equations satisfied by the hypergeometric integrals of type (k, n) is sufficiently simple. I would like to take advantage of this fact. Even in the much simpler case $(2, 4)$, in order to know the local behavior of solutions, the argument in Chapter IV treating integral representations is not enough for studying local properties around the singular points. We also need the argument in Chapter III making a suitable pair of local solutions at each singular point. This requires studying transformations of the differential equation.

Let us see what kind of equations hypergeometric integrals of type (k, n) satisfy. We regard x as variables and α as parameters. In the transformation rules above, put

$$g = I_k + \varepsilon E_{il}, \quad h = I_n + \varepsilon E_{jj},$$

where E_{ij} represents the matrix whose unique non-zero component is 1 at (i, j), and differentiate both sides with respect to ε. We then see that these integrals satisfy the linear equations

$$\sum_{j=1}^{n} x_{lj} \frac{\partial u}{\partial x_{ij}} + \delta_{il} u = 0, \quad 1 \le i, l \le k,$$

$$\sum_{i=1}^{k} x_{ij} \frac{\partial u}{\partial x_{ij}} - (\alpha_j - 1) u = 0, \quad 1 \le j \le n,$$

with unknown u. The following equality can be easily checked:

$$\frac{\partial^2 u}{\partial x_{ip} \partial x_{jq}} - \frac{\partial^2 u}{\partial x_{iq} \partial x_{jp}} = 0 \quad 1 \le i, j \le k, \ 1 \le p, q \le n.$$

Are there any equations which are not obtained from the three kinds of equations above? Good question. We shall answer this question in §5. In any case, let us name the system of three kinds of equations above

the *hypergeometric system $E(k, n; \alpha)$ of type (k, n)*.

2. Domains of Integration, Loaded Cycles

Recall the hypergeometric integrals of type $(2, 4)$:

$$\int_C \prod_{j=1}^{4} (t - x_j)^{\alpha_j - 1} dt,$$

where $x_j \in \mathbb{P}^1$ are real and arranged as

$$x_1 < x_2 < x_3 < x_4,$$

and $\alpha_1 + \cdots + \alpha_4 = 4 - 2 = 2$. In Chapter IV we integrated along two real intervals, $C = I_1 = (x_1, x_2)$ and $I_2 = (x_2, x_3)$. Of course you can choose other paths, but we do not need to take unnecessarily complicated ones. I hope you still remember that (Chapter IV, Proposition 4.1) the integrals along (x_3, x_4) and (x_4, x_1) are linear combinations of those along I_1 and I_2 provided that $\alpha_j \notin \mathbb{Z}$. So if you put $x_4 = \infty$, the two intervals I_1 and I_2 are the bounded chambers in the space $\mathbb{R} - \{x_1, x_2, x_3\}$.

Now please guess what kind of domains of integration we should use in the present case, $(k, n) = (3, 6)$. We choose a base arrangement $\overset{\bullet}{x}$ of six lines satisfying the following conditions:

(1) The lines are in general position, i.e., no three lines meet at a point.

(2) The lines are defined over the real numbers, and the sixth line is at infinity.

(3) The lines are of type O in the sense of §8 of Chapter VII, i.e., they bound a hexagon in $\mathbb{P}^2_{\mathbb{R}}$.

(4) There is a conic tangent to all of these lines.

Though I have no tools to measure rigorously the complexity of arrangements, you will see that this is a simple and convenient arrangement in various respects. Such an arrangement is illustrated in Figure 2.1, where the sixth line H_6 is at infinity, and the lines $H_1, \ldots, H_6$ bound, in this order, the (infinitely large) hexagon. You can see there are six bounded chambers, which we shall use as domains of integration. We must assign the arguments of linear forms L_j, defining the lines H_j, on each chamber to make each a loaded cycle. You can do as you please. You have a great deal of freedom. This probably annoys you. In order to make the story transparent, we use notation involving r and n; please set

$$r = 2(= 3 - 1) \quad \text{and} \quad n = 5(= 6 - 1).$$

Then let $X = X(r + 1, n + 1)$ be the configuration space of $n + 1$ hyperplanes in general position in the r-dimensional projective space $\mathbb{P}^r$.

2.1. The Submanifold Q of X and a Base Arrangement. Let Q be the $(n - 2)$-dimensional submanifold of X consisting of the arrangements for which there is a nonsingular curve of degree r along which the

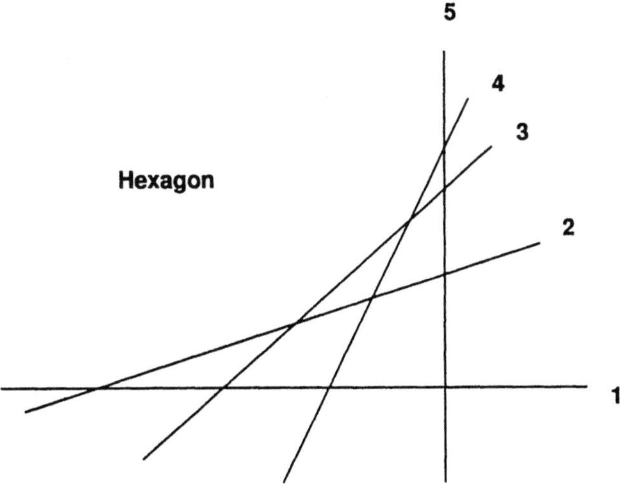

FIGURE 2.1. A base arrangement

$n + 1$ hyperplanes osculate, or equivalently, for which there is a nonsingular curve of degree r in the dual projective space on which the $n + 1$ points dual to the $n + 1$ hyperplanes are located. Since any nonsingular curve of degree r is projectively equivalent to the curve (the *Veronese embedding* of $\mathbb{P}^1$)

$$t_0 = s^r, \ t_1 = s^{r-1}, \ldots, t_{r-1} = s, \ t_r = 1$$

parametrized by $s \in P^1$, the manifold Q can be parametrized by the configuration space $X(2, n + 1)$ of $n + 1$ points on the projective line as follows:

$$\iota: \quad X(2, n+1) \quad \longrightarrow \quad Q$$

$$(\xi_0 = \infty, \xi_1, \ldots, \xi_n) \longmapsto \begin{pmatrix} 1 & (-\xi_1)^r & \cdots & (-\xi_n)^r \\ 0 & (-\xi_1)^{r-1} & \cdots & (-\xi_n)^{r-1} \\ \vdots & \vdots & \cdots & \cdots \\ 0 & -\xi_1 & \cdots & -\xi_n \\ 0 & 1 & 1 & 1 \end{pmatrix}.$$

Here, without loss of generality, we assume that the 0-th point of $\xi \in X(2, n + 1)$ is at infinity and that the remaining n points ξ_j $(1 \leq j \leq n)$ are on the complex affine line S parametrized by s. The image $\iota(\xi) \in Q$ represents the hyperplane H_0 at infinity and n hyperplanes H_j in the complex affine r-space T, with coordinates $t = (t_1, \ldots, t_r)$, defined by

$$L_j(t) := t_r + (-\xi_j)t_{r-1} + \cdots + (-\xi_j)^{r-1}t_1 + (-\xi_j)^r, \quad 1 \le j \le n.$$

Let us fix n real points on S:

$$\xi_1 < \cdots < \xi_j < \cdots < \xi_n, \quad \xi_0 = \xi_{n+1} = \infty.$$

We then define the *base arrangement* $\overset{\bullet}{x} \in X$ as

$$\iota(\xi) = (H_0, \overset{\bullet}{H}_1, \ldots, \overset{\bullet}{H}_n).$$

In the next subsection, for notational simplicity, we drop the "bullets" above the H_j.

2.2. Loaded Cycles. Set

$$U(t) = \prod_{j=1}^{n} L_j(t)^{\alpha_j - 1},$$

where $L_j(t)$ is the linear form in t, determined by ξ_j, just defined above. Fix an orientation on $T_{\mathbb{R}} = T \cap \mathbb{R}^r$ by $(-1)^{r(r-1)/2}dt_1 \wedge \cdots \wedge dt_r$. The reason for choosing this orientation will be made clear below. For a multi-index,

$$P = (p_1 \ldots p_r), \quad 1 \le p_1 < \cdots < p_r \le n - 1,$$

we define loaded cycles $\{D_P\}$ and $\{\check{D}_P\}$ on chambers (see Figure 2.2):

$$D'_P = \{t \in T_{\mathbb{R}} \mid (-1)^{P(j)}L_j(t) > 0, \quad 1 \le j \le n\}.$$

(The orientation here is induced as a domain of $T_{\mathbb{R}}$.) On the chamber D'_P, we assign U and U^{-1}, respectively, with

$$\arg L_j = -P(j)\pi, \quad 1 \le j \le n$$

where

$$P(j) := \text{ cardinality of } \{i \mid p_i < j\}.$$

The values of the $\arg L_j$ for the chambers shown in Figure 2.2 (in the case $(r, n) = (2, 5)$) are as follows:

	D_{12}	D_{13}	D_{14}	D_{23}	D_{24}	D_{34}
$\arg L_1$	0	0	0	0	0	0
$\arg L_2$	$-\pi$	$-\pi$	$-\pi$	0	0	0
$\arg L_3$	-2π	$-\pi$	$-\pi$	$-\pi$	$-\pi$	0
$\arg L_4$	-2π	-2π	$-\pi$	-2π	$-\pi$	$-\pi$
$\arg L_5$	-2π	-2π	-2π	-2π	-2π	-2π

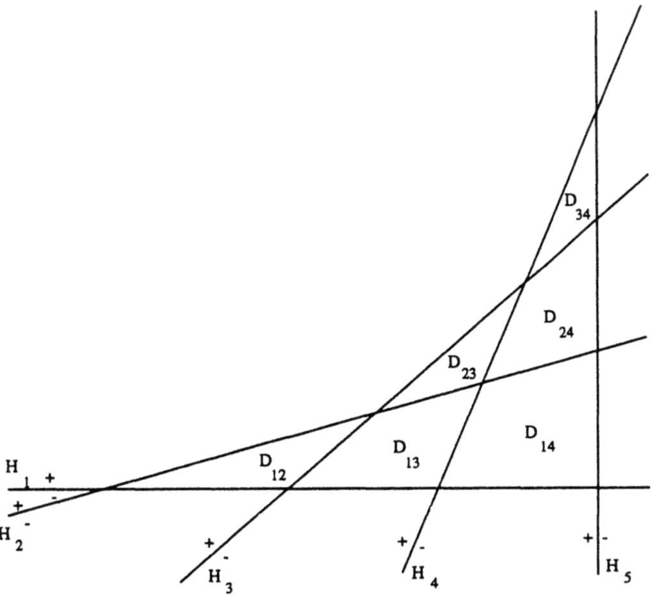

FIGURE 2.2. $r = 2, \quad n = 5$

The following argument will explain the meaning of this assignment. Let

$$\phi : S^r = \overbrace{S \times \cdots \times S}^{r} \ni (s^{(1)}, \ldots, s^{(r)}) \mapsto (t_1, \ldots, t_r) \in T$$

be a holomorphic mapping, where t_j is the elementary symmetric polynomial of degree j in $s^{(1)}, \ldots, s^{(r)}$, as determined by the relation

$$\prod_{i=1}^{r} (s^{(i)} - z) = t_r + (-z)t_{r-1} + \cdots + (-z)^{r-1}t_1 + (-z)^r.$$

Note that this map can be considered as a realization of the quotient space of S^r under the symmetric group S_r. The critical set of ϕ is

$$\bigcup_{i \neq j} \{s \in S^r \mid s^{(i)} = s^{(j)}\};$$

outside this it is locally biholomorphic.

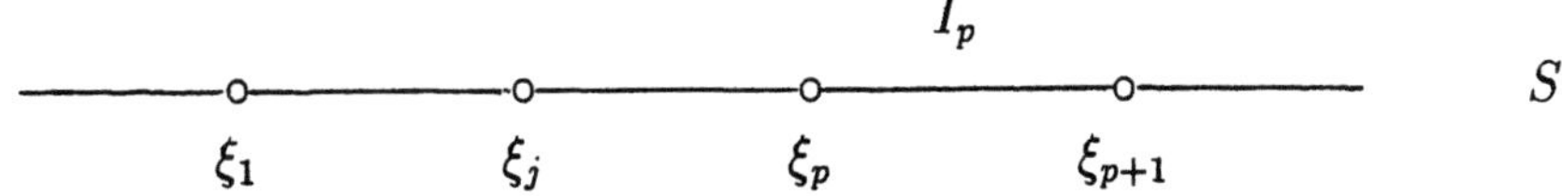

Note that

$$L_j(t) = \prod_{i=1}^{r}(s^{(i)} - \xi_j)$$

and that the map ϕ induces an orientation preserving isomorphism:

$$\phi : I_{p_1} \times \cdots \times I_{p_r} \xrightarrow{\sim} D_P \subset T_{\mathbb{R}}.$$

Indeed the formula

$$\det \frac{\partial(t_1, \ldots, t_r)}{\partial(s^{(1)}, \ldots, s^{(r)})} = \prod_{i<j}(s^{(i)} - s^{(j)})$$

explains the strange orientation we put on $T_{\mathbb{R}}$. Note also that the arguments of $s^{(i)} - \xi_j$ on D_j are compatible with those of $L_j(t)$ on D_P:

$$\arg L_j = \arg \prod_{i=1}^{r}(s^{(i)} - \xi_j)$$

$$= \sum_{i=1}^{r} \arg(s^{(i)} - \xi_j) = -P(j)\pi.$$

Since we assume $1 \leq p_1 < \cdots < p_r \leq n-1$ for $P = (p_1, \ldots, p_r)$, and since ϕ is orientation preserving, it is natural to write

$$D_P = I_{p_1} \wedge \cdots \wedge I_{p_r}.$$

Figure 2.3 illustrates the map

$$\phi : (s^{(1)}, s^{(2)}) \longmapsto (t_1, t_2) = (s^{(1)} + s^{(2)}, s^{(1)}s^{(2)}).$$

The diagonal line $s^{(1)} = s^{(2)}$ in $S^{(1)} \times S^{(2)}$ is mapped to a conic, and the two lines $s^{(1)} - \xi_j = 0$ and $s^{(2)} - \xi_j = 0$ are mapped to the line

$$H_j : L_j = (s^{(1)} - \xi_j)(s^{(2)} - \xi_j) = 0$$

tangent to this conic. In the figure, you might suspect that the map ϕ is orientation reversing; recall the orientation of $T_{\mathbb{R}}$ we chose above!

In general, the domain $D_P \subset T_{\mathbb{R}}$ is bounded by walls $H_j := \{t \in T_{\mathbb{R}} \mid L_j = 0\}$, where

$$j \in \{p_1, p_1 + 1\} \cup \cdots \cup \{p_r, p_r + 1\} \quad \text{(not necessarily disjoint)}.$$

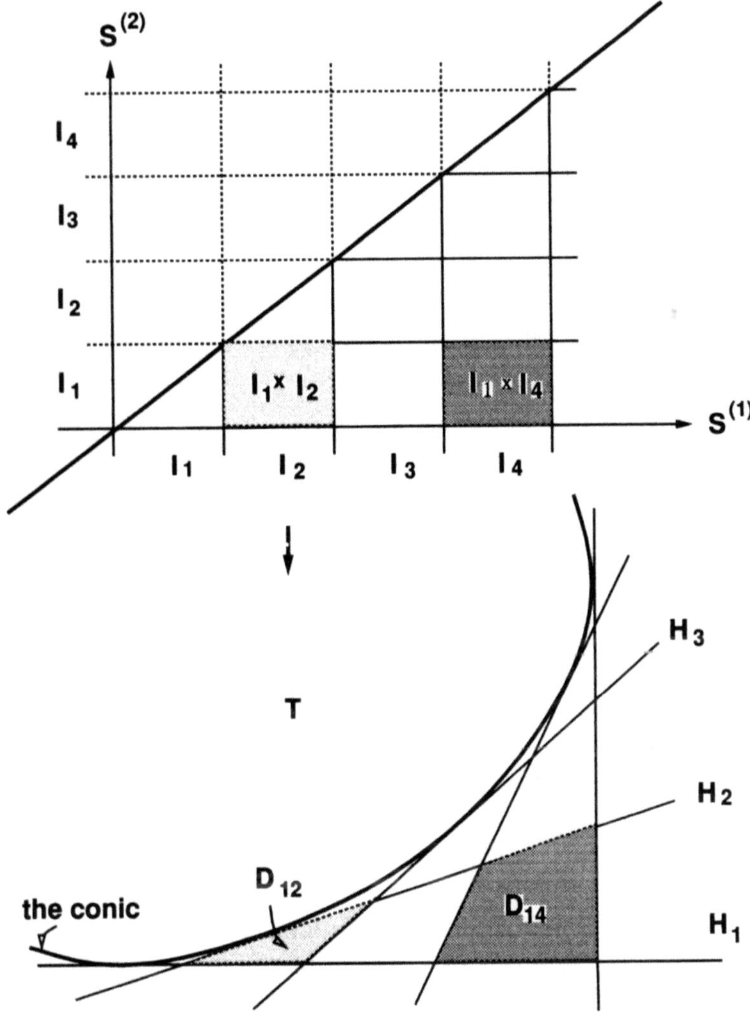

FIGURE 2.3. The map ϕ

The following figure hopefully helps understand why such walls bound the domain D_P.

Thus the set of vertices of D_P is given by

$$V_P = \{\langle j_1, \ldots, j_r\rangle := H_{j_1} \cap \cdots \cap H_{j_r} \mid j_1 < \cdots < j_r,$$
$$j_1 \in \{p_1, p_1 + 1\}, \ldots, j_r \in \{p_r, p_r + 1\}\}.$$

2.3. Regularizations. If you wish to regularize the loaded cycles D_P, you can proceed as follows. Each loaded cycle is locally a direct product of 1-dimensional cycles, so all you need to do is to regularize the 1-dimensional cycles, make the appropriate product, patch them, and load the function accordingly. It is not difficult. The neighborhood of each j-codimensional face of the chamber D_P' is then replaced by the product of j circles and $(k - j)$ segments. For example, for the case $r = 2$, near the vertex $H_1 \cap H_2$, the chamber D_{12}' can be considered as the product $(0, 2\varepsilon) \times (0, 2\varepsilon) \subset I_1' \times I_2'$, which should be replaced by

$$\{\frac{C_1}{d_1} + (\varepsilon, \overrightarrow{2\varepsilon})\} \wedge \{\frac{C_2}{d_2} + (\varepsilon, \overrightarrow{2\varepsilon})\}$$
$$= \frac{C_1 \wedge C_2}{d_1 d_2} + \frac{C_1}{d_1} \wedge (\varepsilon, \overrightarrow{2\varepsilon}) + (\varepsilon, \overrightarrow{2\varepsilon}) \wedge \frac{C_2}{d_2} + (\varepsilon, \overrightarrow{2\varepsilon}) \wedge (\varepsilon, \overrightarrow{2\varepsilon}),$$

where S_1 and S_2 are positively oriented circles with initial point ε (see Figure 2.4).

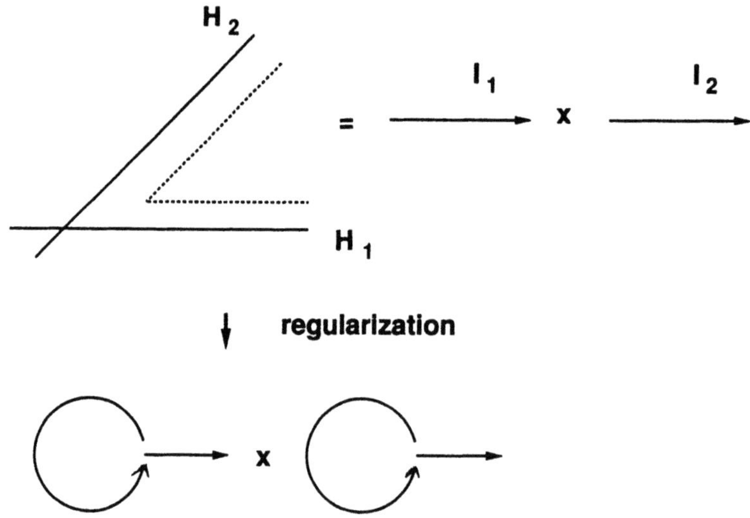

FIGURE 2.4. A regularization

3. Intersections of Loaded Cycles and the Invariant Form

3.1. The Intersection Matrix and the Invariant Form H.

Since the loaded cycle D_P can be thought of as the wedge product

$$I_{p_1} \wedge I_{p_2}, \quad P = (p_1, p_2),$$

whatever definition of intersection numbers is made, it is natural to believe that the intersection $D_P \cdot \check{D}_Q$ of the loaded cycles D_P and $\check{D}_Q$ should be given by

$$D_P \cdot \check{D}_Q = \det(I_{p_i} \cdot \check{I}_{q_j}).$$

We shall rigorously prove this conjecture in the following subsections only in the case $(k, n) = (3, 6)$. Let us evaluate the intersection numbers to support this conjecture. The intersection numbers of the 1-dimensional cycles I_j are known to be (§7 of Chapter IV)

$$Int(6; \alpha) = \begin{pmatrix} I_1 \cdot \check{I}_1 & \cdots & I_1 \cdot \check{I}_4 \\ \vdots & & \vdots \\ I_4 \cdot \check{I}_1 & \cdots & I_4 \cdot \check{I}_4 \end{pmatrix}$$

$$= - \begin{pmatrix} d_{12}/d_1 d_2 & -1/d_2 & 0 & 0 \\ -c_2/d_2 & d_{23}/d_2 d_3 & -1/d_3 & 0 \\ 0 & -c_3/d_3 & d_{34}/d_3 d_4 & -1/d_4 \\ 0 & 0 & -c_4/d_4 & d_{45}/d_4 d_5 \end{pmatrix}.$$

The 2×2-minors of this matrix arranged in the lexicographic order,

$$P = (12), (13), (14), (23), (24), (34),$$

is $(d_1 d_2 d_3 d_4 d_5)^{-1}$ times

$$K =$$

$$\begin{pmatrix} d_{123}d_4d_5 & -d_{12}d_4d_5 & 0 & d_1d_4d_5 & 0 & 0 \\ -d_{12}c_3d_4d_5 & d_{12}d_{34}d_5 & -d_{12}d_3d_5 & -d_1d_{34}d_5 & d_1d_3d_5 & 0 \\ 0 & -d_{12}d_3c_4d_5 & d_{12}d_3d_{45} & d_1d_3c_4d_5 & -d_1d_3d_{45} & 0 \\ d_1c_2c_3d_4d_5 & -d_1c_2d_{34}d_5 & d_1c_2d_3d_5 & d_1d_{234}d_5 & -d_1d_{23}d_5 & d_1d_2d_5 \\ 0 & d_1c_2d_3c_4d_5 & -d_1c_2d_3d_{45} & -d_1d_{23}c_4d_5 & d_1d_{23}d_{45} & -d_1d_2d_{45} \\ 0 & 0 & 0 & d_1d_2c_3c_4d_5 & -d_1d_2c_3d_{45} & d_1d_2d_{345} \end{pmatrix}$$

Note that

$$\det K = (d_1 \cdots d_5 d_{1\ldots 5})^3.$$

(Warning: Do not try to check this yourself. Instead, order your machine to do it.) This implies the six cycles D_P are linearly independent if

$\alpha_j \notin \mathbb{Z}$. In the following subsections, we will show that these 2×2-minors are actually the intersection numbers $D_P \cdot \check{D}_Q$. Let us simply assume this at this time. Therefore we have

$$Int(3,6;\alpha) := \begin{pmatrix} D_{12} \\ \vdots \\ D_{34} \end{pmatrix} (\check{D}_{12}, \cdots, \check{D}_{34}) = (d_1 d_2 d_3 d_4 d_5)^{-1} K.$$

The inverse matrix of $Int(3,6;\alpha)$ is a constant multiple of the following matrix (be warned as above!):

$$H = d_6 \times$$

$$\begin{pmatrix} d_1 d_2 d_{345} & d_1 d_2 d_{45} & d_1 d_2 d_5 & 0 & 0 & 0 \\ d_1 d_2 c_3 d_{45} & d_1 d_{23} d_{45} & d_1 d_{23} d_5 & d_1 d_3 d_{45} & d_1 d_3 d_5 & 0 \\ d_1 d_2 c_3 c_4 d_5 & d_1 d_{23} c_4 d_5 & d_1 d_{234} d_5 & d_1 d_3 c_4 d_5 & d_1 d_{34} d_5 & d_1 d_4 d_5 \\ 0 & d_1 c_2 d_3 d_{45} & d_1 c_2 d_3 d_5 & d_{12} d_3 d_{45} & d_{12} d_3 d_5 & 0 \\ 0 & d_1 c_2 d_3 c_4 d_5 & d_1 c_2 d_{34} d_5 & d_{12} d_3 c_4 d_5 & d_{12} d_{34} d_5 & d_{12} d_4 d_5 \\ 0 & 0 & d_1 c_2 c_3 d_4 d_5 & 0 & d_{12} c_3 d_4 d_5 & d_{123} d_4 d_5 \end{pmatrix}.$$

The actual relation is

$$Int(3,6;\alpha) = d_6 d_{1\ldots 5} H^{-1}, \quad K = d_1 \cdots d_5 d_6 d_{1\ldots 5} H^{-1}$$

so

$$\det H = (d_1 \cdots d_5 d_{12345})^3 (d_6)^6.$$

This matrix $H = H(\alpha)$ will play a very important role in our story (cf. §6 of Chapter IV). Let $\Gamma(\alpha)$ be the monodromy group of the loaded cycles

$$D_{12}, \ D_{13}, \ D_{14}, \ D_{23}, \ D_{24}, \ D_{34}.$$

(A precise definition will be given in the next section.) We will see in the next section that every $M \in \Gamma(\alpha)$ can be defined over the field $\mathbb{Q}(c_1, \ldots, c_n)$, which admits the *involution* $\vee$:

$$\check{c}_j := c_j^{-1}.$$

$\check{M}$ and $\check{H}$ are defined in the obvious manner. We have

$$^t\check{H} = H.$$

Note that if all the α_j are real (or equivalently, $|c_j| = 1$), the operator $\vee$ is the complex conjugation, and the above identity states that H is hermitian.

Since the intersection numbers are invariant under small deformation (you will see this in the following subsections), we have

$$M \; Int(3,6;\alpha) \; {}^t\check{M} = Int(3,6;\alpha) \quad \text{for all} \quad M \in \Gamma(\alpha),$$

which is equivalent to

$$ {}^t\check{M} H M = H \quad \text{for all} \quad M \in \Gamma(\alpha).$$

In this sense we call H the *invariant form*.

REMARK 3.1. We can prove that if the α_j are non-integral, such a form is unique (up to factors of real constants).

REMARK 3.2. Discoveries were made in the other way round compared to what I described above. We found a set of generators of the monodromy group $\Gamma(\alpha)$, obtained the invariant form H through brute force computation by using these generators, guessed that H^{-1} should give the intersection matrix of the loaded cycles, and found a way to evaluate intersection numbers for loaded cycles.

3.2. Deformations of Loaded Cycles. Now our task is to establish a reasonable theory for intersections of loaded cycles. There are elaborate theories to establish the conjecture made at the beginning of §3.1; for these, please consult [KY1] and [IK]. Here I am going to present a down-to-earth method. Do you remember how we computed the intersection numbers of the 1-dimensional loaded cycles? We made a regularization $reg D_j$ of D_j, loaded with the function u, by putting two circles at the edge of the interval and deformed $\check{D}_j$, loaded with u^{-1}, to assume the shape of a sine curve. We then summed up the product of the two functions at the intersection points. Note that the three intersection points are situated near the endpoints of the interval and the midpoint. Thus you can say these are (near) the barycenters of the faces of the interval. In §2.3 we defined regularizations of two dimensional loaded cycles.

So we deform $\check{D}_j$, but how? The sine curve is described in the complex s-line S, the product space of the real s-line and the imaginary s-line (see Figure 3.1). We can represent it as a vector field on the real s-line (see Figure 3.2). If you want to be precise, let Δ be a segment $[a_0, a_1]$ and b

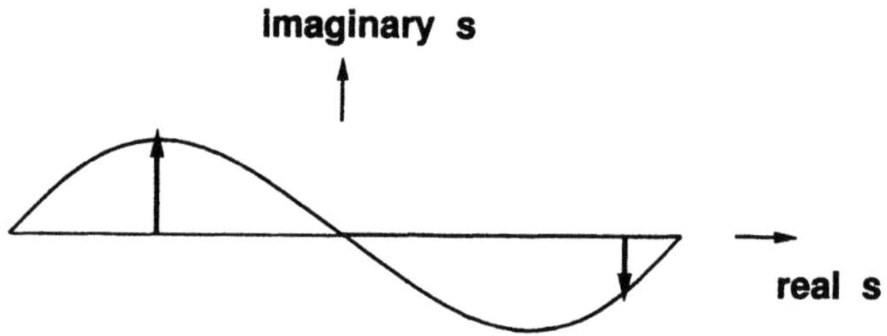

FIGURE 3.1. A sine curve drawn in the complex s-line

FIGURE 3.2. Description in terms of a vector field on Δ

its midpoint. We define a vector field v on Δ by

$$
v(x) = \begin{cases} (\sin \pi \dfrac{a_0 x}{a_0 b})\overrightarrow{a_0 b} & \text{for} \quad x \in [a_0, b] \\[2em] (\sin \pi \dfrac{a_1 x}{a_1 b})\overrightarrow{a_1 b} & \text{for} \quad x \in [b, a_1], \end{cases}
$$

where $a_0 x$ represents the distance between a_0 and x. The sine curve in question can be given by the map

$$
\Delta \ni x \mapsto x + \sqrt{-1}\,v(x) \in S.
$$

Now we are ready to consider the 2-dimensional case.

When $r = 2$, let Δ be a polygon in $T_{\mathbb{R}}$ with vertices $a_1, \ldots, a_m$, and b its barycenter. On each segment $[a_i, a_{i+1}]$ $(a_{m+1} := a_1)$, we define a vector field v as above. If y is in the cone with apex b and base $[a_i, a_{i+1}]$, define $v(y)$ by

$$
v(y) = \frac{by}{bx}v(x) + (\sin \pi \frac{xy}{bx})\overrightarrow{xb},
$$

where x is the intersection of $[a_i, a_{i+1}]$ and the line joining b and y (see Figure 3.3). Since the polygon is the union of such cones, the vector field

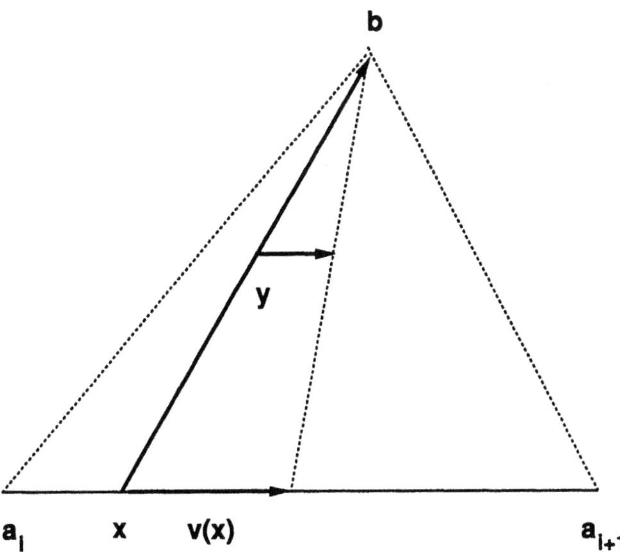

FIGURE 3.3. The vector field v on a cone

v is defined on Δ. It is readily seen that v is continuous and has zeros only at the vertices, the midpoints of the edges, and the barycenter b (see Figure 3.4), and that the corresponding section

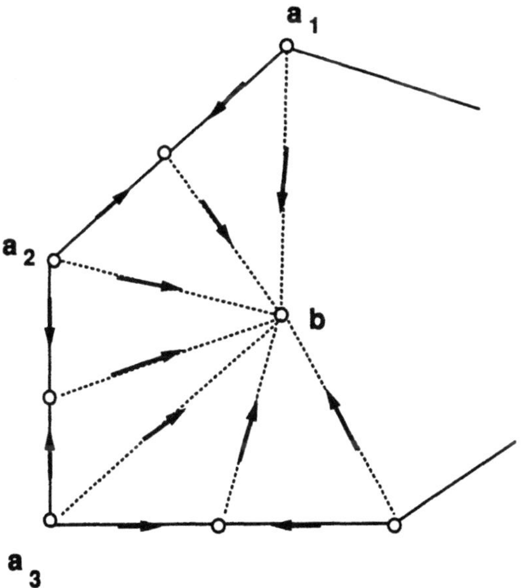

FIGURE 3.4. The vector field v on a polygon

$$\Delta \ni x \mapsto x + \sqrt{-1}v(x) \in T$$

meets the zero-section transversally. When $r \geq 3$, one can inductively continue this construction to obtain its higher dimensional version.

We have thus finished the topological preparation. Now we evaluate intersection numbers.

3.3. Intersections of Loaded Cycles. Let l_j be real linear forms on the r-dimensional affine space $T_{\mathbb{R}}$ defining hyperplanes H_j.

PROPOSITION 3.1. *Let D_1 and D_2 be loaded cycles with supports*

$$|\overline{D}_1| : l_1 \geq 0, \ldots, l_p \geq 0; \quad l_{p+1} \geq 0, \ldots, l_q \geq 0,$$
$$|\overline{D}_2| : l_1 \leq 0, \ldots, l_p \leq 0; \quad l_{p+1} \geq 0, \ldots, l_q \geq 0,$$
$$|\overline{D}_1| \cap |\overline{D}_2| : l_1 = 0, \ldots, l_p = 0; \quad l_{p+1} \geq 0, \ldots, l_q \geq 0$$

(cf. Figure 3.5) on which branches of $\prod_{j=1}^{q} l_j^{\alpha_j}$ are loaded, where $\log l_j$

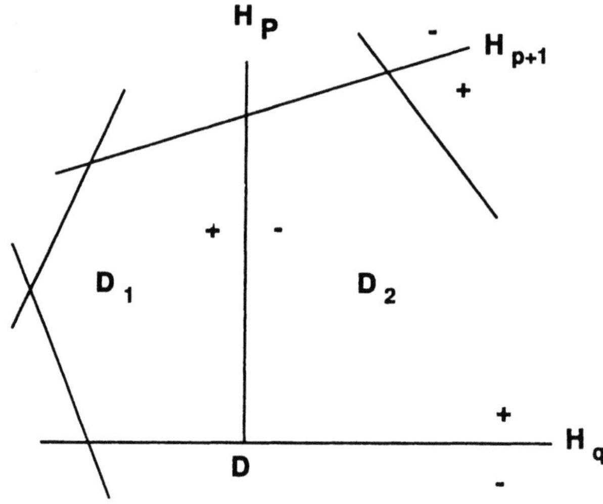

FIGURE 3.5. The two polygons $|D_1|$ and $|D_2|$ touching along $|D|$

are assigned as

$$D_1 : \arg l_1 = \cdots = \arg l_q = 0,$$
$$D_2 : \arg l_1 = \cdots = \arg l_p = -\pi, \quad \arg l_{p+1} = \cdots = \arg l_q = 0.$$

If the hyperplanes $H_j = \{l_j = 0\}(j = 1, \ldots q)$ meet normally along $|\overline{D}_1| \cap |\overline{D}_2|$, then the intersection number $D_1 \cdot \check{D}_2$ is given by

$$D_1 \cdot \check{D}_2 = (-)^{r-p}\frac{c_1 \cdots c_p}{d_1 \cdots d_p}\left(1 + \sum_{m=1}^{r-p} \sum_{p+1 \le j_1 < \cdots < j_m \le q} \frac{1}{d_{j_1} \cdots d_{j_m}}\right),$$

where the summation is taken over $p + 1 \le j_1 < \cdots < j_m \le q$ satisfying the condition

$$|D_1| \cap |D_2| \cap \cap_{s=1}^{m} H_{j_s} \ne \emptyset.$$

PROOF. Recall that the intersection number can be expressed as the sum of the local contributions at the barycenters of the faces of the supports of D_1 and D_2. Let D be the loaded cycle on the $(r - p)$-dimensional space

$$H_P := H_1 \cap \cdots \cap H_p$$

with support

$$|D| : \quad l_1 = 0, \ldots, l_p = 0; \quad l_{p+1} > 0, \ldots, l_q > 0,$$

(the interior of $|\overline{D}_1| \cap |\overline{D}_2|$ in H_P) on which $\prod_{j=p+1}^{q} l_j^{\alpha_j}$ is assigned by

$$\arg l_{p+1} = \cdots = \arg l_q = 0.$$

On the l_j-space $(1 \le j \le p)$, the cycles in question appear as in Figure 3.6 (in which the dotted segment indicates that the branches of $\log l_j$ on

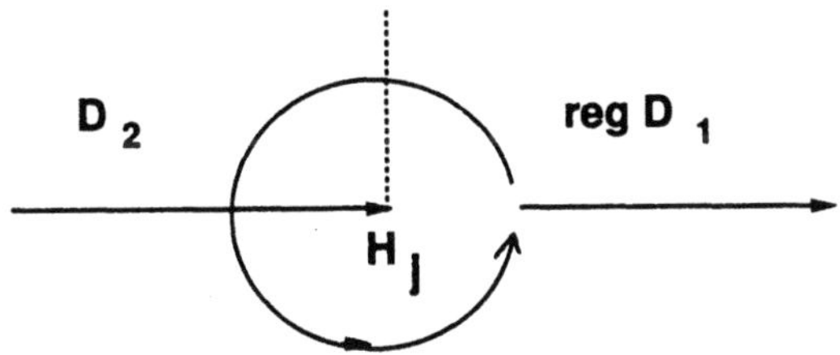

FIGURE 3.6. D_2 and $reg\,D_1$ in the l_j-line $(1 \le j \le p)$

D_1 and D_2 can be continued through the lower half plane). Thus we have

$$D_1 \cdot \check{D}_2 = \frac{c_1 \cdots c_p}{d_1 \cdots d_p} D \cdot \check{D}.$$

Now we evaluate $D \cdot \check{D}$ by making the regularization $reg\,D$ described in §2.3 and the deformation of $\check{D}$ given in the previous subsection. Since $reg\,D$ and $\check{D}$ (deformed) meet near the barycenters of the faces of the

$(r - p)$-dimensional polyhedron $|D|$, and since $|D|$ is bounded by the hyperplanes

$$H_P \cap H_{p+1}, \ldots, H_P \cap H_q$$

in H_P, the intersection number $D \cdot \check{D}$ is equal to the sum of the contributions at the barycenters of all the non-empty faces

$$|\overline{D}| \cap H_{j_1} \cap \cdots \cap H_{j_m}, \quad p+1 \leq j_1 < \cdots < j_m \leq q, \quad 0 \leq m \leq r - p$$

(when $m = 0$, the face is $|D|$ itself). If you reflect on how the deformation of $\check{D}$ was made above, near the barycenter of the face, $regD$ and $\check{D}$ in the full space are found by making the direct product of the 1-dimensional counterparts in the l_j-spaces $(j = j_1, \ldots, j_m)$ illustrated on the left-hand of Figure 3.7, and $(r - p - m)$ times that on the right-hand of this figure. Thus the contribution at this barycenter is equal to

$$\frac{(-)^{r-p}}{d_{j_1} \cdots d_{j_m}}.$$

This completes the proof. $\square$

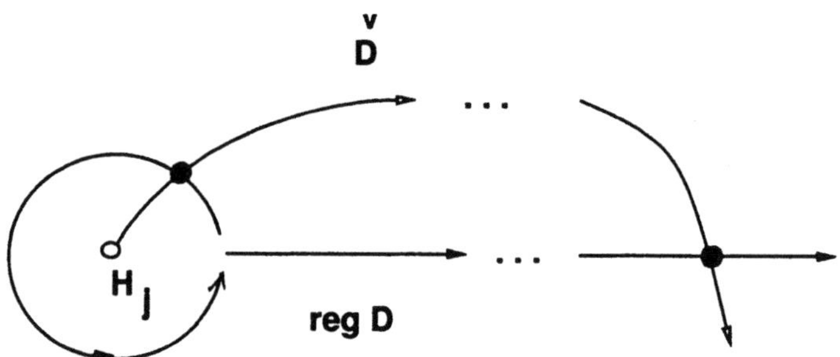

FIGURE 3.7. $regD$ and the deformed $\check{D}$ in the l_j-line $(p + 1 \leq j \leq q)$

REMARK 3.3. When the branches of $\log l_j$ are different from those given above, one has only to multiply D_1 and D_2 by suitable powers of the c_j and make use of the linearity:

$$(c_j D_1) \cdot \check{D}_2 = c_j(D_1 \cdot \check{D}_2), \quad D_1 \cdot (c_j \check{D}_2) = \frac{1}{c_j}(D_1 \cdot \check{D}_2).$$

EXAMPLE 3.1. When $\Delta := |\overline{D}|$ is an r-simplex bounded by $H_1, \ldots, H_{r+1}$, the intersection number $(-)^r D \cdot \check{D}$ is the sum of the following quantities (cf. Figure 3.8):

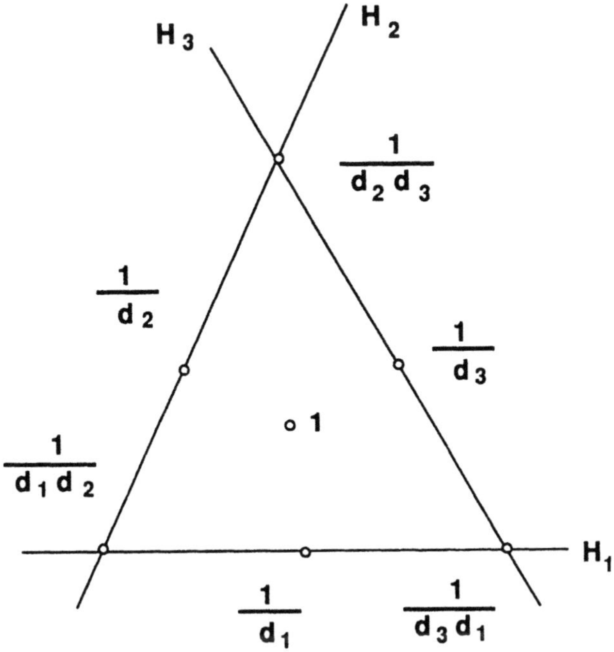

FIGURE 3.8. A triangle with contributions at the barycenters

$$\prod_{i\neq j}\frac{1}{d_i} \qquad \text{near the vertex } A_j = \cap_{i\neq j}H_i,$$

$$\prod_{i\neq j_1,j_2}\frac{1}{d_i} \qquad \text{near the midpoint of the edge } \langle A_{j_1},A_{j_2}\rangle = \Delta\cap_{i\neq j_1,j_2}H_i,$$

$$\vdots$$

$$\prod_{i\neq j_1,\ldots,j_p}\frac{1}{d_i} \qquad \text{near the barycenter of the } (p-1)\text{-dimensional face}$$

$$\langle A_{j_1},\ldots,A_{j_p}\rangle = \Delta\cap_{i\neq j_1,\ldots,j_p}H_i,$$

$$\vdots$$

$$\frac{1}{d_j} \qquad \text{near the barycenter of the hyperface}$$

$$\langle A_i\mid i\neq j\rangle = \Delta\cap H_j,$$

$$1 \qquad \text{at the barycenter of } \Delta.$$

Summing up all these quantities we have

$$D \cdot \check{D} = (-)^r \sum_{p=1}^{r+1} \ \sum_{1 \le j_1 < \cdots < j_p \le r+1} \ \prod_{i \ne j_1,\ldots,j_p} \frac{1}{d_i}$$

$$= (-)^r \frac{d_{1\cdots r+1}}{d_1 \cdots d_{r+1}}$$

(the second equality can be proved by induction). Note that this is just the determinant of the intersection matrix $Int(r+2;\alpha)$ for the I_j.

REMARK 3.4. If the support $|D|$ of D is of the direct product type, then the intersection number $D \cdot \check{D}$ is the product of those direct factors.

EXAMPLE 3.2. When $|D|$ is a rectangle, as in Figure 3.9, we have

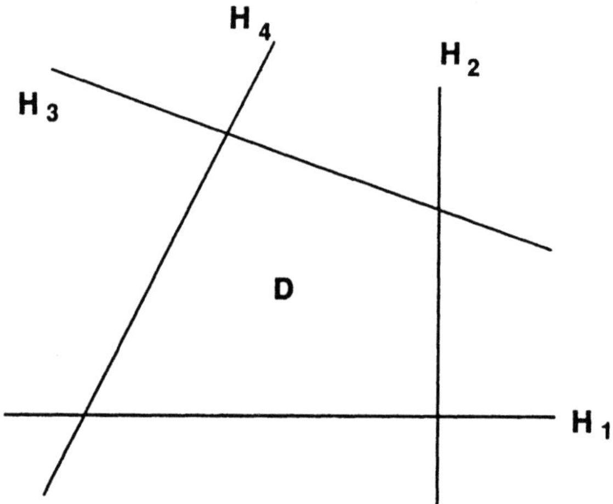

FIGURE 3.9. A rectangle

$$D \cdot \check{D} = \frac{1}{d_1 d_2} + \frac{1}{d_2 d_3} + \frac{1}{d_3 d_4} + \frac{1}{d_4 d_1} + \frac{1}{d_1} + \frac{1}{d_2} + \frac{1}{d_3} + \frac{1}{d_4} + 1$$

$$= \frac{d_{13}}{d_1 d_3} \frac{d_{24}}{d_2 d_4}.$$

3.4. Intersection Numbers for $D_{12}, \ldots, D_{34}$. Since D_{12}, D_{23} and D_{34} are triangles (see Figure 2.2) with D_{12} bounded by H_1, H_2 and H_3, D_{23} bounded by H_2, H_3 and H_4, and D_{34} bounded by H_3, H_4 and H_5, we have (cf. Example 3.1)

$$D_{12} \cdot \check{D}_{12} = \frac{d_{123}}{d_1 d_2 d_3}, \quad D_{23} \cdot \check{D}_{23} = \frac{d_{234}}{d_2 d_3 d_4}, \quad D_{34} \cdot \check{D}_{34} = \frac{d_{345}}{d_3 d_4 d_5}.$$

The remaining chambers are rectangles, so we have (cf. Example 3.2)

$$D_{13} \cdot \check{D}_{13} = \frac{d_{12}}{d_1 d_2} \frac{d_{34}}{d_3 d_4}, \quad D_{14} \cdot \check{D}_{14} = \frac{d_{12}}{d_1 d_2} \frac{d_{45}}{d_4 d_5}, \quad D_{24} \cdot \check{D}_{24} = \frac{d_{23}}{d_2 d_3} \frac{d_{45}}{d_4 d_5}.$$

We shall evaluate

$$D_{12} \cdot \check{D}_{13}, \ D_{13} \cdot \check{D}_{12}, \quad D_{12} \cdot \check{D}_{23}, \ D_{23} \cdot \check{D}_{12}.$$

Here are the assignments for the arguments of L_j we made in §2.2 for the D_{ij} which are of interest right now (see Figure 3.10):

	D_{12}	D_{13}	D_{23}
$\arg L_1$	0	0	0
$\arg L_2$	$-\pi$	$-\pi$	0
$\arg L_3$	-2π	$-\pi$	$-\pi$

The proposition in the previous subsection implies

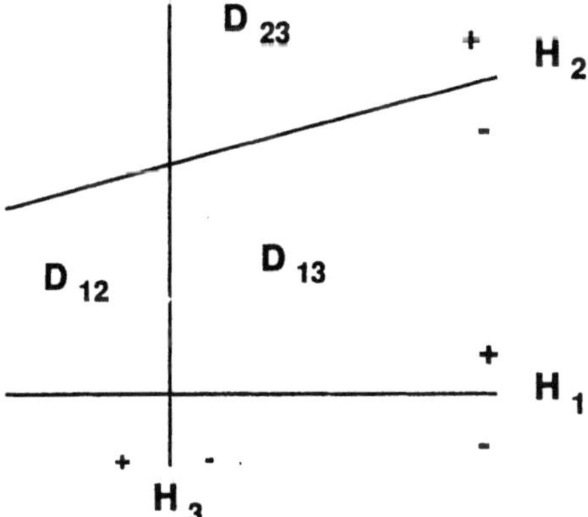

FIGURE 3.10. D_{12}, D_{13} and D_{23}

$$D_{12} \cdot \check{D}_{13} = (-)^{2-1} \frac{1}{d_3} \left(1 + \frac{1}{d_1} + \frac{1}{d_2}\right) = (-)^{2-1} \frac{1}{d_3} \frac{d_{12}}{d_1 d_2},$$

$$D_{13} \cdot \check{D}_{12} = (-)^{2-1} \frac{c_3}{d_3} \frac{d_{12}}{d_1 d_2},$$

$$D_{12} \cdot \check{D}_{23} = (-)^{2-2} \frac{1}{d_2} \frac{1}{d_3},$$

$$D_{23} \cdot \check{D}_{12} = (-)^{2-2} \frac{c_2}{d_2} \frac{c_3}{d_3}.$$

200 VIII. HYPERGEOMETRIC FUNCTIONS OF TYPE (3,6)

Thus the intersection matrix for $D_{12}, D_{13}, D_{14}, D_{23}, D_{24}, D_{34}$ is given by

$$Int(3,6;\alpha) = \begin{pmatrix} D_{12}\cdot\check{D}_{12} & \cdots & D_{12}\cdot\check{D}_{34} \\ \vdots & & \vdots \\ D_{34}\cdot\check{D}_{12} & \cdots & D_{34}\cdot\check{D}_{34} \end{pmatrix}$$

$$= \begin{pmatrix}
\dfrac{d_{123}}{d_1 d_2 d_3} & -\dfrac{d_{12}}{d_1 d_2 d_3} & 0 & \dfrac{1}{d_2 d_3} & 0 & 0 \\[2ex]
-\dfrac{d_{12}c_3}{d_1 d_2 d_3} & \dfrac{d_{12}d_{34}}{d_1 d_2 d_3 d_4} & -\dfrac{d_{12}}{d_1 d_2 d_4} & -\dfrac{d_{34}}{d_2 d_3 d_4} & \dfrac{1}{d_2 d_4} & 0 \\[2ex]
0 & -\dfrac{d_{12}c_4}{d_1 d_2 d_4} & \dfrac{d_{12}d_{45}}{d_1 d_2 d_4 d_5} & \dfrac{c_4}{d_2 d_4} & -\dfrac{d_{45}}{d_2 d_4 d_5} & 0 \\[2ex]
\dfrac{c_2 c_3}{d_2 d_3} & -\dfrac{c_2 d_{34}}{d_2 d_3 d_4} & \dfrac{c_2}{d_2 d_4} & \dfrac{d_{234}}{d_2 d_3 d_4} & \dfrac{d_{23}}{d_2 d_3 d_4} & \dfrac{1}{d_3 d_4} \\[2ex]
0 & \dfrac{c_2 c_4}{d_2 d_4} & -\dfrac{c_2 d_{45}}{d_2 d_4 d_5} & -\dfrac{d_{23}c_4}{d_2 d_3 d_4} & \dfrac{d_{23}d_{45}}{d_2 d_3 d_4 d_5} & -\dfrac{d_{45}}{d_3 d_4 d_5} \\[2ex]
0 & 0 & 0 & \dfrac{c_3 c_4}{d_3 d_4} & -\dfrac{c_3 d_{45}}{d_3 d_4 d_5} & \dfrac{d_{345}}{d_3 d_4 d_5}
\end{pmatrix}.$$

So the prediction we made in the beginning of this section was right! For later use let us evaluate this intersection matrix for the case $\alpha_j = 1/2$ (i.e., $c_j = -1, d_j = -2, d_{jk} = 0, d_{ijk} = -2$):

$$Int(3,6;\tfrac{1}{2}) = \frac{1}{4}\begin{pmatrix} 1 & 0 & 0 & 1 & 0 & 0 \\ 0 & 0 & 0 & 0 & 1 & 0 \\ 0 & 0 & 0 & -1 & 0 & 0 \\ 1 & 0 & -1 & 1 & 0 & 1 \\ 0 & 1 & 0 & 0 & 0 & 0 \\ 0 & 0 & 0 & 1 & 0 & 1 \end{pmatrix}.$$

Note that this matrix is symmetric.

3.5. Higher Dimensional Pochhammer Loops. In the 1-dimensional case, for a loaded cycle I with the segment $(0,1)$ as support, we defined the compact loaded cycle $reg I$ by attaching two circles. The Pochhammer loop (double contour loop) is defined in §2.2 of Chapter IV (cf. Figure 2.1 of Chapter IV). These two are related by

$$(1-c_1)(1-c_2)reg I = (1-c_1-c_2+c_1 c_2)reg I,$$

where we think of the right-hand side in terms of the 4-folded path representing the Pochhammer loop, as we did in §2.2 of Chapter IV.

REMARK 3.5. The Pochhammer loop corresponding to $regI$ is not unique. You can take a path around 1 first in the negative direction (clockwise) and then in the positive direction (counterclockwise), and similarly for 0. So there are $2 \times 2 = 4$ ways.

If you unfold this loop, it becomes a square, or more precisely, a truncated square, as is shown in Figure 3.11 (the four truncated corners correspond to the four circles with slits). The arguments of t (resp. $1 - t$)

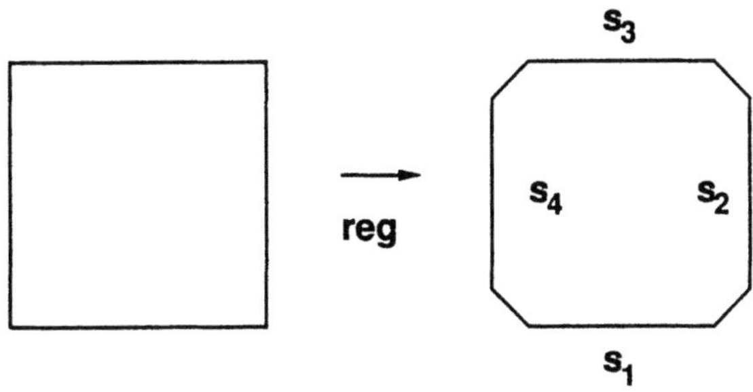

FIGURE 3.11. Unfolded Pochhammer loop

at the starting point and ending point agree.

What do you think will happen for a 2-dimensional loaded cycle D, where $|D|$ is a simplex bounded by H_1, H_2, H_3? We multiply $regD$ by

$$(1 - c_1)(1 - c_2)(1 - c_3) = 1 - c_1 - c_2 - c_3 + c_1c_2 + c_2c_3 + c_3c_1 - c_1c_2c_3.$$

We then see eight triangles glued together to form an octahedron (see Figure 3.12). Precisely speaking, the octahedron should be truncated along the edges and the vertices so that the edges become rectangles and the vertices become squares, each corresponding to

$$\text{(an interval)} \times \text{(a circle with a slit)}$$

and

$$\text{(a circle with a slit)} \times \text{(a circle with a slit)},$$

respectively. At any rate, you will find three square equators in our octahedron, corresponding to H_1, H_2, H_3. Please look at Figure 3.12, where H_1 is emphasized. In the northern hemisphere, which is the union of four simplices including D, assign the value (for example) $\arg l_1 = 0$ and on the southern hemisphere $\arg l_1 = 2\pi$ or -2π (recall Remark 3.5). Then do the same for the hemispheres bounded by H_2 and H_3.

Now you see that for any closed path on the octahedron, the arguments of l_j $(j = 1, 2, 3)$ at the starting point and the ending point agree. So you can think of this as a 2-dimensional version of the Pochhammer loop.

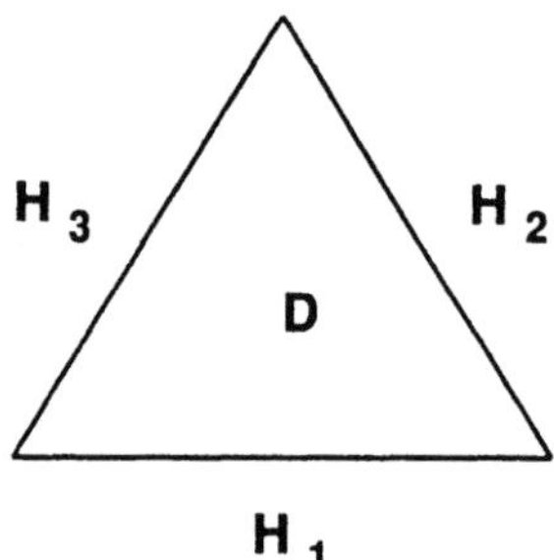

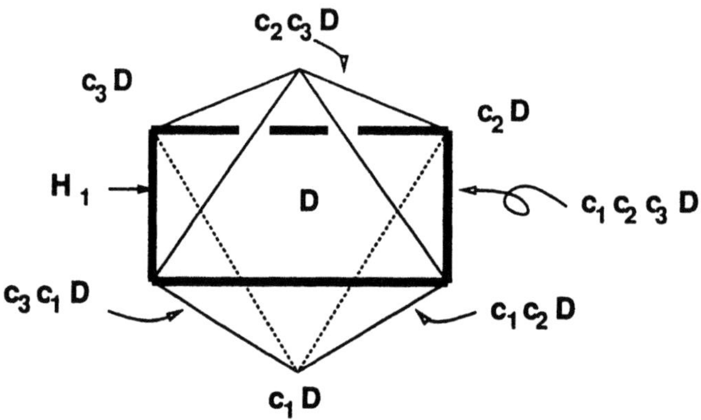

FIGURE 3.12. 2-dimensional unfolded Pochhammer cycle

Before proceeding to higher dimensional cases, let us recall the following fact that everyone should know:

There are infinitely many (1-dimensional) regular polygons (in $\mathbb{R}^2$), obviously. There are five (2-dimensional) regular polyhedra: the boundaries of the tetrahedron, cube, octahedron, dodecahedron and icosahedron. There are six 3-dimensional regular polyhedra. For $k \geq 4$, there are only three k-dimensional regular polyhedra: the boundaries of the simplex, the hypercube and its dual.

For a k-dimensional loaded cycle D, where $|D|$ is a k-simplex, we glue 2^{k+1} simplices corresponding to the terms of the expansion of

$$(1 - c_1) \cdots (1 - c_{k+1}),$$

and we get the dual of a hypercube which has 2^{k+1} faces and $2(k+1)$ vertices. The discussion above for the octahedron can be paraphrased word for word for the present case, and we obtain a k-dimensional version of the Pochhammer loop. Though I omit the details, I strongly recommend to the reader to think about what will happen in the 3-dimensional case. (Hint: Sixteen tetrahedra fill the space. Find the four equators.)

4. Monodromy of Loaded Cycles

4.1. The Circuit Matrix $M(1,\ldots,r+1;\alpha)$.

The chamber $D'_{(1,\ldots,r)}(\overset{\bullet}{x})$ of $T_{\mathbb{R}}$ is a simplex bounded by the $r+1$ hyperplanes $\overset{\bullet}{H}_1,\ldots,\overset{\bullet}{H}_r$ and $\overset{\bullet}{H}_{r+1}$. (See §2.2.) In this subsection, we study the *circuit matrix* $M(1,\ldots,r+1;\alpha)$ of the system $(D_P)_P$ relative to a loop $\rho(1,\ldots,r+1)$ in X with base point $\overset{\bullet}{x}$, i.e., the linear change of the D_P caused by the move of the hyperplanes representing the loop in X. The loop is described as follows: make a parallel displacement of the hyperplane $H_{r+1}(x)$ in $T_{\mathbb{R}}$ so that the simplex becomes small. Then let the hyperplane go once around the intersection point $\overset{\bullet}{H}_1 \cap \cdots \cap \overset{\bullet}{H}_r$ in the complex space T in the positive sense. Finally let it return. During the entire journey, always keep $H_{r+1}(x)$ parallel to $\overset{\bullet}{H}_{r+1}$. Similarly letting another hyperplane move around the intersection point of the remaining r hyperplanes defines a loop homotopic to $\rho(1,\ldots,r+1)$ in X.

PROPOSITION 4.1. *The deformation of the hyperplanes represented by the loop $\rho(1,\ldots,r+1)$ induces the transformation $M(1,\ldots,r+1;\alpha)$ of the loaded cycles D_P as follows:*

$$D_k \longrightarrow D_k + (-1)^{r-k}c_{k+1}\cdots c_{r+1}(1 - c_1 \cdots c_k)D_{r+1}, \quad 1 \le k \le r,$$
$$D_{r+1} \longrightarrow c_1 \cdots c_{r+1}D_{r+1},$$

where $c_j = \exp 2\pi i \alpha_j$ and

$$D_k := D_{(1,\ldots,k-1,k+1,\ldots,r+1)}, \quad 1 \le k \le r+1.$$

The cycles D_P do not change for other P.

Before beginning the proof of the proposition, we give several remarks.

REMARK 4.1. When $r = 1$, the proposition agrees with the full-turn formula obtained in §4 of Chapter IV.

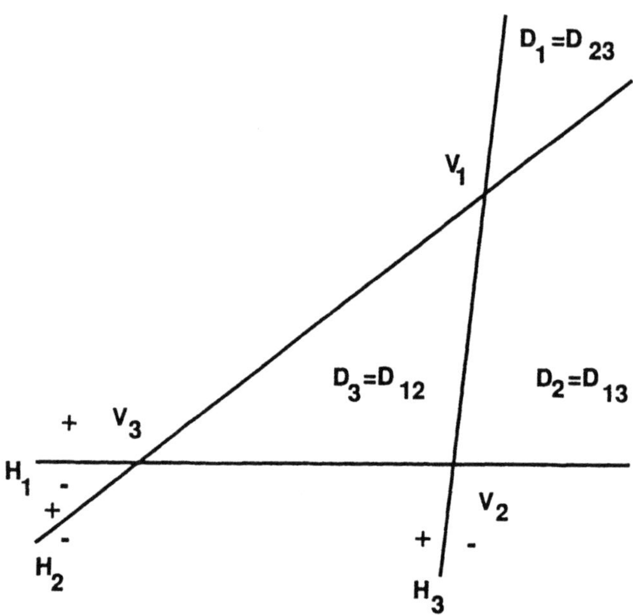

FIGURE 4.1. $r = 2$, the domains touching D_{12}

REMARK 4.2. We have

$$\det M(1,\ldots,r+1;\alpha) = c_1 \cdots c_{r+1}.$$

If, moreover, at least one of the $\alpha_j (j = 1,\ldots,r+1)$ is not an integer,

$$\operatorname{rank}\,[M(1,\ldots,r+1;\alpha) - id_r] = 1.$$

REMARK 4.3. Though we chose the base point $\overset{\bullet}{x}$ carefully in $Q \subset X$, the loop cannot stay within Q; it must go beyond Q in X.

PROOF. We apply the convention in the proposition also to the domains D'_P. The real domain $D'_{r+1} = D'_{(1,\ldots,r)}$ is a simplex bounded by the $r+1$ hyperplanes $\overset{\bullet}{H}_1,\ldots,\overset{\bullet}{H}_{r+1}$. We name its $r+1$ vertices according to

$$V_k = \overset{\bullet}{H}_1 \cap \cdots \cap \overset{\bullet}{H}_{k-1} \cap \overset{\bullet}{H}_{k+1} \cap \cdots \cap \overset{\bullet}{H}_{r+1}, \quad 1 \le k \le r+1.$$

There are exactly r domains D'_P which touch the simplex D'_{r+1} (cf. Figures 4.1 and 4.2); in fact the domains

$$D'_1,\ldots,D'_k,\ldots,D'_r$$

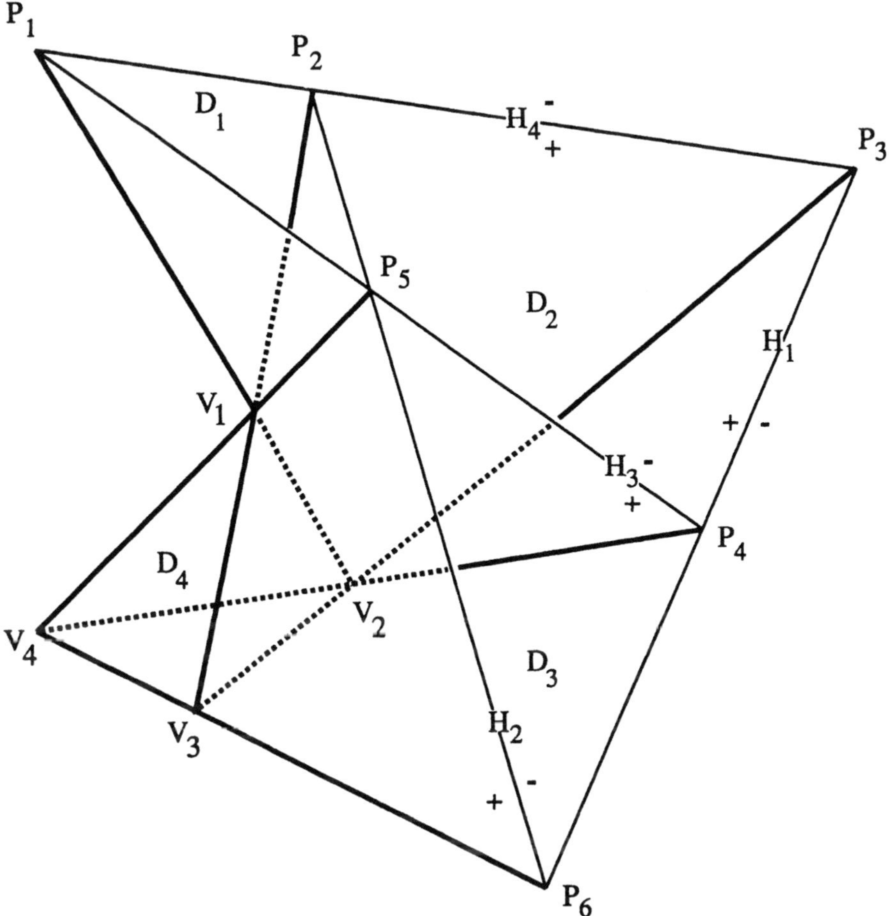

FIGURE 4.2. $r = 3$, the domains touching $D_4 = D_{123}$. Section by
the generic hyperplane $P_1P_2P_3P_4P_5P_6$, which we regard as H_5, is added
in order to show the hyperplanes $H_1,...,H_4$ and the domains $D'_1,...,D'_3$,
where $D'_1=(V_1,P_1,P_2,P_5)$ is a simplex, $D'_2=(V_1,V_2,P_4,P_3,P_2,P_5)$ is a polytope,
$D'_3=(V_1,V_2,V_3,P_6,P_4,P_5)$ is a polytope, and $D'_4=(V_1,V_2,V_3,V_4)$ is a simplex.

touch the simplex D'_{r+1} along the faces

$$V_1,\ldots,(V_1,\ldots,V_k),\ldots,(V_1,\ldots,V_r),$$

respectively, where $(V_1,\ldots,V_k)$ denotes the $(k-1)$-simplex with vertices $V_1,\ldots,V_k$. It is obvious that by moving the arrangement along
$\rho(1,\ldots,r+1)$, only $D_1,\ldots,D_{r+1}$ among the $\binom{n-1}{r}$ cycles D_P change.

   In order to study the change of D_k, we consider the complex line l

passing through a point A in the simplex $(V_1, \ldots, V_k)$ and a point B in the complementary simplex $(V_{k+1}, \ldots, V_{r+1})$ (see Figure 4.5). Figure 4.3 shows the line l as well as the points A, B and the two segments $l \cap D'_{r+1}$ and $l \cap D'_k$.

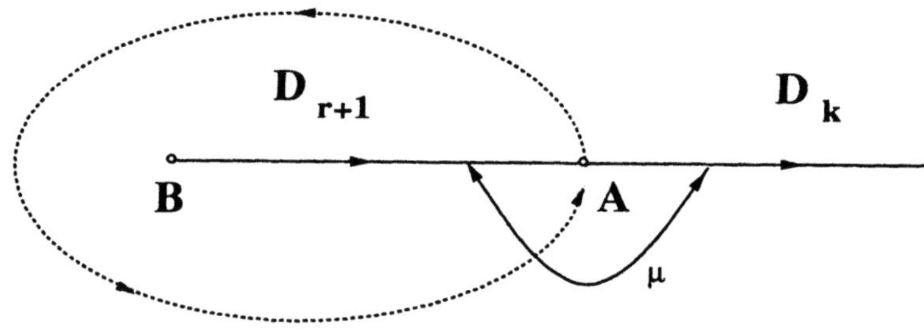

FIGURE 4.3. The line l through A and B

For the assignment of arguments of the L_j, we have

$$\arg L_{k+1} = -k\pi \quad \text{on } D'_{r+1}, \qquad -(k-1)\pi \quad \text{on } D'_k,$$
$$\vdots \qquad\qquad \vdots \qquad\qquad\qquad \vdots$$
$$\arg L_{r+1} = -r\pi \quad \text{on } D'_{r+1}, \qquad -(r-1)\pi \quad \text{on } D'_k.$$

Therefore, for each m $(k+1 \leq m \leq r+1)$, the power function $L_m^{\alpha_m - 1}$ defined on D'_{r+1} and that defined on D'_k are analytic continuations of each other along a path μ in the *lower* half plane of the line l. According to the move along $\rho(1, \ldots, r+1)$, the point A goes once around the point B in the positive direction (see the dotted curve in Figure 4.3). This causes the change of the segment $l \cap D'_k$ shown in Figure 4.4.

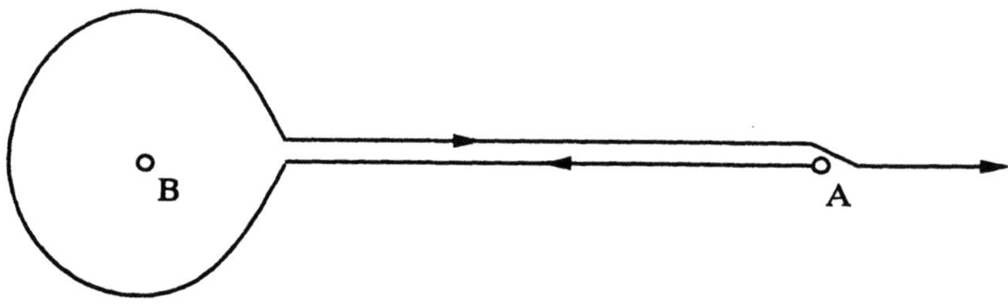

FIGURE 4.4. The transformed curve

Since the transformed curve of $l \cap D'_k$ passes *above* the point A and goes around the point B, we have

$$l \cap D_k \longrightarrow l \cap D_k + c_{k+1} \cdots c_{r+1}(1 - c_1 \cdots c_k)l \cap D_{r+1}, \quad 1 \le k \le r.$$

With the point A fixed and as the point B is moved in $(V_{k+1}, \ldots, V_{r+1})$, we consider the map sending B to its antipodal point relative to A (see Figure 4.5); the map is orientation preserving or reversing if the dimension of the simplex $(V_{k+1}, \ldots, V_{r+1})$, which is equal to $r - k$, is even or odd, respectively. When the point B is fixed and the point A moves in $(V_1, \ldots, V_k)$ in some direction, then $l \cap D'_{r+1}$ and $l \cap D'_k$ move in the same direction. Since D'_{r+1} is the *join* of two simplices, $(V_1, \ldots, V_k)$ and $(V_{k+1}, \ldots, V_{r+1})$, we have

$$D_k \longrightarrow D_k + (-1)^{r-k}c_{k+1} \cdots c_{r+1}(1 - c_1 \cdots c_k)D_{r+1}, \quad 1 \le k \le r.$$

(The join of two subsets V and W of an affine space is defined as the union of segments with one end in V and the other end in W.)

In the course of this procedure the segment $l \cap D'_{r+1}$ travels around the point A as well as the point B. Thus we have

$$D_{r+1} \longrightarrow c_1 \cdots c_{r+1}D_{r+1}.$$

The proof is now complete. $\square$

4.2. The Circuit Matrix $M(123; \alpha)$ as a Quasi-reflection. The circuit matrix $M(1, \ldots, r+1; \alpha)$ obtained in the previous section has one eigenvalue $c_1 \cdots c_{r+1}$, and all others are 1. In particular, for the circuit matrix $M = M(123; \alpha)$ we have

$$M(123; \alpha) = \begin{pmatrix} c_1 c_2 c_3 & 0 & 0 & 0 & 0 & 0 \\ -d_{12}c_3 & 1 & 0 & 0 & 0 & 0 \\ 0 & 0 & 1 & 0 & 0 & 0 \\ d_1 c_2 c_3 & 0 & 0 & 1 & 0 & 0 \\ 0 & 0 & 0 & 0 & 1 & 0 \\ 0 & 0 & 0 & 0 & 0 & 1 \end{pmatrix},$$

which operates on the column vector $^t(D_{12}, \ldots, D_{34})$ from the left. This matrix M has the following specific characters:

(1) It preserves the invariant form H defined in the previous section (recall that $^t\check{H} = H$), i.e.,

$$^t\check{M}HM = H.$$

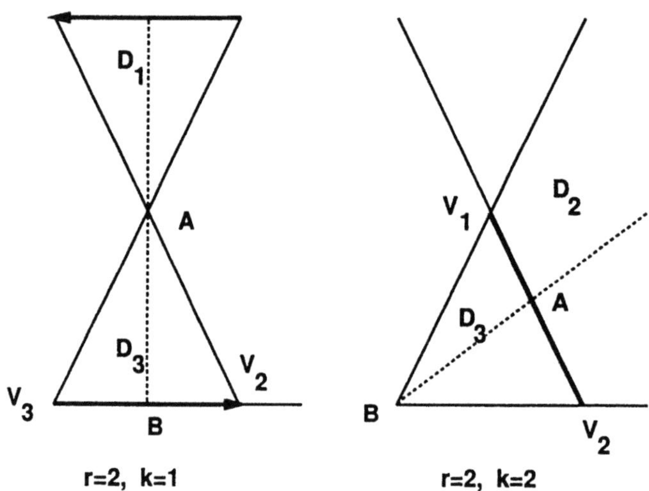

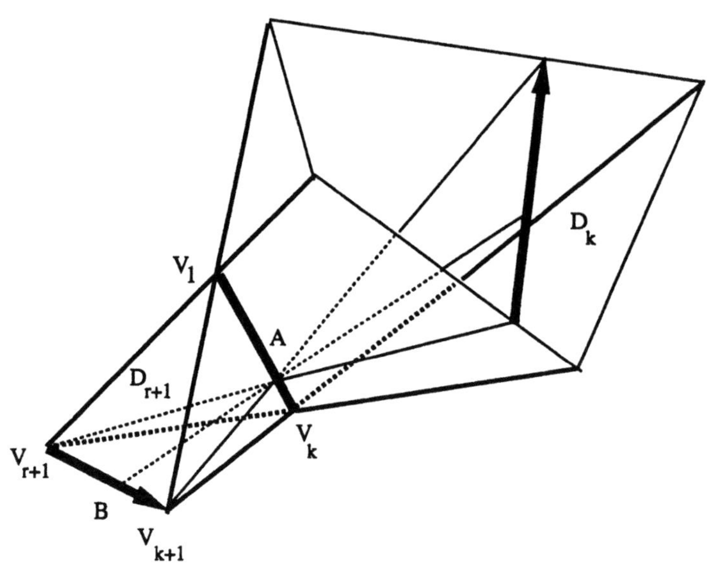

FIGURE 4.5. Sending B to the point antipodal of B relative to A

(2) The rank of $M - I_6$ is 1.

(3) $\det M = c_1 c_2 c_3$.

In general we have

PROPOSITION 4.2. *Let a scalar $\varepsilon(\neq 0)$, a row r-vector b, and a nonsingular $r \times r$-matrix S be defined over $\mathbb{Q}(c_1, \ldots, c_n)$. Assume*

$$\check{\varepsilon} = \varepsilon^{-1}, \quad {}^t\check{S} = S, \quad bS{}^t\check{b} \neq 0,$$

where the operator $\vee$ is defined by $c_j \mapsto c_j^{-1}$. Then

$$R = R_b := I_r - (1 - \varepsilon)\frac{{}^t\check{b}bS}{bS{}^t\check{b}}$$

satisfies

$$ {}^t\check{R}SR = S.$$

PROOF.

$$
\begin{aligned}
{}^t\check{R}SR &= \left(I_r - (1 - \check{\varepsilon})\frac{S{}^t\check{b}b}{bH{}^t\check{b}}\right) S \left(I_r - (1 - \varepsilon)\frac{{}^t\check{b}bS}{bS{}^t\check{b}}\right) \\
&= S - \frac{1 - \check{\varepsilon} + 1 - \varepsilon}{bS{}^t\check{b}} S{}^t bbS + (1 - \check{\varepsilon})(1 - \varepsilon)\frac{S{}^t bbS{}^t bbS}{(bS{}^t\check{b})^2} \\
&= S,
\end{aligned}
$$

since $1 - \check{\varepsilon} + 1 - \varepsilon = (1 - \check{\varepsilon})(1 - \varepsilon)$. $\square$

We call this R_b a *quasi-reflection* belonging to S with *root* b and *determinant* ε. Note that if b is replaced by b times a monomial in $\mathbb{Q}[\ldots, c_j, c_j^{-1}, \ldots]$, the resulting quasi-reflection is the same. In our case, the circuit matrix $M(123; \alpha)$ is the quasi-reflection belonging to $H = H(\alpha)$ with root

$$a(123) = (d_{123}, -d_{12}, 0, d_1, 0, 0)$$

and with determinant $c_1 c_2 c_3$, i.e.,

$$M(123; \alpha) = I_6 + d_{123}\frac{{}^t\check{a}(123)a(123)H}{a(123)H{}^t\check{a}(123)}.$$

4.3. Circuit Matrices $M(j_1, \ldots, j_{r+1}; \alpha)$. Let

$$\rho(j_1, \ldots, j_{r+1}), \quad 1 \leq j_1 < \cdots < j_{r+1} \leq n + 1$$

be a loop in X with base point $\overset{\bullet}{x}$ which is described by the following move of hyperplanes H_j $(0 \leq j \leq n)$. Let us first choose one index among $j_1, \ldots, j_{r+1}$ and call it j'. All the hyperplanes but $H_{j'}$ remain fixed. The hyperplane approaches sufficiently close to the intersection point of the r remaining hyperplanes among the $r + 1$ hyperplanes $H_{j_1}, \ldots, H_{j_{r+1}}$, goes once around the point in the positive sense, and then travels back to the original position, tracing over the previous route. Note that the choice of the loop$\rho(j_1,\ldots,j_{r+1})$ is by no means unique.

The circuit matrix $M(1,\ldots,r+1;\alpha)$ is the linear change of the basis caused by the loop $\rho(1,\ldots,r+1)$ defined in the beginning of this section. In order to describe the change of the basis caused by another loop $\rho(1,\ldots,r,r+2)$, we first exchange the two hyperplanes H_{r+1} and H_{r+2}, which can be done by a move of x in Q, apply $M(1,\ldots,r+1;\alpha)$ (of course α_{r+1} must be considered as α_{r+2}), and then again exchange the two hyperplanes H_{r+1} and H_{r+2}. Since the process of exchange can be done *inside* Q, by virtue of the map ϕ in §2.2, giving the identity

$$D_P = I_{p_1} \wedge \cdots \wedge I_{p_r}, \quad P = (p_1,\ldots,p_r),$$

we can describe it in terms of the change of 1-dimensional cycles. By successively applying the interchanging process, we can find circuit matrices $M(j_1,\ldots,j_{r+1};\alpha)$ corresponding to the other loops $\rho(j_1,\ldots,j_{r+1})$ by

$$M(j_1,\ldots,j_{r+1};\alpha) = W M(1,\ldots,n+1;\alpha')W^{-1},$$

where α' is a permutation of $\alpha = (\alpha_1,\ldots,\alpha_{n+1})$, and W is the matrix describing the successive exchanges. This implies that the unique eigenvalue (of multiplicity 1) of the matrix $M(j_1,\ldots,j_{r+1};\alpha)$ distinct from 1 is $e(\alpha_{j_1} + \cdots + \alpha_{j_{r+1}})$.

Let us carry out the program for $r = 2, n = 5$. Arrange the P in the lexicographic order,

$$(12), \ (13), \ (14), \ (23), \ (24), \ (34).$$

Then think of $\{D_P\}_P$ as a column vector, with $M(ijk;\alpha)$ the 6×6 matrix operating on the vector from the left. ($M(123;\alpha)$ is given in the previous subsection.) Next we try $M(124;\alpha)$. If you would like to let the line H_4 go around the point $H_1 \cap H_2$, for instance, you can first *exchange* H_4 and H_3 so that H_1, H_2 and H_4 bound a triangle. Next let H_4 go around the point $H_1 \cap H_2$, and then exchange H_4 and H_3 again. You can perform this interchanging process in Q. In other words, you can exchange ξ_3 and ξ_4 in the s-space, as in Figure 5.1 in Chapter IV. Denote this move by s_3. We already saw what happens for the loaded cycles I_j after the exchange in the half-turn formula in §5 of Chapter IV. Since the loaded cycles D_P can be thought of as exterior products of the I_{p_j}, we know the change $W(s_3, \alpha)$ of the cycles D_P along the move s_3 is

$$W(s_3, \alpha) = \begin{pmatrix} 1 & c_4^{-1} & 0 & 0 & 0 & 0 \\ 0 & -c_4^{-1} & 0 & 0 & 0 & 0 \\ 0 & 1 & 1 & 0 & 0 & 0 \\ 0 & 0 & 0 & -c_4^{-1} & 0 & 0 \\ 0 & 0 & 0 & 1 & 1 & c_4^{-1} \\ 0 & 0 & 0 & 0 & 0 & -c_4^{-1} \end{pmatrix}.$$

Thus we have

$$M(124; \alpha) = W(s_3, \alpha) M(123; \alpha_1, \alpha_2, \alpha_4, \alpha_3, \alpha_5, \alpha_6) W(s_3, \alpha)^{-1}$$

$$= \begin{pmatrix} 1 + c_1 c_2 d_4 & c_1 c_2 d_4 & 0 & 0 & 0 & 0 \\ d_{12} & c_1 c_2 & 0 & 0 & 0 & 0 \\ -d_{12} c_4 & -d_{12} c_4 & 1 & 0 & 0 & 0 \\ -d_1 c_2 & -d_1 c_2 & 0 & 1 & 0 & 0 \\ d_1 c_2 c_4 & d_1 c_2 c_4 & 0 & 0 & 1 & 0 \\ 0 & 0 & 0 & 0 & 0 & 1 \end{pmatrix}.$$

If we put

$$a(124) = (d_4, c_4 d_{12}, -d_{12}, -d_1 c_4, d_1, 0),$$

then $M(124; \alpha)$ can be expressed as follows:

$$M(124; \alpha) = I_6 + d_{124} \frac{{}^t \breve{a}(124) a(124) H}{a(124) H {}^t \breve{a}(124)}.$$

By performing the "exchanges" the necessary number of times, we can determine the changes $M(ijk; \alpha)$ to the cycles D_P when the line H_i goes around the point $H_j \cap H_k$. The actual computation can be done (an economical procedure is shown in Figure 4.6), and we have all of them.

Since every matrix $M(ijk; \alpha)$ can be written as

$$M(ijk; \alpha) = I_6 + d_{ijk} \frac{{}^t \breve{a}(ijk) a(ijk) H}{a(ijk) H {}^t \breve{a}(ijk)},$$

I tabulate the roots $a(ijk)$. In the table below, as usual, the $c...$ and the $d...$ represent

$$c_j = e(\alpha_j) = \exp(2\pi i \alpha_j), \quad c_{jk...} = c_j c_k \cdots,$$
$$d_j = c_j - 1, \qquad\qquad d_{jk...} = c_{jk...} - 1, \quad 1 \le j, k, \cdots \le 6.$$

$$
\begin{aligned}
a(123) &= (\quad d_{123}, \quad -d_{12}, \quad 0, \quad d_1, \quad 0, \quad 0) \\
a(124) &= (\quad d_4, \quad d_{12}c_4, \quad -d_{12}, \quad -d_1c_4, \quad d_1, \quad 0) \\
a(125) &= (\quad -d_5, \quad 0, \quad -d_{12}c_5, \quad 0, \quad d_1c_5, \quad 0) \\
a(126) &= (\quad 1, \quad 0, \quad 0, \quad 0, \quad 0, \quad 0) \\
a(134) &= (\quad c_3d_4, \quad -d_{34}, \quad d_3, \quad -d_1c_{34}, \quad d_1c_3, \quad -d_1) \\
a(135) &= (\quad c_3d_5, \quad -d_5, \quad -d_3c_5, \quad 0, \quad -d_1c_{35}, \quad d_1c_5) \\
a(136) &= (\quad d_2c_3, \quad -d_2, \quad 0, \quad 0, \quad 0, \quad 0) \\
a(145) &= (\quad 0, \quad c_4d_5, \quad -d_{45}, \quad 0, \quad 0, \quad -d_1c_{45}) \\
a(146) &= (\quad 0, \quad c_4, \quad -1, \quad 0, \quad 0, \quad 0) \\
a(156) &= (\quad 0, \quad 0, \quad 1, \quad 0, \quad 0, \quad 0) \\
a(234) &= (-c_{23}d_4, \quad c_2d_{34}, \quad -c_2d_3, \quad -d_{234}, \quad d_{23}, \quad -d_2) \\
a(235) &= (-c_{23}d_5, \quad c_2d_5, \quad c_2d_3c_5, \quad -d_5, \quad -d_{23}c_5, \quad d_2c_5) \\
a(236) &= (\quad c_{23}, \quad -c_2, \quad 0, \quad 1, \quad 0, \quad 0) \\
a(245) &= (\quad 0, \quad -c_{24}d_5, \quad c_2d_{45}, \quad c_4d_5, \quad -d_{45}, \quad -d_2c_{45}) \\
a(246) &= (\quad 0, \quad c_2c_4, \quad -c_2, \quad -c_4, \quad 1, \quad 0) \\
a(256) &= (\quad 0, \quad 0, \quad c_2, \quad 0, \quad -1, \quad 0) \\
a(345) &= (\quad 0, \quad 0, \quad 0, \quad -c_{34}d_5, \quad c_3d_{45}, \quad -d_{345}) \\
a(346) &= (\quad 0, \quad 0, \quad 0, \quad c_{34}, \quad -c_3, \quad 1) \\
a(356) &= (\quad 0, \quad 0, \quad 0, \quad 0, \quad c_3, \quad -1) \\
a(456) &= (\quad 0, \quad 0, \quad 0, \quad 0, \quad 0, \quad 1)
\end{aligned}
$$

4.4. Monodromy of the Loaded Cycles. It is believed that the loops $\rho(j_1, \ldots, j_{r+1})$ generate the fundamental group of X. However, I must confess that I have no rigorous proof which works in general for $r + 1 \geq 3$. What I can prove is that when $(r + 1, n + 1) = (3, 6)$, the monodromy group of the loaded cycles $D_{12}, \ldots, D_{34}$ is generated by $M(ijk; \alpha)$. I have only one proof ([MSTY 1, 425–426]), which I do not like. If you have any idea, please let me know.

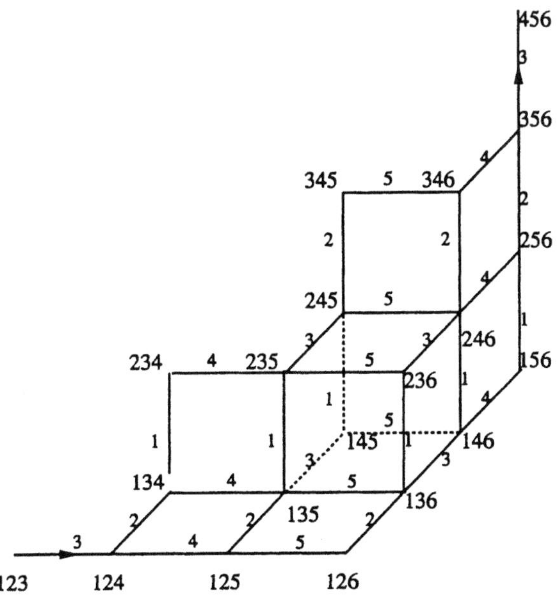

FIGURE 4.6. An economical method to obtain circuit matrices $M(ijk;\alpha)$. The numeral j beside a segment indicates the exchange of j and $j+1$.

5. The Hypergeometric System $E(k,n;\alpha)$

In this and the next sections we study basic facts about the hypergeometric system of type $(3,6)$.

For natural numbers $k < n$ and a set of parameters

$$\alpha = (\alpha_1, \ldots, \alpha_n), \quad \sum_{j=1}^{n} \alpha_j = n - k,$$

the hypergeometric integral

$$\int \prod_{j=1}^{n} (\sum_{i=1}^{k} x_{ij} t^i)^{\alpha_j - 1} dt,$$

as a function on $M^*(k,n)$ satisfies (cf. §1) the system $E(k,n;\alpha)$ of linear differential equations

$$\sum_{j=1}^{n} x_{lj}\frac{\partial u}{\partial x_{ij}} + \delta_{il}u = 0, \quad 1 \le i, l \le k,$$

$$\sum_{i=1}^{k} x_{ij}\frac{\partial u}{\partial x_{ij}} - (\alpha_j - 1)u = 0, \quad 1 \le j \le n,$$

$$\frac{\partial^2 u}{\partial x_{ip}\partial x_{jq}} - \frac{\partial^2 u}{\partial x_{iq}\partial x_{jp}} = 0, \quad 1 \le i, j \le k,\ 1 \le p, q \le n.$$

This is called the *hypergeometric system of type* (k, n).

Instead of modifying this system into a $GL(k) \times H_n$-invariant system so that it may be defined on

$$X(k, n) = GL(k)\backslash M^*(k, n)/H_n,$$

we will "restrict" the system on the "subset" of $M^*(k, n)$ consisting of the matrices with the following form:

$$\begin{pmatrix} 1 & 0 & \cdots & 0 & 1 & 1 & \cdots & 1 \\ 0 & 1 & \cdots & 0 & 1 & x_{2\ k+2} & \cdots & x_{2n} \\ \vdots & & \ddots & & \vdots & \vdots & & \vdots \\ 0 & 0 & \cdots & 1 & 1 & x_{k\ k+2} & \cdots & x_{kn} \end{pmatrix}.$$

Note that any element of $M^*(k, n)$ can be put into this form under the $GL(k) \times H_n$-action. The transformation formulae given in §1 imply in particular that the "ratio" of solutions is invariant under the $GL(k) \times H_n$-action. Since we are interested in the ratio of solutions, the restricted system works as well as the invariant system.

The restricted system in

$$x_{ip}, \quad 2 \le i \le k, \quad k + 1 \le p \le n$$

can be written as follows:

$$(\alpha - 1 + \theta)\theta_{jq}u = x_{jq}(\theta^q - \alpha_q + 1)(\theta_j + \alpha_j)u, \quad \begin{aligned} &2 \leq j \leq k, \\ &k + 2 \leq q \leq n; \end{aligned}$$

$$x_{jp}(\theta^p - \alpha_p + 1)\theta_{jq}u = x_{jq}(\theta^q - \alpha_q + 1)\theta_{jq}u, \quad \begin{aligned} &2 \leq j \leq k, \\ &k + 2 \leq p < q \leq n; \end{aligned}$$

$$x_{iq}(\theta_i + \alpha_i)\theta_{jq}u = x_{jq}(\theta_j + \alpha_j)\theta_{iq}u, \quad \begin{aligned} &2 \leq i < j \leq k, \\ &k + 2 \leq q \leq n; \end{aligned}$$

$$x_{iq}x_{jp}\theta_{ip}\theta_{jq}u = x_{ip}x_{jq}\theta_{iq}\theta_{jp}u, \quad \begin{aligned} &2 \leq i < j \leq k, \\ &k + 2 \leq p < q \leq n, \end{aligned}$$

where

$$\alpha = \sum_{j=2}^{k+1}\alpha_j, \qquad \theta_{ip} = x_{ip}\frac{\partial}{\partial x_{ip}},$$

$$\theta_i = \sum_{p=k+2}^{n}\theta_{ip}, \qquad \theta^p = \sum_{i=2}^{k}\theta_{ip},$$

$$\theta = \sum_{i=2}^{k}\theta_i = \sum_{p=k+2}^{n}\theta^p = \sum_{i=2}^{k}\sum_{p=k+2}^{n}\theta_{ip}.$$

This can be checked by a straightforward computation, which we omit (cf. [MSY1]). Of course we expect that the system is the original hypergeometric differential equation when $k = 2$ and $n = 4$; let us check. In this case we have

$$z = \begin{pmatrix} 1 & 0 & 1 & 1 \\ 0 & 1 & 1 & x \end{pmatrix} \in M^*(2,4),$$

$$\alpha = \alpha_2 + \alpha_3,$$

$$\theta = \theta_i = \theta^p = \theta_{ip} = x\frac{d}{dx}.$$

Only the first equation (among the above system of four equations) survives:

$$(\alpha - 1 + \theta)\theta u = x(\theta - \alpha_4 + 1)(\theta + \alpha_2)u.$$

This can be written as

$$(\theta + 1 - \alpha_4)(\theta + \alpha_2)u = (\alpha + \theta)(\theta + 1)\frac{1}{x}u,$$

which is the hypergeometric equation $E(1 - \alpha_4, \alpha_2, \alpha_2 + \alpha_3)$.

PROPOSITION 5.1. *The system $E(k,n;\alpha)$ admits a solution expressed by the power series*

$$\sum A(n)x^n, \quad x = (x_{ip}), \ n = (n_{ip}), \ x^n = \prod_i \prod_p (x_{ip})^{n_{ip}},$$

where

$$A(n) = \frac{\prod_p(1 - \alpha_p, \sum_i n_{ip}) \prod_i(\alpha_i, \sum_p n_{ip})}{(\alpha, \sum_i \sum_p n_{ip}) \prod_i \prod_p(1, n_{ip})},$$

$$\prod_i = \prod_{2 \le i \le k}, \quad \prod_p = \prod_{k+2 \le p \le n},$$

$$\sum_i = \sum_{2 \le i \le k}, \quad \sum_p = \sum_{k+2 \le p \le n}$$

and the summation is taken over $n_{ip} = 0$ to ∞ for all $2 \le i \le k$ and $k + 2 \le p \le n$.

PROOF. Recall the computation in §1 of Chapter III. Substituting the expression $u = \sum A(n)x^n$ into the first set of equations of the system, we have

$$\left(\alpha + \sum_i \sum_p n_{ip} \right)(n_{jq} + 1)A(n_{jq} + 1)$$

$$= \left(1 - \alpha_p + \sum_i n_{iq} \right)\left(\alpha_j + \sum_p n_{jp} \right)A(n),$$

where

$$A(n_{jq} + 1) = A(n_{2\ k+2}, \ldots, n_{jq} + 1, \ldots, n_{kn}).$$

Therefore if we set $A(0) = 1$, the above recurrence formulae lead to the desired power series. It is easy to see that the series satisfies the recurrence formulae corresponding to the remaining three equations of the system. $\square$

Let us consider the case $k = 3, n = 6$ and introduce a system of local coordinates $x = (x^1, \ldots, x^4)$ on $X = X(3,6)$ by normalizing z as follows:

$$z = (z_{ij}) = \begin{pmatrix} 1 & 0 & 0 & 1 & 1 & 1 \\ 0 & 1 & 0 & 1 & x^1 & x^2 \\ 0 & 0 & 1 & 1 & x^3 & x^4 \end{pmatrix}.$$

For these variables $x^1,\ldots,x^4$ the system $E(3,6;\alpha)$ can be written as follows:

$$(\alpha_{234} - 1 + \theta_{1234})\theta_1 u = x^1(\theta_{13} + 1 - \alpha_5)(\theta_{12} + \alpha_2)u,$$

$$(\alpha_{234} - 1 + \theta_{1234})\theta_2 u = x^2(\theta_{24} + 1 - \alpha_6)(\theta_{12} + \alpha_2)u,$$

$$(\alpha_{234} - 1 + \theta_{1234})\theta_3 u = x^3(\theta_{13} + 1 - \alpha_5)(\theta_{34} + \alpha_3)u,$$

$$(\alpha_{234} - 1 + \theta_{1234})\theta_4 u = x^4(\theta_{24} + 1 - \alpha_6)(\theta_{34} + \alpha_3)u,$$

$$x^1(\alpha_5 - 1 - \theta_{13})\theta_2 u = x^2(\alpha_6 - 1 - \theta_{24})\theta_1 u,$$

$$x^3(\alpha_5 - 1 - \theta_{13})\theta_4 u = x^4(\alpha_6 - 1 - \theta_{24})\theta_3 u,$$

$$x^1(-\alpha_2 - 1 - \theta_{12})\theta_3 u = x^3(-\alpha_3 - 1 - \theta_{34})\theta_1 u,$$

$$x^2(-\alpha_2 - 1 - \theta_{12})\theta_4 u = x^4(-\alpha_3 - 1 - \theta_{34})\theta_2 u,$$

$$x^2 x^3 \theta_1 \theta_4 u = x^1 x^4 \theta_2 \theta_3 u,$$

where $\theta_i = x^i \partial/\partial x^i$, $\alpha_{ijk} = \alpha_i + \alpha_j + \alpha_k$, $\theta_{ij} = \theta_i + \theta_j$, $\theta_{1234} = \theta_1 + \cdots + \theta_4$. The system admits a solution u expressed by the following power series in x:

$$u = \sum \frac{(1 - \alpha_5, n_{13})(1 - \alpha_6, n_{24})(\alpha_2, n_{12})(\alpha_3, n_{34})}{(\alpha_{234}, n_{1234})(1, n_1) \cdots (1, n_4)}(x^1)^{n_1} \cdots (x^4)^{n_4},$$

where $n_{jk} = n_j + n_k$, $n_{1234} = n_1 + \cdots + n_4$, and the summation is taken over $n_1,\ldots,n_4 = 0$ to ∞.

The above system can be written in the following form:

$$\frac{\partial^2 u}{\partial x^i \partial x^j} = G_{ij}\frac{\partial^2 u}{\partial x^1 \partial x^4} + \sum_{k=1}^{4} A_{ij}^k \frac{\partial u}{\partial x^k} + A_{ij}^0 u, \quad (1 \le i,j \le 4).$$

This implies that the rank (= dimension of the space of solutions at a generic point) is not greater than 6. Indeed at a fixed point x, all the derivatives of u at x can be determined by the six quantities:

$$u(x), \; \frac{\partial u}{\partial x^1}(x), \; \ldots, \; \frac{\partial u}{\partial x^4}(x), \; \frac{\partial^2 u}{\partial x^1 x^4}(x).$$

Since we already know that there are six linearly independent loaded cycles, we can conclude that the rank is exactly six. Strictly speaking, we must prove that the hypergeometric integrals over the six cycles are linearly independent solutions (cf. [Kit2]).

The coefficients $G_{ij} = G_{ji}$ of the principal part appearing in the above equation for u are independent of the α_j and are given as follows:

$$G_{12} = \frac{x^4 - x^3}{x^1 - x^2}, \quad G_{13} = \frac{x^4 - x^2}{x^1 - x^3}, \quad G_{14} = 1,$$

$$G_{23} = 1, \quad G_{24} = \frac{x^3 - x^1}{x^2 - x^4}, \quad G_{34} = \frac{x^2 - x^1}{x^3 - x^4},$$

$$G_{11} = \frac{x^2 x^3 - x^4}{x^1(1 - x^1)} - \frac{x^3(x^4 - x^2)}{x^1(x^1 - x^3)} - \frac{x^2(x^4 - x^3)}{x^1(x^1 - x^2)},$$

$$G_{22} = \frac{x^1 x^4 - x^3}{x^2(1 - x^2)} - \frac{x^1(x^3 - x^4)}{x^2(x^2 - x^1)} - \frac{x^4(x^3 - x^1)}{x^2(x^2 - x^4)},$$

$$G_{33} = \frac{x^1 x^4 - x^2}{x^3(1 - x^3)} - \frac{x^1(x^2 - x^4)}{x^3(x^3 - x^1)} - \frac{x^4(x^2 - x^1)}{x^3(x^3 - x^4)},$$

$$G_{44} = \frac{x^2 x^3 - x^1}{x^4(1 - x^4)} - \frac{x^3(x^1 - x^2)}{x^4(x^4 - x^3)} - \frac{x^2(x^1 - x^3)}{x^4(x^4 - x^2)}.$$

We do not write down the coefficients A_{ij}^k and A_{ij}^0 explicitly. But if you are patient enough, you can calculate these and can furthermore check (maybe with a machine) that the system is *integrable*, which proves again that the rank is six. Here the term "integrable" means that the coefficients

$$G_{ij}, \ A_{ij}^k = A_{ji}^k, \ A_{ij}^0 = A_{ji}^0$$

satisfy the system of (non-linear) differential equations derived from

$$\frac{\partial}{\partial x^k}\left(\frac{\partial^2 u}{\partial x^i x^j}\right) = \frac{\partial}{\partial x^j}\left(\frac{\partial^2 u}{\partial x^i x^k}\right).$$

REMARK 5.1. We will see in §2 of Chapter IX that six linearly independent solutions of the system $E(3,6;\alpha)$ are quadratically related when $\alpha = (1/2,\dots,1/2)$. A differential geometric aspect of this fact can be found in [SY3] and [MSY].

6. Local Properties of $E(3,6;\alpha)$

We study local properties of the system $E(3,6;\alpha)$ around a point in $\overline{X}$ where the divisors $X^{ijk} := \overline{X}_3^{ijk}$ intersect normally.

Using hypergeometric integrals, we can know the local properties of the system $E(3,6;\alpha)$ to a certain extent. Consider, for example, the integral

$$F(x) = \int_\Delta \prod_{j=1}^{6}(L_j)^{\alpha_j-1}ds \wedge dt,$$

where

$$L_1 = s, \quad L_2 = t, \quad L_3 = 1, \quad L_4 = s + t + 1,$$
$$L_5 = s + x^1 t + x^3, \quad L_6 = s + x^2 t + x^4,$$

and Δ is the triangle bounded by the three lines defined by $L_1 = 0, L_2 = 0$ and $L_6 = 0$. Transforming the variables of integration as $s \to x^4 s, t \to x^4 t/x^2$ and applying the *Dirichlet formula* (a generalization of the beta function formula presented in §2.1 of Chapter IV)

$$\int_{s,t\geq 0,\ s+t\leq 1} s^{p-1}t^{q-1}(1 - s - t)^{r-1}dsdt = \frac{\Gamma(p)\Gamma(q)\Gamma(r)}{\Gamma(p + q + r)},$$

we obtain the following expression of the integral F around the divisor $x^4 = 0$. Let x^1, x^2 and x^3 be fixed and let x^4 tend to 0. Then F can be expressed as

$$(x^4)^{\alpha_1+\alpha_2+\alpha_6-1}f(x^1, x^2, x^3, x^4),$$

where $f(x^1, x^2, x^3, 0)$ does not vanish identically. Since the divisor defined by $x^4 = 0$ corresponds to the divisor X_3^{126}, thanks to the S_6-symmetry of the system $E(3,6;\alpha)$, we conclude that around a generic point of the divisor X_3^{ijk} the system admits a singular solution with exponent $\alpha_i + \alpha_j + \alpha_k - 1$. (The corresponding fact can be obtained for general (k, n).) To have a more detailed information for the local property of the system, you need to choose different domains of integration very carefully. This is not as simple as in the case of $(k, n) = (2, n)$. Therefore, instead of manipulating hypergeometric integrals, I honestly study the hypergeometric system itself. The results given here can be obtained through a routine computation which is however very clumsy (only machines can bear it).

Let us see which part of X the local coordinate $x = (x^1, \ldots, x^4)$ covers. Let U be the affine 4-space with coordinate x and $D(ijk)$ be the minors of the matrix z. Then define

$$X_U^{ijk} := \{x \in U \mid D(ijk) = 0\}, \quad X_U^{***} := \cup X_U^{ijk}.$$

Note that only 14 among the twenty X_U^{ijk} are non-empty. These are given by

$$D(135) = x^1, \quad D(136) = x^2, \quad D(345) = x^1 - 1, \quad D(346) = x^2 - 1,$$
$$D(125) = x^3, \quad D(126) = x^4, \quad D(245) = x^3 - 1, \quad D(246) = x^4 - 1,$$
$$D(145) = x^1 - x^3, \quad D(146) = x^2 - x^4,$$
$$D(256) = x^3 - x^4, \quad D(356) = x^1 - x^2,$$
$$D(156) = x^1 x^4 - x^2 x^3, \quad D(456) = (x^1 - 1)(x^4 - 1) - (x^2 - 1)(x^3 - 1).$$

We know that $U - X_U^{***}$ is naturally isomorphic to X, that, of course, X_U^{ijk} corresponds (if it is non-empty) to the divisor X^{ijk} of X, and that

$$U - \cup_{|\{i,j,k\} \cap \{p,q,r\}|=2} X_U^{ijk} \cap X_U^{pqr}$$

is isomorphic to an open set of $\overline{X}'$.

We work around the point $x = (0, 1, 1, 0)$ where only the following four divisors meet:

$$D(135) = x^1 = 0, \quad D(346) = x^2 - 1 = 0,$$
$$D(126) = x^4 = 0, \quad D(245) = x^3 - 1 = 0.$$

Note that this point corresponds to the point

$$X_{0b}^{125;346;245;126} \in \overline{X}'.$$

We present two methods.

6.1. Transforming the System into a Pfaffian Form. We transform the system $E(3, 6; \alpha)$ written in the form (in §5)

$$\frac{\partial^2 u}{\partial x^i \partial x^j} = G_{ij} \frac{\partial^2 u}{\partial x^1 \partial x^4} + \sum_{k=1}^{4} A_{ij}^k \frac{\partial u}{\partial x^k} + A_{ij}^0 u \quad (1 \leq i, j \leq 4)$$

into a system of equations of first order with an unknown vector. Putting

$$\vec{u} = {}^t(u, \frac{\partial u}{\partial x^1}, \ldots, \frac{\partial u}{\partial x^4}, \frac{\partial^2 u}{\partial x^1 \partial x^4}),$$

we can find a matrix valued 1-form ω satisfying

$$d\vec{u} = \omega \vec{u},$$

where d is the exterior differentiation with respect to $x = (x^1, \ldots, x^4)$. The entries of ω can be written in terms of the coefficients G_{ij}, A_{ij}^k and A_{ij}^0 given in §5. The 1-form ω admits the expression

$$\omega = \omega(135)\frac{dD(135)}{D(135)} + \omega(346)\frac{dD(346)}{D(346)}$$
$$+ \omega(245)\frac{dD(245)}{D(245)} + \omega(126)\frac{dD(126)}{D(126)} + \omega',$$

where each $\omega(pqr)$ is a matrix whose entries are polynomials in x independent of $D(pqr)$, and ω' is a matrix valued 1-form whose entries have no poles along the four divisors. Using an explicit form of each $\omega(pqr)$, we can prove Proposition 6.1 in the next subsection.

6.2. Expanding Solutions in Power Series. We change variables as

$$y^1 = x^1, \quad y^2 = x^2 - 1, \quad y^3 = x^3 - 1, \quad y^4 = x^4,$$

and find power-series solutions of the system in the form

$$u = (y^1)^{\rho_1} \cdots (y^4)^{\rho_4} \sum_{n_1,\ldots,n_4=0}^{\infty} A(n_1,\ldots,n_4)(y^1)^{n_1} \cdots (y^4)^{n_4}, \quad A(0) = 1.$$

Substitute this expression to the system $E(3,6;\alpha)$ written in the form (in §5)

$$(\alpha_{234} - 1 + \theta_{1234})\theta_1 u = x^1(\theta_{13} + 1 - \alpha_5)(\theta_{12} + \alpha_2)u,$$

$$\cdots$$

$$x^2 x^3 \theta_1 \theta_4 u = x^1 x^4 \theta_2 \theta_3 u,$$

and transform them into a system of difference equations for $A(n)$. It turns out that the system of exponents $(\rho_1, \ldots, \rho_4)$ must be one of

$$(\alpha_{135}-1,0,0,0), \quad (0,\alpha_{346}-1,0,0), \quad (0,0,\alpha_{245}-1,0), \quad (0,0,0,\alpha_{126}-1)$$

and $(0,0,0,0)$, where $\alpha_{ijk} := \alpha_i + \alpha_j + \alpha_k$. In this way we get

PROPOSITION 6.1. *Assume* $\alpha_{135}, \ldots, \alpha_{126} \notin \mathbb{Z}$. *Around* $y = 0$, *the system has two linearly independent holomorphic solutions (one is non-vanishing) and four solutions with the four exponents consisting of one of the systems* $(\rho_1, \ldots, \rho_4)$ *given above.*

Thanks to the S_6-symmetry of the hypergeometric system, we have

PROPOSITION 6.2. *Assume $\alpha_{ijk} \notin \mathbb{Z}$ for all $1 \leq i < j < k \leq 6$. Let U be a sufficiently small neighborhood in $\overline{X}'$ of a point*

$$p := X_{0b}^{ijk;klm;mni;jln} = X^{ijk} \cap X^{klm} \cap X^{mni} \cap X^{jln},$$

and $f_1, \ldots, f_4$ holomorphic functions on U with simple zeros along X^{ijk}, $\ldots, X^{jln}$, respectively. Then there is a projective solution (the ratio of six linearly independent solutions) on U of $E(3,6;\alpha)$ in the following form:

$$f_1^{\alpha_{ijk}-1} h_1 : \cdots : f_4^{\alpha_{jln}-1} h_4 : h_5 : h_6,$$

where the h_j are holomorphic in U, and $h_j(p) = 1$.

REMARK 6.1. Since this result is so simple-looking, I have long been wondering whether there is a fancy theory which makes the situation transparent. I was informed that the recent preprint [SST] does.

7. The Duality of $E(3,6;\alpha)$

Roughly speaking, by the involution $*$ on $X(3,6)$, the hypergeometric system $E(3,6;\alpha)$ is transformed into $E(3,6;1-\alpha)$, where $1-\alpha = (1-\alpha_1, \ldots, 1-\alpha_6)$. Note that when $\alpha_1 = \cdots = \alpha_6 = 1/2$, the system is invariant under $*$. In this section we shall prove

PROPOSITION 7.1. *By the change of variables*

$$x = (x_1, x_2) \mapsto \xi = (\xi_1, \xi_2) = ({}^t x_1^{-1}, {}^t x_2^{-1}) \in GL(3) \times GL(3) \supset M^*(3,6),$$

where $x \in M^*(3,6)$, and the change of the unknown

$$u \longrightarrow v = |x_1||x_2|u,$$

the system $E(3,6;\alpha)$ is transformed into the system $E(3,6;1-\alpha)$.

DEFINITION 7.1. Two systems are said to be *projectively equivalent* if one is obtained from the other by replacing the unknown by its product with a non-zero function.

REMARK 7.1. By the change of the variables introduced in the proposition above, the system $E(3,6;\alpha)$ is transformed to a system projectively equivalent to $E(3,6;1-\alpha)$. In [MSY1] and [KM], you can find another aspect of the "duality" of the hypergeometric systems.

The system $E(3,6;\alpha)$ can be written as follows:

$$^{x}E_{ij}u = -\delta_{ij}u, \quad 1 \le i,j \le 3$$
$$^{x}E^{p}u = (\alpha_p - 1)u, \quad 1 \le p \le 6$$
$$\frac{\partial^2 u}{\partial x_{ip}\partial x_{jq}} = \frac{\partial^2 u}{\partial x_{iq}\partial x_{jp}}, \quad 1 \le i,j \le 3,\ 1 \le p,q \le 6,$$

where ^{x}E are Euler operators

$$^{x}E_{ij} = \sum_{q=1}^{6} x_{iq}\frac{\partial}{\partial x_{jq}}, \quad ^{x}E^{p} = \sum_{i=1}^{3} x_{ip}\frac{\partial}{\partial x_{ip}}.$$

Define $^{\xi}E_{ij}$ and $^{\xi}E^{p}$ by the formulae above using ξ in place of x. These Euler operators are related as

$$^{\xi}E_{ij} = -^{x}E_{ij}, \quad ^{\xi}E^{p} = -^{x}E^{p}.$$

Thus we have

$$^{\xi}E_{ij}v = -^{x}E_{ij}(|x_1||x_2|u) = -(2-1)\delta_{ij}v = -\delta_{ij}v,$$
$$^{\xi}E^{p}v = -^{x}E^{p}(|x_1||x_2|u) = -(1+\alpha_p-1)v = ((1-\alpha_p)-1)v.$$

Our task is to show

$$\frac{\partial^2 v}{\partial \xi_{ip}\partial \xi_{jq}} = \frac{\partial^2 v}{\partial \xi_{iq}\partial \xi_{jp}}, \quad 1 \le i,j \le 3,\ 1 \le p,q \le 6.$$

These equalities can be shown by a straightforward computation; I will give a sketch of the proof for the case $i = p = 2, j = q = 1$. I do not dare give proofs of the lemmas below. For notational simplicity, we write x for x_1, and ξ for ξ_1, because the computation does not involve x_2 nor ξ_2. Since x is $|x| = |\xi|^{-1}$ times the cofactor matrix of ξ, and since

$$|\xi|\frac{\partial|x|}{\partial \xi_{ij}} = -x_{ij},$$

we have

$$\frac{\partial}{\partial \xi_{ij}} = |x|P_{ij} - x_{ij}E, \quad E = \sum_{a,b=1}^{3} x_{ab}\frac{\partial}{\partial x_{ab}},$$

where P_{ij} is the product of two vectors P_{ij}^{ξ} and P_{ij}^{∂}, where we have, for example,

$$P_{11}^{\xi} = (\xi_{33}, -\xi_{32}, -\xi_{23}, \xi_{22}), \quad P_{11}^{\partial} = (\partial/\partial x_{22}, \partial/\partial x_{23}, \partial/\partial x_{32}, \partial/\partial x_{33}),$$

$$P_{22}^{\xi} = (\xi_{33}, -\xi_{31}, -\xi_{13}, \xi_{11}), \quad P_{22}^{\partial} = (\partial/\partial x_{11}, \partial/\partial x_{13}, \partial/\partial x_{31}, \partial/\partial x_{33}),$$

$$P_{12}^{\xi} = (-\xi_{33}, \xi_{31}, \xi_{23}, -\xi_{21}), \quad P_{12}^{\partial} = (\partial/\partial x_{21}, \partial/\partial x_{23}, \partial/\partial x_{31}, \partial/\partial x_{33}),$$

$$P_{21}^{\xi} = (-\xi_{33}, \xi_{32}, \xi_{13}, -\xi_{12}), \quad P_{21}^{\partial} = (\partial/\partial x_{12}, \partial/\partial x_{13}, \partial/\partial x_{32}, \partial/\partial x_{33}).$$

LEMMA 7.2.

$$x_{11}P_{22} + x_{22}P_{11} - x_{12}P_{21} - x_{21}P_{12} = \xi_{33}E + \frac{\partial}{\partial x_{33}}.$$

COROLLARY 7.3.

$$\frac{\partial|x|}{\partial\xi_{11}}\frac{\partial}{\partial x_{22}} + \frac{\partial|x|}{\partial\xi_{22}}\frac{\partial}{\partial x_{11}} - \frac{\partial|x|}{\partial\xi_{12}}\frac{\partial}{\partial x_{21}} - \frac{\partial|x|}{\partial\xi_{21}}\frac{\partial}{\partial x_{12}}$$
$$= |x|^2\{\xi_{33}E - \frac{\partial}{\partial x_{33}}\}.$$

The third equation of the system $E(3,6;\alpha)$ for u, together with the formula

$$Eu = (\alpha. - 3)u, \quad \alpha. = \alpha_1 + \alpha_2 + \alpha_3,$$

leads to

LEMMA 7.4.

$$(P_{11}^{\xi}P_{22}P_{11}^{\partial} - P_{12}^{\xi}P_{21}P_{12}^{\partial})u = |\xi|(\alpha. - 4)u_{33}, \quad u_{33} = \partial u/\partial x_{33}.$$

Since we have

$$\frac{\partial}{\partial\xi_{22}}P_{11} = P_{11}^{\xi}\frac{\partial}{\partial\xi_{22}}P_{11}^{\partial} + \frac{\partial}{\partial x_{33}}$$
$$= |x|P_{11}^{\xi}P_{22}P_{11}^{\partial} - x_{22}P_{11}(E-1) + \frac{\partial}{\partial x_{33}}$$

and similarly

$$\frac{\partial}{\partial\xi_{21}}P_{12} = |x|P_{12}^{\xi}P_{21}P_{12}^{\partial} - x_{21}P_{12}(E-1) - \frac{\partial}{\partial x_{33}},$$

we have

$$\frac{\partial}{\partial \xi_{22}} P_{11} - \frac{\partial}{\partial \xi_{21}} P_{12} = |x|(P_{11}^{\xi} P_{22} P_{11}^{\partial} - P_{12}^{\xi} P_{21} P_{12}^{\partial})$$
$$- (x_{22} P_{11} - x_{21} P_{12})(E - 1) + 2\frac{\partial}{\partial x_{33}},$$

so that

$$\left(\frac{\partial}{\partial \xi_{22}} P_{11} - \frac{\partial}{\partial \xi_{21}} P_{12}\right) u = (\alpha. - 2)u_{33} - (\alpha. - 4)(x_{22} P_{11} - x_{21} P_{12})u.$$

Since

$$\frac{\partial u}{\partial \xi_{11}} = |x| P_{11} - (\alpha. - 3)x_{11}u,$$

we have

$$\frac{\partial^2 u}{\partial \xi_{22} \partial \xi_{11}} = \frac{\partial |x|}{\partial \xi_{22}} P_{11} u + |x|\frac{\partial}{\partial \xi_{22}} P_{11} u$$
$$- (\alpha. - 3)(|x| P_{22} - x_{22} E)x_{11} u$$
$$= -|x| x_{22} P_{11} u + |x|\frac{\partial}{\partial \xi_{22}} P_{11} u$$
$$- (\alpha. - 3)\{|x|(x_{11} P_{22} + \xi_{33}) - (\alpha. - 2)x_{11}x_{22}\}u,$$

and similarly

$$\frac{\partial^2 u}{\partial \xi_{21} \partial \xi_{12}} = -|x| x_{21} P_{12} u + |x|\frac{\partial}{\partial \xi_{21}} P_{12} u$$
$$- (\alpha. - 3)\{|x|(x_{12} P_{21} - \xi_{33}) - (\alpha. - 2)x_{12}x_{21}\}u.$$

Now we are ready. We finally have

$$\frac{\partial^2 |x|u}{\partial \xi_{22} \partial \xi_{11}} - \frac{\partial^2 |x|u}{\partial \xi_{21} \partial \xi_{12}}$$
$$= \left(\frac{\partial^2 |x|}{\partial \xi_{22} \partial \xi_{11}} - \frac{\partial^2 |x|}{\partial \xi_{21} \partial \xi_{12}}\right) u$$
$$+ \left(\frac{\partial |x|}{\partial \xi_{11}}\frac{\partial u}{\partial x_{22}} + \frac{\partial |x|}{\partial \xi_{22}}\frac{\partial u}{\partial x_{11}} - \frac{\partial |x|}{\partial \xi_{12}}\frac{\partial u}{\partial x_{21}} - \frac{\partial |x|}{\partial \xi_{21}}\frac{\partial u}{\partial x_{12}}\right)$$
$$+ |x|\left(\frac{\partial^2 u}{\partial \xi_{22} \partial \xi_{11}} - \frac{\partial^2 u}{\partial \xi_{21} \partial \xi_{12}}\right)$$

$$= 0 + |x|^2 \{(\alpha. - 3)\xi_{33}u - u_{33}\}$$

$$- |x|^2 (x_{22}P_{11} - x_{21}P_{12})u + |x|^2 \left(\frac{\partial}{\partial \xi_{22}}P_{11} - \frac{\partial}{\partial \xi_{21}}P_{12}\right) u$$

$$- |x|(\alpha. - 3)\{|x|(x_{11}P_{22} - x_{12}P_{21} + 2\xi_{33})$$

$$- (\alpha. - 2)(x_{11}x_{22} - x_{12}x_{21})\}u$$

$$= |x|^2 \{(\alpha. - 3)\xi_{33}u - u_{33}\}$$

$$- |x|^2 (\alpha. - 3)(x_{22}P_{11} - x_{21}P_{12})u + |x|^2 (\alpha. - 2)u_{33}$$

$$- (\alpha. - 3)|x|\{|x|(x_{11}P_{22} - x_{12}P_{21}) - 2|x|\xi_{33} - (\alpha. - 2)|x|\xi_{33}\}u$$

$$= 0.$$

CHAPTER IX

Modular Interpretation of the Configuration Space $X(3,6)$

In this chapter we present a modular interpretation of the configuration space $X(3,6)$ of six lines on the projective plane. It is an evolution of the story of the configuration space $X(2,4)$ presented in Part 1, which can be summarized as follows:

$$
\begin{array}{ccc}
& \text{hypergeometric functions} & \\
X(2,4) & \longrightarrow & \mathbb{H}/\Gamma(2) \\
\text{democratic embedding} \searrow & & \nearrow \text{ theta functions} \\
& Y(2,4) &
\end{array}
$$

Roughly speaking, we are going to tell a story which can be illustrated as

$$
\begin{array}{ccc}
& \text{hypergeometric functions} & \\
X(3,6) & \longrightarrow & \mathbb{H}_2/\Gamma \\
\text{democratic embedding} \searrow & & \nearrow \text{ theta functions} \\
& Y(3,6) &
\end{array}
$$

We already studied the space $X(3,6)$, together with its democratic embedding, and the hypergeometric functions of type $(3,6)$. So we are going to answer the questions

- What is the domain $\mathbb{H}_2$?
- What is the group Γ?
- What kind of theta functions do we consider?

1. A Family of K3 Surfaces

In this section, we describe algebro-geometric aspects of the hypergeometric integrals of type $(3,6)$ with parameters $\alpha_j = 1/2$ $(j = 1,\ldots,6)$. Since this book is not intended to be a textbook in algebraic geometry, the treatment will be sketchy. But do not worry. Only the Riemann equality and the Riemann inequality given in §2 will be used later.

Just as the hypergeometric integrals of type $(2,4)$,

$$\int \frac{dt}{\sqrt{(t - x_1)\cdots(t - x_4)}},$$

can be regarded as periods of the family of elliptic curves

$$s^2 = (t - x_1)\cdots(t - x_4)$$

presented as a double cover of $\mathbb{P}^1$, our hypergeometric integrals of type $(3,6)$ can be regarded as periods of a family of surfaces.

Let

$$l_j = \{(t^1, t^2, t^3) \in \mathbb{P}^2 \mid x_{1j}t^1 + x_{2j}t^2 + x_{3j}t^3 = 0\}, \quad (1 \leq j \leq 6)$$

be six lines in general position on the projective plane $\mathbb{P}^2$ with homogeneous coordinates $t^1 : t^2 : t^3$, and denote by l the system of the six lines ordered. Recall that six lines are said to be in general position if no three of them intersect at a point. Consider the double cover $S'(l)$ of $\mathbb{P}^2$ branching along the union of the six lines and $S(l)$ the smooth surface obtained from $S'(l)$ by replacing $6 \cdot 5/2 = 15$ singular points by 15 non-singular rational curves. Then let $\pi : S(l) \to \mathbb{P}^2$ be the projection. The pull-back $\pi^*\eta(l)$ of the (2-valued) 2-form on $\mathbb{P}^2$ gives a nowhere-vanishing holomorphic 2-form on $S(l)$, where

$$\eta(l) = \prod_{j=1}^{6}(x_{1j}t^1 + x_{2j}t^2 + x_{3j}t^3)^{-\frac{1}{2}}dt,$$

and here

$$dt = t^1 dt^2 \wedge d^3 + t^3 dt^1 \wedge d^2 + t^2 dt^3 \wedge dt^1.$$

By the way, a non-singular surface is called a *K3 surface* if it admits a nowhere-vanishing holomorphic 2-form and its Euler characteristic is 24.

The Euler characteristic $\chi(S)$ of $S = S(l)$ can be computed as follows:

$$\chi(S) = \chi(S') + 15\{\chi(\mathbb{P}^1) - \chi(\text{point})\} = \chi(S') + 15,$$
$$\chi(S') = 2\{\chi(\mathbb{P}^2) - \chi(6 \text{ lines in general position})\}$$
$$+ \chi(6 \text{ lines in general position})$$
$$= 2\chi(\mathbb{P}^2) - \chi(6 \text{ lines in general position})$$
$$= 2\chi(\mathbb{P}^2) - \{6\chi(\mathbb{P}^1) - 15\chi(\text{point})\},$$

so we have

$$\chi(S) = 2 \cdot 3 - \{6 \cdot 2 - 15\} + 15 = 24.$$

Thus our surface $S(l)$ is a K3 surface, and we have a 4-dimensional family of K3 surfaces on $X(3,6)$:

$$\cup_{l \in X(3,6)} S(l) \to X(3,6).$$

There are many studies of K3 surfaces. We now state some basic facts. A K3 surface S is simply connected, which implies that the second homology group $L = H_2(S, \mathbb{Z})$ is free of rank 22. By counting the number of intersections of two cycles, we have the symmetric bi-linear form

$$L \times L \to \mathbb{Z},$$

called the *intersection form*. It is known that this form is non-degenerate and of signature $(2+, 20-)$.

A *period* of $S(l)$ is defined as an integral of the 2-form $\pi^*\eta(l)$ on a cycle in $L(l) = H_2(S(l), \mathbb{Z})$. Hence a hypergeometric integral of type $(3,6)$ with $\alpha_j = 1/2$ can be interpreted as a period of the K3 surface $S(l)$.

If the cycle is represented by an algebraic curve in $S(l)$, then the integral vanishes, so we are interested in transcendental (= non-algebraic) cycles. We can see 16 algebraic curves on $S(l)$: 15 curves coming from the desingularization and the pull-back of a generic line in $\mathbb{P}^2$. Let us call $AC(l)$ the sublattice of the homology group $L(l)$ spanned by the 16 algebraic curves. We can show that $AC(l)$ is free of rank 16, and we can find two systems of 2-cycles,

$$c_1(l), \ldots, c_6(l) \quad \text{and} \quad c_1'(l), \ldots, c_6'(l),$$

depending continuously on l, such that the cycles $c_j'(l)$ form a $\mathbb{Z}$-basis of the orthogonal complement of $AC(l)$ in $L(l)$, the cycles $c_i(l)$ are dual to the $c_j'(l)$ (i.e., $c_i \cdot c_j' = \delta_{ij}$), and that the intersection matrix $(c_i' \cdot c_j')_{ij}$ is given by

$$A = (A_{ij}) = 2\{U \oplus U \oplus (-I_2)\}, \quad U = \begin{pmatrix} 0 & 1 \\ 1 & 0 \end{pmatrix}.$$

Note that the signature of A is $(2+, 4-)$.

These cycles are explicitly constructed in [MSY1] by using an elliptic fibration of the surface $S(l)$. Their monodromy behavior (for moves of l) was obtained by tracing these cycles. This procedure was an enormous labor, far from clear-cut.

Our idea to avoid this brute-force business is to consider the loaded cycles $D_{12}, \ldots, D_{34}$ on $\mathbb{P}^2$ instead of the cycles $c_1, \ldots, c_6$ and $c_1', \ldots, c_6'$ on $S(l)$. There is a natural way to do this (cf. [CY]), via the double covering $\pi : S(l) \to \mathbb{P}^2$ branching along $\cup l_j$, making a 2-cycle on $S(l)$ from a loaded cycle on $\mathbb{P}^2 - \cup_{j=1}^6 l_j$ loaded with the function

$$\prod_{j=1}^6 (x_{1j}t^1 + x_{2j}t^2 + x_{3j}t^3)^{-\frac{1}{2}}.$$

The intersection number of the loaded cycles, defined in §3 of Chapter VIII, and that of the corresponding cycles on $S(l)$ differ only by a negative constant.

Indeed, the signature of the intersection matrix $Int(3, 6; 1/2) = {}^t(D_{12}, \ldots) \cdot (\check{D}_{12}, \ldots)$ is $(4+, 2-)$, while that of $c_1', \ldots, c_6'$ is $(2+, 4-)$. This difference in signature is caused by the orientation of the 2-dimensional complex space. For a holomorphic local coordinate $(z_1, z_2) = (x_1 + iy_1, x_2 + iy_2)$, the usual orientation is given by $dx_1 dy_1 dx_2 dy_2$ (of course, also when defining the intersection form $L \times L \to \mathbb{Z}$ on $S(l)$). When evaluating the intersection numbers of loaded cycles, we have used the orientation $dx_1 dx_2 dy_1 dy_2$. We have

$$dx_1 dy_1 dx_2 dy_2 = -dx_1 dx_2 dy_1 dy_2.$$

2. The Riemann Equality and the Riemann Inequality

The following is a standard argument for K3 surfaces (cf. [Kod]): Let S be a K3 surface, $\{\gamma_j \mid 1 \leq j \leq 22\}$ a basis of the homology group $H_2(S, \mathbb{Z})$,

$$b_{ij} := \gamma_i \cdot \gamma_j,$$

and ψ a holomorphic 2-form on S. Then the periods

$$\lambda_j = \int_{\gamma_j} \psi, \quad (1 \le j \le 22)$$

satisfy the *Riemann equality*

$$\sum_{i,j=1}^{22} a_{ij}\lambda_i\lambda_j = 0$$

and the *Riemann inequality*

$$\sum_{i,j=1}^{22} a_{ij}\lambda_i\overline{\lambda}_j > 0,$$

where (a_{ij}) is the inverse of the intersection matrix (b_{ij}). These can be proved by computing the right-hand sides of the obvious equality and inequality

$$0 = \int_S \psi \wedge \psi \quad \text{and} \quad 0 < \int_S \psi \wedge \overline{\psi}.$$

We apply this argument to $S = S(l)$ and the six cycles corresponding to the loaded cycles $D_{12}, \ldots, D_{34}$. (Note that the period along an algebraic curve vanishes.) We now give our conclusion.

Let $u^1(l), \ldots, u^6(l)$ be the hypergeometric integrals of type $(3,6)$ with parameters $\alpha_j = 1/2$ along the cycles $D_{12}, \ldots, D_{34}$, and $H = H(1/2) = (h_{ij})$ be the inverse of the intersection matrix given in §3.4 of Chapter VIII. We have the equality

$$\sum_{i,j=1}^{6} h_{ij}u^i(l)u^j(l) = 0,$$

and the inequality

$$-\sum_{i,j=1}^{6} h_{ij}u^i(l)\overline{u}^j(l) > 0,$$

which will be called the *Riemann equality* and the *Riemann inequality*, respectively. (They can be proved (see [CH] and [HY]) without using geometry of K3 surfaces.)

3. The Monodromy Group MG as a Reflection Group

We showed in §4.2 of Chapter VIII that each generator $M(ijk;\alpha)$ of the monodromy group $\Gamma(\alpha)$ is a quasi-reflection belonging to the invariant form $H(\alpha) = {}^t\breve{H}(\alpha)$, with root $a(ijk;\alpha)$ and eigenvalue $c_i c_j c_k$. Hereafter we set

$$\alpha_j = \frac{1}{2}, \quad j = 1,\ldots,6.$$

A quasi-reflection is called a *reflection* if it is of order 2. Our generators $M(ijk) = M(ijk;1/2)$ turn out to be reflections belonging to the "real symmetric" matrix $H = H(1/2)$ with "real" roots $a(ijk) = a(ijk;1/2)$, i.e.,

$$M(ijk) = R_{a(ijk)} = I_6 - 2\frac{{}^t aaH}{(a,a)_H},$$

where the *norm* of a is defined as

$$(a,a)_H := aH{}^t a.$$

So the monodromy group $\Gamma(1/2)$ is a *reflection group*, that is, a group generated by reflections.

3.1. A Cosmetic Treatment. Putting $\alpha_j = 1/2$ $(j = 1,\ldots,6)$ into $H(\alpha)$ and $a(ijk;\alpha)$, we have

$$H(\tfrac{1}{2}) = 2^4 \times \begin{pmatrix} 1 & 0 & 1 & 0 & 0 & 0 \\ 0 & 0 & 0 & 0 & 1 & 0 \\ 1 & 0 & 1 & -1 & 0 & 1 \\ 0 & 0 & -1 & 0 & 0 & 0 \\ 0 & 1 & 0 & 0 & 0 & 0 \\ 0 & 0 & 1 & 0 & 0 & 1 \end{pmatrix},$$

and

$$
\begin{aligned}
a(123) &= (-2,0,0,-2,0,0), & a(124) &= (-2,0,0,-2,-2,0), \\
a(125) &= (2,0,0,0,2,0), & a(126) &= (1,0,0,0,0,0), \\
a(134) &= (2,0,-2,2,2,2), & a(135) &= (2,2,-2,0,2,2), \\
a(136) &= (2,2,0,0,0,0), & a(145) &= (0,2,0,0,0,2), \\
a(146) &= (0,-1,-1,0,0,0), & a(156) &= (0,0,1,0,0,0),
\end{aligned}
$$

$$a(234) = (2,0,-2,2,0,2), \quad a(235) = (2,2,-2,2,0,2),$$
$$a(236) = (1,1,0,1,0,0), \quad a(245) = (0,2,0,2,0,2),$$
$$a(246) = (0,1,1,1,1,0), \quad a(256) = (0,0,-1,0,-1,0),$$
$$a(345) = (0,0,0,2,0,2), \quad a(346) = (0,0,0,1,1,1),$$
$$a(356) = (0,0,0,0,-1,-1), \quad a(456) = (0,0,0,0,0,1).$$

These do not look nice. We shall perform some cosmetics on them. Our strategy is the following:

(1) Perform a $\mathbb{Z}$-linear change of the basis $D_{12}, \ldots, D_{34}$ of loaded cycles in order to make H look nicer.
(2) Replace a root a, with negative entries or even entries, by $-a$ or $a/2$.
(3) Replace a generating reflection R_a by its conjugate with respect to another generating reflection.

First we choose $P \in GL(6, \mathbb{Z})$, for example, as follows:

$$P = \begin{pmatrix} 0 & 0 & 0 & -1 & 1 & 0 \\ 1 & 0 & 0 & 0 & 0 & 0 \\ 0 & 0 & 0 & 1 & 0 & 0 \\ 0 & 0 & 1 & 0 & 0 & 1 \\ 0 & -1 & 0 & 0 & 0 & 0 \\ 0 & 0 & 0 & 0 & 0 & 1 \end{pmatrix}, \quad P^{-1} = \begin{pmatrix} 0 & 1 & 0 & 0 & 0 & 0 \\ 0 & 0 & 0 & 0 & -1 & 0 \\ 0 & 0 & 0 & 1 & 0 & -1 \\ 0 & 0 & 1 & 0 & 0 & 0 \\ 1 & 0 & 1 & 0 & 0 & 0 \\ 0 & 0 & 0 & 0 & 0 & 1 \end{pmatrix},$$

so that

$$^tPHP = -2^4A, \quad A = U \oplus U \oplus (-I_2), \quad U = \begin{pmatrix} 0 & 1 \\ 1 & 0 \end{pmatrix}.$$

Recall that for a real symmetric matrix H and a real root a, the corresponding reflection R_a belonging to H is defined by

$$R_a = I_6 - 2\frac{^taaH}{aH^ta}.$$

Since we have

$$P^{-1}R_aP = I_6 - 2\frac{P^{-1t}aa^tP^{-1t}PHP}{a^tP^{-1t}PHPP^{-1t}a}$$
$$= I_6 - 2\frac{P^{-1t}aa^tP^{-1}A}{a^tP^{-1}AP^{-1t}a}$$
$$= I_6 - 2\frac{^tbbA}{bA^tb}, \quad b = a^tP^{-1}, \quad ^tPHP = -2^4A,$$

the matrix $P^{-1}R_a P$ is the reflection belonging to A with the root $b = a^t P^{-1}$. I am lazy enough to write the reflection belonging to A with a root b as R_b. Hereafter, except in §11, reflections are always assumed to belong to A. (Do you think I should write $R_b^{(A)} = P^{-1}R_a^{(H)}P$? The notation would smother you!) The following is the list of roots $b(ijk) = a(ijk)^t P^{-1}$ (when all the entries of a given $b(ijk)$ are divisible by 2, $b(ijk)/2$ is shown, and when all the entries of a $b(ijk)$ are negative, $-b(ijk)$ is shown; I am again lazy):

$$b(123) = (0,0,1,0,1,0), \qquad b(124) = (0,1,-1,0,-1,0),$$
$$b(125) = (0,-1,0,0,1,0), \qquad b(126) = (0,0,0,0,1,0),$$
$$b(134) = (0,-1,0,-1,0,1), \quad b(135) = (1,-1,-1,-1,0,1),$$
$$b(136) = (1,0,0,0,1,0), \qquad b(145) = (1,0,-1,0,0,1),$$
$$b(146) = (1,0,0,1,1,0), \qquad b(156) = (0,0,0,1,1,0),$$

$$b(234) = (0,0,0,-1,0,1), \qquad b(235) = (1,0,0,-1,0,1),$$
$$b(236) = (1,0,1,0,1,0), \qquad b(245) = (1,0,0,0,0,1),$$
$$b(246) = (1,-1,1,1,1,0), \qquad b(256) = (0,1,0,-1,-1,0),$$
$$b(345) = (0,0,0,0,0,1), \qquad b(346) = (0,-1,0,0,0,1),$$
$$b(356) = (0,1,1,0,0,-1), \qquad b(456) = (0,0,-1,0,0,1).$$

Now the roots $b(ijk)$ became fairly simple. In particular, their norm

$$(b,b) := bA^t b$$

is -1. I want to apply a little more cosmetic surgery to the roots $b(ijk)$ with negative entries.

Incidentally, a reflection $R_{b(ijk)}$ with root $b(ijk)$ represents a linear change of the loaded cycles caused by a loop in X which starts from $\overset{\bullet}{x}$, travels along a path γ near a point on the divisor X_3^{ijk}, goes once around the divisor, and travels back along γ. The (homotopy class of the) path γ is not unique. If I make use of another path γ', then the result is the reflection $C^{-1}R_{b(ijk)}C$ conjugate to $R_{b(ijk)}$, where C is the circuit matrix for the loop $\gamma' \circ \gamma^{-1}$.

Note that (since reflections are of order 2)

$$R_c R_b R_c = R_{b^t R_c},$$

and that we have the formula

$$b^t R_c = b + 2(b,c)c,$$

where $(b,c) = bA^t c$ and $(c,c) = -1$. This transformation of b will be expressed as

$$b \xrightarrow{c} b + 2(b,c)c.$$

For example,

$$b(124) = (0,1,-1,0,-1,0) \xrightarrow{b(125)} -(0,1,1,0,-1,0) \xrightarrow{b(126)} -(0,1,1,0,1,0),$$

$$b(125) = (0,-1,0,0,1,0) \xrightarrow{b(126)} -(0,1,0,0,1,0),$$

$$b(134) = (0,-1,0,-1,0,1) \xrightarrow{b(345)} -(0,1,0,1,0,1),$$

$$b(135) = (1,-1,-1,-1,0,1) \xrightarrow{b(134)} (1,1,-1,1,0,1) \xrightarrow{b(345)} (1,1,-1,1,0,1).$$

In this way, we transform the roots $b(ijk)$ (some of them do not change) and obtain the following (this time I am not lazy, I change b into α)

$$
\begin{aligned}
\alpha(126) &= (0,0,0,0,1,0), & \alpha(246) &= (1,1,1,-1,1,0), \\
\alpha(136) &= (1,0,0,0,1,0), & \alpha(236) &= (1,0,1,0,1,0), \\
\alpha(125) &= (0,1,0,0,1,0), & \alpha(146) &= (1,0,0,1,1,0), \\
\alpha(123) &= (0,0,1,0,1,0), & \alpha(124) &= (0,1,1,0,1,0), \\
\alpha(156) &= (0,0,0,1,1,0), & \alpha(256) &= (0,1,0,1,1,0),
\end{aligned}
$$

$$
\begin{aligned}
\alpha(345) &= (0,0,0,0,0,1), & \alpha(135) &= (1,1,1,-1,0,1), \\
\alpha(245) &= (1,0,0,0,0,1), & \alpha(145) &= (1,0,1,0,0,1), \\
\alpha(346) &= (0,1,0,0,0,1), & \alpha(235) &= (1,0,0,1,0,1), \\
\alpha(456) &= (0,0,1,0,0,1), & \alpha(356) &= (0,1,1,0,0,1), \\
\alpha(234) &= (0,0,0,1,0,1), & \alpha(134) &= (0,1,0,1,0,1).
\end{aligned}
$$

NOTATION. We shall refer to the reflection group $P^{-1}\Gamma(1/2)P$, generated by twenty reflections $R_{\alpha(ijk)}$, as the *monodromy group* and shall denote it by *MG*.

3.2. The Geometric Meaning of the Reflections. Let us consider the geometric meaning of the reflections. For notational simplicity, we write (z, w) in place of $(z, w)_A = zA^t w$ for $z, w \in \mathbb{C}^6$. Consider the quadratic hypersurface $Quad$ in $\mathbb{P}^5$ defined by

$$Quad : (z, z) = 2z^1 z^2 + 2z^3 z^4 - (z^5)^2 - (z^6)^2 = 0.$$

Its group of holomorphic automorphisms is given by

$$Aut(Quad) = \{Y \in GL(6, \mathbb{R}) \mid {}^t Y A Y = A\}/ \pm .$$

The inequality

$$(z, \bar{z}) = 2\Re z^1 \bar{z}^2 + 2\Re z^3 \bar{z}^4 - |z^5|^2 - |z^6|^2 > 0$$

defines an open subset of the quadratic hypersurface $Quad$. This open subset has two connected components. Indeed, by putting $z^1 = 1$, eliminating z^2 in the quadric inequality $(z, \bar{z}) > 0$, and using the quadric equality $(z, z) = 0$, we have

$$\begin{aligned}
0 &< 2\Re z^1 \bar{z}^2 + 2\Re z^3 \bar{z}^4 - |z^5|^2 - |z^6|^2 \\
&= -\Re\{2z^3 z^4 - (z^5)^2 - (z^6)^2\} + 2\Re z^3 \bar{z}^4 \\
&\quad - \{(\Re z^5)^2 + (\Im z^5)^2 + (\Re z^6)^2 + (\Im z^6)^2\} \\
&= 4\Im z^3 \cdot \Im z^4 - 2(\Im z^5)^2 - 2(\Im z^6)^2.
\end{aligned}$$

This implies that $\Im z^3/z^1$ and $\Im z^4/z^1$ are non-zero and of the same sign, and that if we put

$$y_3 = \Im z^3/z^1, \ldots, y_6 = \Im z^6/z^1,$$

the open set in question is the direct product of the real affine 4-space and the imaginary bi-cone defined by $2y_3 y_4 - (y_5)^2 - (y_6)^2 > 0$ (see Figure 3.1). Since an automorphism of the quadric preserves the inner product, it maps the two connected components to themselves. Let us call one of these $\mathbb{D}$, say

$$\mathbb{D} = \{z = (z^1, \ldots, z^6) \in \mathbb{P}^5 \mid (z, z) = 0, (z, \bar{z}) > 0, \Im \frac{z^3}{z^1} > 0\}.$$

The group $Aut(\mathbb{D})$ of holomorphic automorphisms of $\mathbb{D}$ is therefore a subgroup of the above group of index 2. Though it is possible to write down a condition $\mathrm{Con}\mathbb{D}$ for $g \in Aut(Quad)$ to be in $Aut(\mathbb{D})$, I do not because we do not use it explicitly.

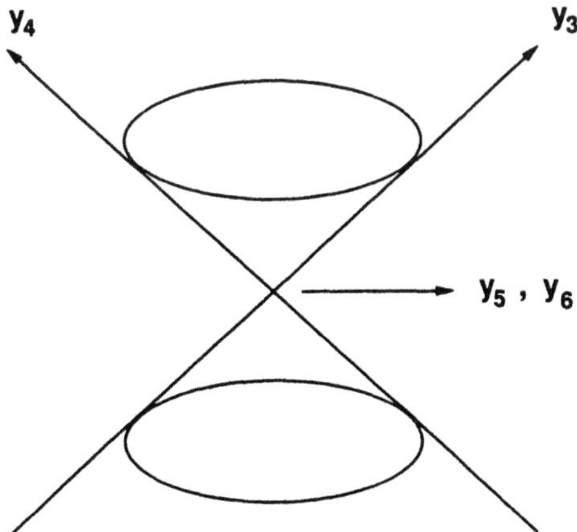

FIGURE 3.1. The cone $2y_3y_4 - (y_5)^2 - (y_6)^2 > 0$

A reflection R_α considered as an automorphism of *Quad* fixes the hypersurface (the intersection of a hyperplane in $\mathbb{P}^5$ and *Quad*) called the *mirror*:

$$\alpha^\perp = \{x \in Quad|\, (x, \alpha) = 0\}.$$

By considering the inner product (γ, γ), we can determine whether the mirror $\gamma^\perp$ passes through $\mathbb{D}$, touches $\mathbb{D}$ or does neither. We have

PROPOSITION 3.1. *For a non-zero real vector* γ,

$$if\ (\gamma, \gamma) < 0\ then\ \gamma^\perp \cap \mathbb{D} \neq \emptyset,$$
$$if\ (\gamma, \gamma) = 0\ then\ \gamma^\perp \cap \mathbb{D} = \emptyset,\ \gamma^\perp \cap \partial\mathbb{D} \neq \emptyset,$$
$$if\ (\gamma, \gamma) > 0\ then\ \gamma^\perp \cap (\mathbb{D} \cup \partial\mathbb{D}) = \emptyset.$$

(See Figure 3.2.)

PROOF. In the three cases above, by an automorphism of *Quad*, the vector γ can be transformed into a constant times $(0, \ldots, 0, 1), (0, 0, 1, 1, 1, 1)$ and $(0, 0, 1, 1, 0, 0)$, respectively. The mirrors of these transformed vectors $\gamma^\perp$ are given by $z^6 = 0, z^3 + z^4 - z^5 - z^6 = 0$ and $z^3 - z^4 = 0$, which define hyperplanes in $(y_3, \ldots, y_6)$-space passing through, tangent to, and outside the cone defined by $2y_3y_4 - (y_5)^2 - (y_6)^2 > 0$. $\square$

Since our root $\alpha(ijk)$ is of norm -1, the mirror $\alpha(ijk)^\perp$ passes through $\mathbb{D}$. Thus the reflection $M(ijk)$ is an automorphism of $\mathbb{D}$.

By considering the inner product of two roots, we can determine how the two mirrors are situated in $\mathbb{D}$. We have

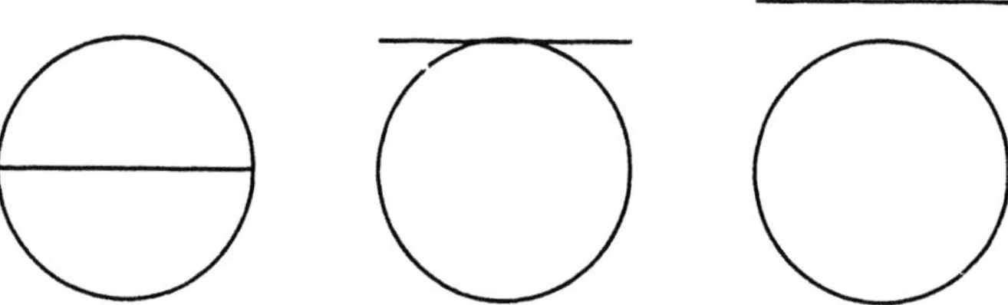

FIGURE 3.2. The three cases $(\gamma, \gamma) < 0$, $(\gamma, \gamma) = 0$, and $(\gamma, \gamma) > 0$

PROPOSITION 3.2. *For two integral 6-vectors α and β of norm -1, the following three cases occur:*

$$(\alpha, \beta) = 0 \iff \alpha^{\perp} \text{ intersects } \beta^{\perp} \text{ orthogonally in } \mathbb{D}$$

$$|(\alpha, \beta)| = 1 \iff \alpha^{\perp} \text{ does not intersect } \beta^{\perp} \text{ in } \mathbb{D}, \text{ but meets in } \partial\mathbb{D}$$

$$|(\alpha, \beta)| \geq 2 \iff \alpha^{\perp} \text{ does not intersect } \beta^{\perp} \text{ in } \mathbb{D} \cup \partial\mathbb{D}.$$

(See Figure 3.3.) The order of $R_{\alpha} R_{\beta}$ is 2 in the first case and infinite otherwise.

FIGURE 3.3. The three cases $(\alpha, \beta) = 0$, $|(\alpha, \beta)| = 1$, and $|(\alpha, \beta)| \geq 2$

PROOF. Since the orthogonal projection $\mathbb{D} \to \alpha^{\perp} \cap \mathbb{D}$ is given by

$$x \mapsto x|_{\alpha}^{\perp} := x - (\alpha, x)\alpha/(\alpha, \alpha),$$

the formula

$$(\beta|_{\alpha}^{\perp}, \beta|_{\alpha}^{\perp}) = \{(\alpha, \alpha)(\beta, \beta) - (\alpha, \beta)^2\}/(\alpha, \alpha)$$
$$= (\alpha, \beta)^2 - 1,$$

together with the proposition above (applied to $\alpha^{\perp} \cap \mathbb{D}$), proves the proposition. $\square$

4. The Monodromy Group as a Congruence Subgroup on $\mathbb{D}$

There are many ways to describe a group. Giving a set of generators in the form of matrices is one of the worst ways. The reason I say this is that such a description tells you little about what the group is. Consider, for example, the group generated by two matrices,

$$\begin{pmatrix} 1 & 2 \\ 0 & 1 \end{pmatrix} \quad \text{and} \quad \begin{pmatrix} 1 & 0 \\ 2 & 1 \end{pmatrix}.$$

This appears very simple, but if you do not know that this group is projectively equivalent to the congruence subgroup $\Gamma(2)$, then your knowledge of it will be very restricted.

In our case the situation is slightly better; the group in question is a reflection group, and we know the roots of the generating reflections. Thus there is a way to see the action of the group geometrically. But this is not enough.

In this section we give a global description of MG as a congruence subgroup of the modular group belonging to A. We shall also see that any reflection belonging to A with integral root of norm -1 is an element of MG.

Let us define the *full modular group* belonging to A by

$$\Gamma_A = \{Y \in GL(6, \mathbb{Z}) \mid {}^tYAY = A,\ \text{Con}\mathbb{D}\},$$

and the *congruence subgroup* of level 2 by

$$\Gamma_A(2) = \{Y \in \Gamma_A \mid Y \equiv I_6 \mod 2\}.$$

Let us then define two sets of integral 6-vectors

$$R_A(-1) = \{\alpha \in \mathbb{Z}^6 \mid (\alpha, \alpha) = -1\},$$
$$R_A(-1, -2) = \{\alpha \in \mathbb{Z}^6 \mid (\alpha, \alpha) = -1, -2\},$$

and two reflection groups generated by reflections with these roots

$$GR_A(-1) = \text{ group generated by } \{R_\alpha \mid \alpha \in R_A(-1)\},$$
$$GR_A(-1, -2) = \text{ group generated by } \{R_\alpha \mid \alpha \in R_A(-1, -2)\}.$$

Since the reflection with root α is defined by

$$R_\alpha = I_6 - 2\frac{{}^t\alpha\alpha A}{(\alpha, \alpha)},$$

we have

$$MG \subset GR_A(-1) \subset \Gamma_A(2), \quad GR_A(-1,-2) \subset \Gamma_A.$$

I am going to tell you that the above inclusions are all equal:

THEOREM 4.1.

$$MG = GR_A(-1) = \Gamma_A(2), \quad GR_A(-1,-2) = \Gamma_A.$$

You might ask why. If I knew a reason why this should be, I would be willing to tell you, as in the case of the elliptic modular group $\Gamma(2)$ (it is a group preserving the 2-division points, so ...). I must confess, however, I do not have a good reason. You worked on Exercise 5 in Chapter II, where you were asked to prove that $\Gamma(2)$ is generated by the above two matrices and $-I_2$. Probably you used a kind of the euclidean algorithm. In the present case, matrices are 6×6, and therefore the situation is not so simple. However, if you are patient enough, you can manage. So please believe it, or better, find a good reason and tell me; I would really appreciate it. If you doubt the statement of Theorem 4.1 see [MSY1 pp. 97–109].

Of course $\Gamma_A(2)$ is a normal subgroup of Γ_A; we have

PROPOSITION 4.2.

$$\Gamma_A/\Gamma_A(2) \cong S_6 \times \mathbb{Z}/2\mathbb{Z},$$

where S_6 is the symmetric group.

The geometric meaning of this fact will become clear later (cf. §10). This proposition can be proved by considering the homomorphism

$$\Gamma_A \ni Y \mapsto Y \mod 2 \in \{Y \in GL(6, \mathbb{Z}/6\mathbb{Z}) \mid {}^t Y A Y \equiv A \mod 2\},$$

whose kernel is exactly $\Gamma_A(2)$. You have only to determine its image.

In the course of the proof of Theorem 4.1 which I omitted, one gets the following two propositions.

PROPOSITION 4.3. *The group Γ_A acts transitively on the set $R_A(-1)$.*

PROPOSITION 4.4. *The set $R_A(-1)$ of roots is the union of twenty $\Gamma_A(2)$-orbits. The twenty $\alpha(ijk)$ given in §3 form a set of representatives.*

The second of these propositions suggests the following definition of the integer (R, R') for two classes R and R' of $R_A(-1)$ modulo $\Gamma_A(2)$:

$$(R, R') := \min\{|(\alpha, \alpha')|; \alpha \in R, \alpha' \in R'\}.$$

The value of this integer is 0 or 1. Denoting the class of $\alpha(ijk)$ by $R(ijk)$, we have

PROPOSITION 4.5.

$$(R(ijk), R(pqr)) = \begin{cases} 0 & \text{if } \#\{i,j,k\} \cap \{p,q,r\} = 0,1 \\ 1 & \text{if } \#\{i,j,k\} \cap \{p,q,r\} = 2,3. \end{cases}$$

PROOF. Set $\alpha = \alpha(ijk)$ and $\beta = \alpha(pqr)$. Notice that if $\gamma \in R_A(-1)$, we have

$$(\alpha, R_\gamma\beta) \equiv (\alpha, \beta) \qquad \mod 2,$$

since we have

$$(\alpha, R_\gamma\beta) = (\alpha, \beta) + 2(\gamma, \beta)(\alpha, \gamma).$$

One can check that $|(\alpha, \beta)| = 0, 1$, or 2. Thus we can conclude that if $|(\alpha, \beta)| = 1$ then $(R(ijk), R(pqr)) = 1$. In case $|(\alpha, \beta)| = 2$, one can easily find an element $\gamma \in R_A(-1)$ such that $(\alpha, R_\gamma\beta) = 0$. $\square$

This corresponds to the fact that the intersection of the divisors $\overline{X}_3^{ijk}$ and $\overline{X}_3^{pqr}$ of $\overline{X}$ is 2-dimensional if $\#\{i, j, k\} \cap \{p, q, r\} = 0, 1$, and 1-dimensional if $\#\{i, j, k\} \cap \{p, q, r\} = 2$.

5. The Map $\phi: X \to \mathbb{D}$ and Its Extension to $\overline{X}'$

Let us define the (multi-valued) map

$$\phi: X \ni l \to u^1(l) : \cdots : u^6(l) \in \mathbb{D} \subset \mathbb{P}^5$$

by

$$u^j(l) = \int_{D_j(l)} \eta(l), \qquad (1 \le j \le 6),$$

where $D_1, \ldots, D_6$ are linear combinations (depending on P in §3.1) of the loaded cycles $D_{12}, \ldots, D_{34}$. Its image is in

$$\mathbb{D} = \{z = (z^1, \ldots, z^6) \in \mathbb{P}^5 \,|\, (z, z) = 0, (z, \overline{z}) > 0, \Im\frac{z^3}{z^1} > 0\},$$

as determined by the Riemann equality and the Riemann inequality (§2). The monodromy group of the multi-valued vector function $(u^1(l), \ldots, u^6(l))$

is the principal congruence subgroup $\Gamma_A(2)$, so the multi-valued map ϕ induces the single-valued map

$$\varphi: \ X \to \mathbb{D}/\Gamma_A(2).$$

Let us extend the domain of definition of φ to the smooth part of $\overline{X}$,

$$\overline{X}' = X \cup X_3 \cup X_{2a} \cup X_{2b} \cup X_{1b} \cup X_{0b},$$
$$= \overline{X} - \cup X_{1a} \cup X_{0a}.$$

In $\overline{X}'$, the divisors $\overline{X}_3^{ijk}$ cross normally. By the local behavior of the solutions of $E(3,6;1/2)$ discussed in §6 of Chapter VIII, for any $p \in \overline{X}'$, there is a neighborhood $U \subset \overline{X}'$ of p such that ϕ has a single-valued inverse. Thus φ can be extended to $\overline{X}'$ as a locally finite branching map which ramifies along $\cup_{i,j,k}\overline{X}_3^{ijk}$. Denote the extended map by the same symbol φ. Since the inverse map φ^{-1} is locally single-valued, and since $\mathbb{D}$ is simply connected, we have the isomorphisms

$$\begin{array}{ccc}
\varphi: \ \overline{X}' & \xrightarrow{\ \sim\ } & \mathbb{D}/\Gamma_A(2) \\
\cup & & \cup \\
\varphi: \ X & \xrightarrow{\ \sim\ } & \mathbb{D}_{reg}/\Gamma_A(2).
\end{array}$$

Here, $\mathbb{D}_{reg}$ is the open dense subset of $\mathbb{D}$ on which the group $\Gamma_A(2)$ acts freely; the complement of $\mathbb{D}_{reg}$ in $\mathbb{D}$ is the union of the mirrors of the reflections with roots of norm -1.

I think you have probably already noticed the correspondence under ϕ between the strata of $\overline{X}'$ and the (intersections of) mirrors of the integral roots of norm -1. Denoting by $R(ijk)^\perp$ the $\Gamma_A(2)$-orbit of $\alpha(ijk)^\perp \cap \mathbb{D}$, of course we have

$$\begin{aligned}
\varphi: X_3^{ijk} &\longrightarrow R(ijk)^\perp, \\
X_{2\alpha}^{ijk;lmn} &\longrightarrow R(ijk)^\perp \cap R(lmn)^\perp, \\
X_{2\beta}^{ijk;imn} &\longrightarrow R(ijk)^\perp \cap R(imn)^\perp, \\
X_{1\beta}^{ijk;klm;mni} &\longrightarrow R(ijk)^\perp \cap R(klm)^\perp \cap R(mni)^\perp, \\
X_{0\beta}^{ijk;klm;mni;jln} &\longrightarrow R(ijk)^\perp \cap R(klm)^\perp \cap R(mni)^\perp \cap R(jln)^\perp.
\end{aligned}$$

The isomorphism φ is the first step in realizing the modular interpretation of the configuration space $X = X(3,6)$. Compare this with the isomorphism $X(4) \cong \mathbb{H}/\Gamma(2)$ stated in Proposition 8.1 of Chapter III.

In the next section, we will compactify the space $\mathbb{D}/\Gamma_A(2)$ so that φ may be extended (in §7) to the isomorphism of $\overline{X}$. To make a compactification of the 4-dimensional space $\mathbb{D}/\Gamma_A(2)$ is not as easy as that of the 1-dimensional space $\mathbb{H}/\Gamma(2)$, so please be patient.

6. Boundary Components

Recall the situation in which $\Gamma = SL(2,\mathbb{Z})$ (or a subgroup of finite index, for example $\Gamma(2)$) acts on $\mathbb{H}$. As we saw in Chapter III, the quotient space $\mathbb{H}/\Gamma$ is isomorphic to $\mathbb{P}^1 - \{$a point$\}$. Since it is 1-dimensional, its compactification is unique; we simply fill the pinhole with a point. We denoted this compactification by $\overline{\mathbb{H}/\Gamma}$.

In higher dimensional cases, e.g., $\mathbb{D}/\Gamma_A$, $\mathbb{D}/\Gamma_A(2),\ldots$, compactifications are not unique. Therefore we must specify what we patch along the holes. Before turning to these more complicated cases, let us consider again the 1-dimensional case above. Though the topological closure $\overline{\mathbb{H}}$ of $\mathbb{H}$ in $\mathbb{P}^1$ is

$$\overline{\mathbb{H}} = \{\tau \in \mathbb{C} \mid \Im\tau \geq 0\} \cup \{\infty\} = \mathbb{H} \cup \mathbb{P}^1_{\mathbb{R}},$$

and the group Γ acts on $\overline{\mathbb{H}}$, the quotient space $\overline{\mathbb{H}}/\Gamma$ is not identical to $\overline{\mathbb{H}/\Gamma}$. It is too big. In order to get the right compactification, we should add only the Γ-orbit of $\infty \in \mathbb{P}^1_{\mathbb{R}}$, or equivalently $\mathbb{P}^1_{\mathbb{Q}}(\subset \mathbb{P}^1_{\mathbb{R}})$, which we will call the *rational boundary*. The space

$$\tilde{\mathbb{H}} := \mathbb{H} \cup \mathbb{P}^1_{\mathbb{Q}}$$

will be called the *rational closure* of $\mathbb{H}$. We then have the isomorphism $\overline{\mathbb{H}/\Gamma} \cong \tilde{\mathbb{H}}/\Gamma$. The compactification $\overline{\mathbb{H}/\Gamma(2)}$ of the quotient space $\mathbb{H}/\Gamma(2) \cong \mathbb{P}^1 - \{$three points$\}$ is given by $\tilde{\mathbb{H}}/\Gamma(2)$.

Let us now consider the situation in which the groups Γ_A, $\Gamma_A(2),\ldots$ act on $\mathbb{D}$. A maximal analytic subset of the topological boundary $\partial\mathbb{D}$ is called a (*boundary*) *component* of $\mathbb{D}$. The group $Aut(\mathbb{D})$ acts on the set of such components. There are two orbits, one consisting of the set of 1-dimensional components and the other of 0-dimensional components. Their representatives are given respectively as follows:

$$F1 := \{(z^1, 0, z^3, 0, 0, 0) \mid \Im(z^3/z^1) > 0\} \cong \mathbb{H},$$
$$F0 := \{(1, 0, 0, 0, 0, 0)\}.$$

The rational closure $\tilde{\mathbb{D}}$ of $\mathbb{D}$ is defined by

$$\tilde{\mathbb{D}} := \mathbb{D} \cup \Gamma_A \cdot F1 \cup \Gamma_A \cdot F0;$$

the Γ_A-orbits $\Gamma_A \cdot F1$ and $\Gamma_A \cdot F0$ are called respectively the *rational boundaries* of dimension 1 and 0. A boundary component is called rational if it is in $\Gamma_A \cdot F1$ or in $\Gamma_A \cdot F0$.

Terminology. The quotients $\bar{\mathbb{D}}/\Gamma_A, \bar{\mathbb{D}}/\Gamma_A(2), \ldots$ are often referred to as the *Satake compactifications* of $\mathbb{D}/\Gamma_A, \mathbb{D}/\Gamma_A(2), \ldots$ They are, in a sense, the smallest compactifications. You surely ask, "Isn't it nonsense to speak about compactness without giving a topology?" I do not bother to answer this question. Please just think that "the Satake compactification of $\mathbb{D}/\Lambda$" is shorter than "the quotient of the rational closure of $\mathbb{D}$ under the group Λ." Sorry.

We have another description of the rational boundary components in terms of the mirrors of the integral roots of norm -1. Roughly speaking, a boundary component is rational if and only if infinitely many such mirrors pass through it. Thanks to this description, we can see that each the 1-dimensional rational boundary and the 0-dimensional rational boundary is the union of fifteen $\Gamma_A(2)$-orbits. Indeed, for example, through $F1$ pass eight mirrors of the twenty $\alpha(ijk)^{\perp}$; their roots are

$$\begin{aligned}
\alpha(126) &= (0,0,0,0,1,0), &\quad \alpha(136) &= (1,0,0,0,1,0), \\
\alpha(123) &= (0,0,1,0,1,0), &\quad \alpha(236) &= (1,0,1,0,1,0), \\
\alpha(345) &= (0,0,0,0,0,1), &\quad \alpha(245) &= (1,0,0,0,0,1), \\
\alpha(456) &= (0,0,1,0,0,1), &\quad \alpha(145) &= (1,0,1,0,0,1).
\end{aligned}$$

Since their labels can be written as

$$\begin{aligned}
(pqr) &: \{p,q,r\} \subset \{1,\ldots,6\} - \{4,5\}, \\
(45s) &: s \in \{1,\ldots,6\} - \{4,5\},
\end{aligned}$$

$F1$ should be labeled by $\{45\}$; let us denote its $\Gamma_A(2)$-orbit by F_1^{45}. Note that the reflections with these eight roots form a infinite group, and therefore there are infinitely many mirrors passing through $F1$. In this way you can find, by using mirrors, representatives of the fifteen $\Gamma_A(2)$-orbits F_1^{ij} in the 1-dimensional rational boundary.

Through $F0$ pass twelve mirrors of the twenty, the eight mirrors given above and the mirrors of the following four roots:

$$\begin{aligned}
\alpha(234) &= (0,0,0,1,0,1), &\quad \alpha(235) &= (1,0,0,1,0,1), \\
\alpha(156) &= (0,0,0,1,1,0), &\quad \alpha(146) &= (1,0,0,1,1,0).
\end{aligned}$$

These twelve mirrors are labeled by

$$(23s) : s \in \{1, \ldots, 6\} - \{2, 3\},$$
$$(45s) : s \in \{1, \ldots, 6\} - \{4, 5\},$$
$$(61s) : s \in \{1, \ldots, 6\} - \{6, 1\}.$$

Thus $F0$ should be labeled by $\{23; 45; 61\}$; let us denote its $\Gamma_A(2)$-orbit by $F_0^{23;45,61}$. In this way you can find representatives of the fifteen $\Gamma_A(2)$-orbits $F_0^{ij;kl;mn}$ in the 0-dimensional rational boundary. Note that

$$(0, 0, 1, 0, 0, 0) \in F_0^{12;45;36}, \quad (1, 0, 1, 0, 0; 0) \in F_0^{13;45;26},$$

and $F0 \in F_0^{23;45;61}$ are on the rational boundary of $F1 \in F_1^{45}$. The infinite reflection group generated by the reflections with these twelve roots will be studied in §11.

Naturally, we are expecting, under an extension of φ, the correspondence

$$X_{1\alpha}^{ij} \longrightarrow F_1^{ij},$$
$$X_{0\alpha}^{ij;kl;mn} \longrightarrow F_0^{ij;kl;mn}.$$

7. The Map ϕ along the Strata $X_{2\alpha}$ and $\varphi : \overline{X} \to \overline{\mathbb{D}}/\Gamma_A(2)$

Let us coordinatize the 2-dimensional stratum $X_{2\alpha}^{135;246}$ as follows:

$$\begin{pmatrix} 1 & 0 & 0 & 1 & 1 & 1 \\ 0 & 1 & 0 & 1 & 0 & z \\ 0 & 0 & 1 & 1 & y & 1 \end{pmatrix}, \quad y, z \in \mathbb{P}^1 - \{0, 1\}.$$

(In terms of the coordinates in §5 of Chapter VIII, $y = x^3, z = x^2$; see the expression for $D(ijk)$ in §7 of Chapter VIII.) The corresponding hypergeometric integral becomes

$$\int s^{\alpha_2 - 1} t^{\alpha_3 - 1} (s + t + 1)^{\alpha_4 - 1} (yt + 1)^{\alpha_5 - 1} (zs + t + 1)^{\alpha_6 - 1} ds \wedge dt$$

$$= \int s^{\alpha_2 - 1} t^{\alpha_3 - 1} (t + 1)^{\alpha_4 - 1} \left(\frac{s}{t + 1} + 1 \right)^{\alpha_4 - 1} (t + 1)^{\alpha_6 - 1}$$

$$\times \left(z \frac{s}{t + 1} + 1 \right)^{\alpha_6 - 1} (yt + 1)^{\alpha_5 - 1} ds \wedge dt$$

$$= \int r^{\alpha_2 - 1} (r + 1)^{\alpha_4 - 1} (zr + 1)^{\alpha_6 - 1} dr$$

$$\times \int t^{\alpha_3 - 1} (t + 1)^{\alpha_2 + \alpha_4 + \alpha_6 - 2} (yt + 1)^{\alpha_5 - 1} dt,$$

where we put $s = r(t + 1)$. Note that this is the product of two hyper-geometric integrals of type $(2,4)$. Note also that when $\alpha_j = 1/2$ these two hypergeometric integrals turn out to be an elliptic integral (because $\alpha_2 + \alpha_4 + \alpha_6 - 2 = -1/2$). In Figure 7.1, you see that two out of the six

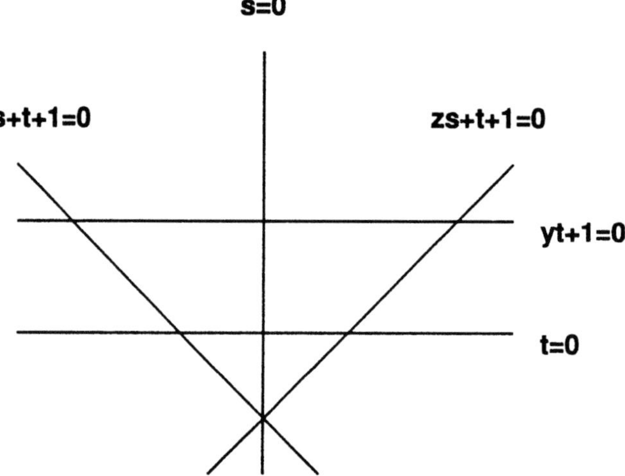

FIGURE 7.1. Four compact chambers

loaded cycles (with bounded support) collapse, and you have four loaded cycles (with bounded support). The map ϕ restricted to

$$X_{2\alpha}^{135;246} \cong (\mathbb{P}^1 - \{\text{three points}\})^2$$

has values in $\alpha^\perp \cap \beta^\perp \cap \mathbb{D} \cong \mathbb{H}^2$, where α and β are conjugate to $\alpha(135)$ and $\alpha(246)$, respectively. The above computation shows that this map is just the direct product of the map

$$\mathbb{P}^1 - \{\text{three points}\} \longrightarrow \mathbb{H}$$

we studied in Chapter III, i.e., the map defined by projective solutions of $E(2,4;1/2)$, in which the monodromy group is conjugate to $\Gamma(2) \subset SL(2,\mathbb{Z})$.

Thus the map ϕ, restricted to $X_{2\alpha}^{135;246}$, can be naturally extended to

$$(\mathbb{P}^1 \times \mathbb{P}^1 \cong) \; \overline{X}_{2\alpha}^{135;246} \longrightarrow \alpha^\perp \cap \beta^\perp \cap \overline{\mathbb{D}}.$$

The group $MG = \Gamma_A(2)$ should induce an action on $\alpha^\perp \cap \beta^\perp \cap \mathbb{D}$, which is isomorphic to the action of $\Gamma(2) \times \Gamma(2)$ on $\mathbb{H} \times \mathbb{H}$.

In order to see directly the action of MG on $\alpha^\perp \cap \beta^\perp \cap \mathbb{D}$, we first introduce some terminology. Let a group Γ act on a space Y. For a subset S of Y, the *isotropy subgroup* Γ_S is the subgroup of Γ consisting of elements fixing S as a set. The group $Id(\Gamma)_S$ is the normal subgroup of Γ_S consisting of elements fixing S pointwise. The quotient group $\Gamma_S/Id(\Gamma)_S$, which can be considered as a group acting on S, is called the *restriction of* Γ *on* S and is denoted by $\Gamma|_S$. We have the exact sequence

$$1 \longrightarrow Id(\Gamma)_S \longrightarrow \Gamma_S \stackrel{\pi}{\longrightarrow} \Gamma|_S \longrightarrow 1.$$

We set $\Gamma := \Gamma_A(2)$ and $Y := \mathbb{D}$. In this section, the term "mirror" refers to the mirror of a reflection belonging to the reflection group $\Gamma_A(2)$. We have the following facts.

PROPOSITION 7.1. *When $S = \alpha^\perp \cap \mathbb{D}$ is a mirror,*

$$Id(\Gamma)_S = \langle R_\alpha \rangle \cong \mathbb{Z}/2\mathbb{Z},$$

and the isotropy subgroup Γ_S is the centralizer of R_α, i.e.,

$$\Gamma_S = \{g \in \Gamma_A(2) \mid gR_\alpha = R_\alpha g\}.$$

Γ_S is the direct product of $\langle R_\alpha \rangle$ and the reflection group

$$\langle R_\beta \mid \beta \in R_A(2),\ (\alpha, \beta) = 0 \rangle.$$

PROPOSITION 7.2. *When $S := \alpha^\perp \cap \beta^\perp \cap \mathbb{D} \cong \mathbb{H}^2$ is the intersection of two mirrors such that*

$$\alpha \in R(ijk), \quad \beta \in R(lmn), \quad \{i, j, k\} \cap \{l, m, n\} = \emptyset,$$

then we have

$$Id(\Gamma)_S = \langle R_\alpha, R_\beta \rangle \cong (\mathbb{Z}/2\mathbb{Z})^2,$$

and the restriction $\Gamma|_S$ is isomorphic to $(\overline{\Gamma}(2))^2$ acting on $\mathbb{H}^2$, where $\overline{\Gamma}(2) = \Gamma(2)/\pm$.

In fact when

$$\alpha = \alpha(345) = (0, 0, 0, 0, 0, 1), \quad \beta = \alpha(126) = (\overline{X}, 0, 0, 0, 1, 0),$$

the isotropy group Γ_S is the direct product of $Id(\Gamma)_S = \langle R_\alpha, R_\beta \rangle$ and the subgroup of $\Gamma_A(2)$ generated by the following four elements:

$$H_1 := R_{\alpha(156)} R_{\alpha(126)} R_{\alpha(136)} R_{\alpha(146)} = \begin{pmatrix} \begin{matrix} 1 & 0 & -2 & 0 \\ 0 & 1 & 0 & 0 \\ 0 & 0 & 1 & 0 \\ 0 & 2 & 0 & 1 \end{matrix} & \\ & I_2 \end{pmatrix},$$

$$H_2 := R_{\alpha(136)} R_{\alpha(126)} R_{\alpha(156)} R_{\alpha(146)} = \begin{pmatrix} \begin{matrix} 1 & 0 & 0 & 0 \\ 0 & 1 & 0 & 2 \\ -2 & 0 & 1 & 0 \\ 0 & 0 & 0 & 1 \end{matrix} & \\ & I_2 \end{pmatrix},$$

$$H_3 := R_{\alpha(123)} R_{\alpha(126)} R_{\alpha(136)} R_{\alpha(236)} = \begin{pmatrix} \begin{matrix} 1 & 0 & 0 & -2 \\ 0 & 1 & 0 & 0 \\ 0 & 2 & 1 & 0 \\ 0 & 0 & 0 & 1 \end{matrix} & \\ & I_2 \end{pmatrix},$$

$$H_4 := R_{\alpha(136)} R_{\alpha(126)} R_{\alpha(123)} R_{\alpha(236)} = \begin{pmatrix} \begin{matrix} 1 & 0 & 0 & 0 \\ 0 & 1 & 2 & 0 \\ 0 & 0 & 1 & 0 \\ -2 & 0 & 0 & 1 \end{matrix} & \\ & I_2 \end{pmatrix}.$$

The space $\alpha^\perp \cap \beta^\perp \cap \mathbb{D}$ is expressed in the z-coordinate as $z^5 = z^6 = 0$,

$$z^1 z^2 + z^3 z^4 = 0, \quad \Im \frac{z^3}{z^1} > 0, \quad z^1 \bar{z}^2 + \bar{z}^1 z^2 + z^3 \bar{z}^4 + \bar{z}^3 z^4 > 0.$$

Eliminating z^4 and putting $\tau^1 = -z^1/z^3$ and $\tau^2 = -z^2/z^3$, we have $\Im \tau^1 > 0$ and $\Im \tau^2 > 0$, because

$$\frac{1}{|z^3|^2}(z^1 \bar{z}^2 + \bar{z}^1 z^2 + z^3 \bar{z}^4 + \bar{z}^3 z^4) = -(\tau_1 - \bar{\tau}_1)(\tau_2 - \bar{\tau}_2).$$

The transformations $H_1, \ldots, H_4$ can be expressed in terms of the coordinates $(\tau^1, \tau^2) \in \mathbb{H}^2$ as follows:

$$H_1 : (\tau^1, \tau^2) \longmapsto (\tau^1 - 2, \tau^2), \quad H_2 : (\tau^1, \tau^2) \longmapsto \left(\frac{\tau_1}{1 - 2\tau^1}, \tau^2 \right),$$

$$H_4 : (\tau^1, \tau^2) \longmapsto (\tau^1, \tau^2 + 2), \quad H_3 : (\tau^1, \tau^2) \longmapsto \left(\tau^1, \frac{\tau_2}{1 + 2\tau^2} \right).$$

Therefore the extended map ϕ induces the isomorphism

$$\overline{X}_{2\alpha}^{345;126} \xrightarrow{\sim} \alpha^{\perp} \cap \beta^{\perp} \cap \overline{\mathbb{D}}/\Gamma_A(2)|_{\alpha^{\perp} \cap \beta^{\perp} \cap \overline{\mathbb{D}}}$$
$$\| \qquad\qquad\qquad\qquad \|$$
$$\mathbb{P}^1 \times \mathbb{P}^1 \xrightarrow{\sim} \qquad \overline{\mathbb{H}}/\Gamma(2) \times \overline{\mathbb{H}}/\Gamma(2).$$

REMARK 7.1. You may wonder why in the former part of this section we worked on $X_{2\alpha}^{135;246}$ and in the latter on $X_{2\alpha}^{345;126}$. Well, this is only a matter of notational convenience; the argument is symmetric with respect to permutations of the six numerals.

We summarize the above arguments in the following

THEOREM 7.3. *The map* $\varphi : X \to \mathbb{D}/\Gamma_A(2)$ *extends naturally to the isomorphism*

$$\varphi : \overline{X} \xrightarrow{\sim} \overline{\mathbb{D}}/\Gamma_A(2).$$

As expected, the strata correspond as follows:

$$X_{1\alpha}^{ij} \longrightarrow F_1^{ij},$$
$$X_{0\alpha}^{ij;kl;mn} \longrightarrow F_0^{ij;kl;mn}.$$

This theorem is the main achievement of Part 3. We will paraphrase this theorem and obtain its final form in §10. In §8, we find an involution on $\overline{\mathbb{D}}/\Gamma_A(2)$ which is equivariant with the involution $*$ on $\overline{X}$, under the isomorphism φ. In order to see the role of the involution on $\overline{\mathbb{D}}/\Gamma_A(2)$ in a nice way (in §10) we will transform the domain $\mathbb{D}$ into another space, $\mathbb{H}_2$. This transformation will be given in §9.

8. The Relation between the Involution $*$ on $\overline{X}$ and the Map φ

We want to know the image of $Q \subset X$ under the map $\phi : X \to \mathbb{D}$, defined by a projective solution of the system $E(3,6;1/2)$ on X.

Since the involution $*$ on X leaves $E(3,6;1/2)$ projectively invariant (§7 of Chapter VIII), there is a system, say $E(3,6;1/2)/\langle *\rangle$, on $X/\langle *\rangle$ such that its pull-back under the projection

$$p_r : X \to X/\langle *\rangle \subset Z$$

is $E(3,6;1/2)$. Thus if we connect a point $x \in X$ and $*x$ by a path $C \subset X$, and denote by $\phi_C(*x)$ the analytic continuation of $\phi(x)$ along C (see Figure 8.1), $\phi(x)$ and $\phi(*x)$ are related as

$$\phi_C(*x) = M_C \phi(x),$$

where $M_C \in PGL(6,\mathbb{C})$ is the projectivized circuit matrix of the system $E(3,6;1/2)/\langle * \rangle$ with respect to the loop $p_r(C) \subset X/\langle * \rangle$ with base $p_r(x) = p_r(*x)$. It is clear that M_C preserves $\mathbb{D}$, i.e., $M_C \in Aut(\mathbb{D})$. We choose a path C so that $p_r(C)$ is a simple loop around Q in $X/\langle * \rangle$. Since Q is the set of fixed points of $*$, M_C is an involution which leaves the intersection H_C of $\mathbb{D}$ and a hyperplane in $\mathbb{P}^5$ unchanged. Thus there exists a 6-vector $q = (q_1, \ldots, q_6)$ satisfying $(q, \bar{q}) < 0$ such that M_C is the reflection R_q with root q and that $H_C = q^{\perp} \cap \mathbb{D}$.

In order to see H_C, we make use of the intersection pattern of Q and $X_3^{ijk} \subset X$ and the pattern of the mirrors $\alpha^{\perp}$ for $\alpha \in R(-1)$: The variety $\overline{Q}$ passes through the intersection $\overline{X}_3^{ijk} \cap \overline{X}_3^{lmn}$ ($\{i, \ldots, n\} = \{1, \ldots, 6\}$), and two such mirrors intersect orthogonally if they meet inside $\mathbb{D}$.

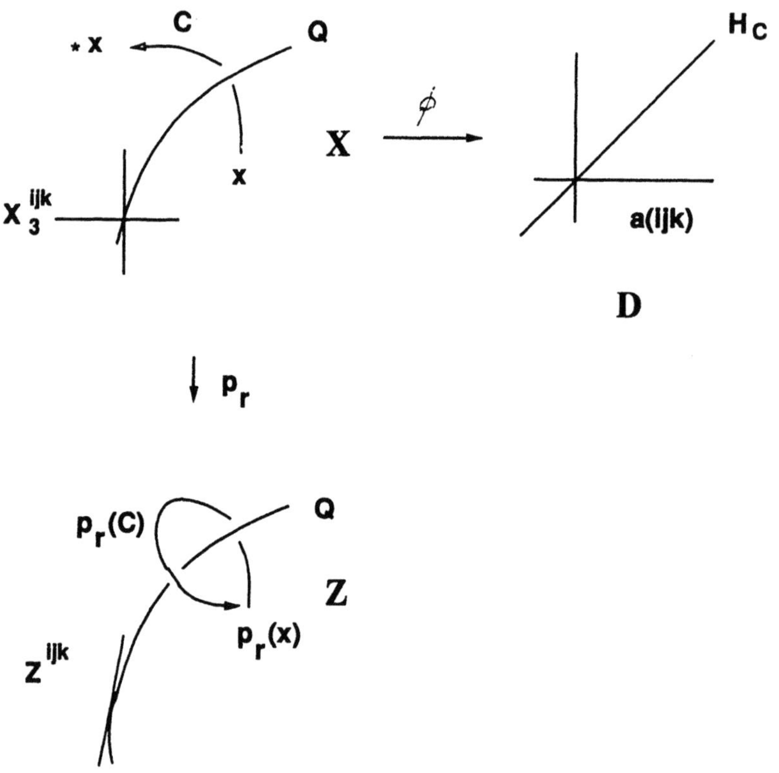

FIGURE 8.1. The path $C \subset X$, the loop $p_r(C) \subset Z$, and the local image H_C of Q under ϕ around C

Note that the set of H_C for such C forms an orbit of MG. Thus without loss of generality, we can assume that $q^{\perp}$ passes through the intersection of the mirrors of the two perpendicular roots

$$\alpha = (0,\dots,0,1) \in R(346) \quad \text{and} \quad \beta = (0,\dots,0,1,0) \in R(125).$$

Thus q must be of the form $(0,\dots,0,q_5,q_6)$. Since the reflection R_q interchanges $\alpha^\perp$ and $\beta^\perp$, we must have $q_5 = q_6$. Therefore we proved

PROPOSITION 8.1. *The image of $Q \subset X$ under $\phi : X \to \mathbb{D}$ is the $\Gamma_A(2)$-orbit of $q^\perp \cap \mathbb{D}$, where $q = (0,\dots,0,1,1) \in R(-2)$.*

COROLLARY 8.2. $\varphi : X \to \mathbb{D}/\Gamma_A(2)$ *satisfies*

$$\varphi(*x) = R_q\varphi(x).$$

COROLLARY 8.3. ϕ *induces the isomorphism*

$$Z = \overline{X}/\langle * \rangle \xrightarrow{\sim} \bar{\mathbb{D}}/\langle \Gamma_A(2), R_q \rangle.$$

REMARK 8.1. The restriction

$$\phi|_Q : Q \longrightarrow q^\perp \cap \mathbb{D}$$

will be studied in §10.

9. The Symmetric Domain $\mathbb{H}_2$

In this section, we transform the space $\mathbb{D}$ into another space, $\mathbb{H}_2$. We first give an isomorphism $\mathbb{D} \dashrightarrow \mathbb{H}_2$, then embed $\mathbb{H}_2$ into the Grassmannian $Gr(4,2)$, and finally give an isomorphism $\iota : \mathbb{H}_2 \xrightarrow{\sim} \mathbb{D}$ in terms of the Plücker embedding $Gr(4,2) \to \mathbb{P}^5$.

9.1. The Isomorphism $\mathbb{D} \dashrightarrow \mathbb{H}_2$. Recall that the space $\mathbb{D} \subset \mathbb{P}^5$ is defined by

$$\frac{1}{2}(z,z) = 2z^1z^2 + 2z^3z^4 - (z^5)^2 - (z^6)^2 = 0,$$

$$\frac{1}{2}(z,\bar{z}) = 2\Re z^1\bar{z}^2 + 2\Re z^3\bar{z}^4 - |z^5|^2 - |z^6|^2 > 0,$$

$$\Im\frac{z^3}{z^1} > 0, \quad z = z^1 : \cdots : z^6 \in \mathbb{P}^5.$$

I am going to now transform this space into a space which has an expression very similar to the upper half plane

$$\mathbb{H} = \{\tau \in \mathbb{C} \mid \Im\tau = \frac{\tau - \bar{\tau}}{2i} > 0\},$$

used in the classical model I presented in Chapter II. Consider the domain

$$\mathbb{H}_2 := \{W \in M(2,2) \mid \frac{W - W^*}{2i} > 0\},$$

where $M(p,q)$ denotes the space of complex $p \times q$-matrices, $*$ stands for the transposed complex conjugate, and 'a matrix > 0' means that the matrix is positive definite hermitian. Since $z^1 z^2 z^3 z^4 \neq 0$ in $\mathbb{D}$, we can define a map $\mathbb{D} \to M(2,2)$ by

$$(z^1, \ldots, z^6) \mapsto W = \begin{pmatrix} w_1 & (w_3 + iw_4)/(1+i) \\ (w_3 - iw_4)/(1-i) & w_2 \end{pmatrix},$$

where

$$w_1 = \frac{z^3}{z^1}, \ w_2 = \frac{z^4}{z^1}, \ w_3 = \frac{z^5}{z^1}, \ w_4 = \frac{z^6}{z^1}.$$

Since

$$\frac{W - W^*}{2i} = \begin{pmatrix} \Im w_1 & \{\Im w_3 + i\Im w_4\}/(1+i) \\ \{\Im w_3 - i\Im w_4\}/(1-i) & \Im w_2 \end{pmatrix},$$

and

$$\det \frac{W - W^*}{2i} = \Im w_1 \cdot \Im w_4 - \frac{1}{2}\{(\Im w_3)^2 + (\Im w_4)^2\}$$

$$= \Im z^3 \cdot \Im z^4 - \frac{1}{2}\{(\Im z^5)^2 + (\Im z^6)^2\}, \quad z^1 = 1$$

(recall from §3.2 that $(z, \bar{z}) > 0$, together with $(z, z) = 0$, is equivalent to $2\Im z^3 \cdot \Im z^4 - (\Im z^5)^2 - (\Im z^6)^2 > 0$), we conclude that the image of this map is in $\mathbb{H}_2$. Conversely, for a given $W \in \mathbb{H}_2$, one can find $w_1, \ldots, w_4$, and so, with $z^1 = 1$, one can also find $z^3, \ldots, z^6$ by the above formulae (define z^2 by the quadric equation). Then the above computation shows that $z^1 : \cdots : z^6$ is in $\mathbb{D}$. Hence the correspondence above gives the isomorphism between $\mathbb{D}$ and $\mathbb{H}_2$.

9.2. The Isomorphism $\iota_1 : \mathcal{D} \xrightarrow{\sim} \mathbb{H}_2$. Note that the upper half plane $\mathbb{H}$ can be written in the homogeneous coordinates $\tau_1 : \tau_2$ on $\mathbb{P}^1$ as

$$\left\{ \begin{pmatrix} \tau_1 \\ \tau_2 \end{pmatrix} \in M(2,1) \mid (\bar{\tau}_1, \bar{\tau}_2) \begin{pmatrix} 0 & i \\ -i & 0 \end{pmatrix} \begin{pmatrix} \tau_1 \\ \tau_2 \end{pmatrix} > 0 \right\} /GL(1, \mathbb{C}) \subset \mathbb{P}^1.$$

We will obtain a homogeneous expression of $\mathbb{H}_2$ similar to the above. First set

$$H(4,2) = \{X \in M(4,2) \mid X^*(iJ)X > 0\},$$

$$S(4,2) = \{X \in H(4,2) \mid {}^t XJX = 0\},$$

where $J = \begin{pmatrix} 0 & I_2 \\ -I_2 & 0 \end{pmatrix}$. The group $GL(2,\mathbb{C})$ acts on $H(4,2)$ and $S(4,2)$ from the right; define

$$\mathcal{D} = H(4,2)/GL(2,\mathbb{C}),$$
$$\mathcal{F} = S(4,2)/GL(2,\mathbb{C}).$$

Before giving an isomorphism between $\mathcal{D}$ and $\mathbb{H}_2$, we study the group of automorphisms of $\mathcal{D}$ which will make clear the nature of $\mathcal{F}$. The groups

$$U(2,2) := \{g \in GL(4,\mathbb{C}) \mid g^*Jg = J\},$$
$$Sp(2,\mathbb{R}) := \{g \in GL(4,\mathbb{R}) \mid {}^tgJg = J\}$$

act naturally on $H(4,2)$ and $S(4,2)$, respectively, from the left, since we have

$$(gX)^*(iJ)(gX) = X^*\{g^*(iJ)g\}X = X^*(iJ)X > 0, \quad X \in H(4,2),$$
$${}^t(gX)J(gX) = {}^tX({}^tgJg)X = {}^tXJX = 0, \quad X \in S(4,2).$$

Note that for any element $X = \begin{pmatrix} X_1 \\ X_2 \end{pmatrix}$, the nonsingularity of X_1 and X_2 follows from the positivity of $X^*(iJ)X = i(X_1^*X_2 - X_2^*X_1)$. Therefore we can define the involution

$$T : \begin{pmatrix} X_1 \\ X_2 \end{pmatrix} \mapsto \begin{pmatrix} {}^tX_2^{-1} \\ {}^tX_1^{-1} \end{pmatrix},$$

which acts on $H(4,2)$, since we have

$$T(X)^*(iJ)T(X) = (\overline{X}_2^{-1t}X_1^{-1})^{*t}\{X^*(iJ)X\}(\overline{X}_2^{-1t}X_1^{-1}) > 0.$$

Then, since we have

$$g(Xh) = (gX)h, \quad T(Xh) = T(X)^t h^{-1}$$

for

$$g \in U(2,2), \quad X \in H(4,2), \quad h \in GL(2,\mathbb{C}),$$

the group $U(2,2)$ and the involution T naturally act on $\mathcal{D}$. Note that $\mathcal{F}$ is the set of fixed points of T and that $Sp(2,\mathbb{R})$ acts on $\mathcal{F}$.

By using the action of $GL(2,\mathbb{C})$, we can represent any element of $\mathcal{D}$ by a unique element of the form $\begin{pmatrix} W \\ I_2 \end{pmatrix}$, where W belongs to $\mathbb{H}_2$. Thus we get the isomorphism $\iota_1 : \mathcal{D} \xrightarrow{\sim} \mathbb{H}_2$. By restricting ι_1 on $\mathcal{F}$, we also find an isomorphism between $\mathcal{F}$ and the *Siegel upper half space*

$$\mathfrak{S}_2 = \{W \in \mathbb{H}_2 \mid {}^t W = W\}$$

of degree 2. Therefore we have the diagram

$$\iota_1 : \quad \begin{array}{ccc} \mathcal{D} & \xrightarrow{\sim} & \mathbb{H}_2 \\ \cup & & \cup \\ \mathcal{F} & \xrightarrow{\sim} & \mathfrak{S}_2 \end{array}$$

The isomorphism ι_1 induces actions of $U(2,2)$ and T on $\mathbb{H}_2$ and those of $Sp(2,\mathbb{R})$ on $\mathfrak{S}_2$:

$$g \cdot W = (AW + B)(CW + D)^{-1}, \quad T \cdot W = {}^t W, \quad W \in \mathbb{H}_2,$$

where $W \in \mathbb{H}_2$ or $\mathfrak{S}_2$, and

$$g = \begin{pmatrix} A & B \\ C & D \end{pmatrix} \in U(2,2) \text{ or } Sp(2,\mathbb{R}),$$

respectively.

9.3. The Isomorphism $\iota_2 : \mathcal{D} \xrightarrow{\sim} \mathbb{D}$. The domain $\mathcal{D}$ can be regarded as part of the *Grassmannian* manifold

$$Gr(4,2) = \{X \in M(4,2) \mid \exists d_{jk}(X) \neq 0\}/GL(2,\mathbb{C}),$$

where

$$d_{jk}(X) = \begin{vmatrix} x_{j1} & x_{j2} \\ x_{k1} & x_{k2} \end{vmatrix}.$$

By using the Plücker embedding of $Gr(4,2)$, we now give realizations of $\mathcal{D}$ and $\mathcal{F}$ in a projective space. We define the map

$$d : M(4,2) \to \mathbb{C}^6$$

by

$$X \mapsto (z_1, \ldots, z_6) = (d_{34}, d_{12}, d_{14}, d_{23}, d_{13}, d_{24}), \quad d_{jk} = d_{jk}(X).$$

The *Plücker relation* reads

$$d_{12}d_{34} - d_{13}d_{24} + d_{14}d_{23} = 0.$$

The map d transforms the condition $X^*(iJ)X > 0$ for $X \in H(2,4)$ into

$$\Re\{d_{12}\overline{d}_{34} + d_{13}\overline{d}_{24} + d_{14}\overline{d}_{23}\} > 0,$$

$$\Im d_{14}\overline{d}_{34} > 0.$$

By the property

$$d_{jk}(Xh) = d_{jk}(X)\det(h), \quad h \in GL(2,\mathbb{C}), \ X \in M(4,2),$$

the map d induces a map

$$d : Gr(4,2) \to \mathbb{P}^5,$$

called the *Plücker embedding* of $Gr(4,2)$. The image $d(Gr(4,2))$ is a quadratic hypersurface of $\mathbb{P}^5$. The image $d(\mathcal{D})$ is a subdomain of the quadratic hypersurface given by the two inequalities above.

The group $U(2,2)$ and the involution T act naturally on $d(\mathcal{D})$ as follows:

$$g \cdot (\ldots, d_{jk}(X), \ldots) := (\ldots, d_{jk}(gX), \ldots),$$
$$T \cdot (\ldots, d_{jk}(X), \ldots) := (\ldots, d_{jk}(T \cdot X), \ldots).$$

An elementary calculation shows

$$d_{jk}(gX) = \sum_{1 \leq p < q \leq 4} \begin{vmatrix} g_{jp} & g_{jq} \\ g_{kp} & g_{kq} \end{vmatrix} d_{pq}(X), \quad g = (g_{pq}) \in U(2,2),$$

and

$$d_{jk}(T \cdot X) = cd_{jk}(X) \text{ for } (jk) \in \{(12),(14),(23),(34)\},$$
$$d_{13}(T \cdot X) = -cd_{24}(X), \quad d_{24}(T \cdot X) = -cd_{13}(X),$$

where $c = (d_{12}(X)d_{34}(X))^{-1}$. That is, $g \in U(2,2)$ corresponds to the $\binom{4}{2}$-matrix whose entries are the 2-minors of g, and T to the matrix $c(I_4 \oplus (-U))$, where $U = \begin{pmatrix} 0 & 1 \\ 1 & 0 \end{pmatrix}$. In this way we get a homomorphism from $\langle U(2,2), T \rangle$ to $PGL(6,\mathbb{C})$. Since $\mathcal{F}$ is the set of fixed points of T in $\mathcal{D}$, the image $d(\mathcal{F})$ is the subset of $d(\mathcal{D})$ given by the equation $z^5 + z^6 = 0$.

To connect with our story, we need one more coordinate change,

$$^t(z_1, \ldots, z_6) = P^t(z^1, \ldots, z^6).$$

This is provided by the matrix

$$P = \operatorname{diag}(1,-1,-1,1) \oplus \begin{pmatrix} -1/(1-i) & -i/(1-i) \\ -1/(1+i) & i/(1+i) \end{pmatrix},$$

which takes $d(\mathcal{D})$ onto $\mathbb{D}$. With this coordinate change and the Plücker embedding d, we obtain the isomorphism

$$\iota_2 : \mathcal{D} \xrightarrow{\sim} \mathbb{D},$$

under which the set $\mathcal{F}$ is transformed into the subset of $\mathbb{D}$ defined by $z_5 - z_6 = 0$.

9.4. The Isomorphism $\iota : \mathbb{H}_2 \xrightarrow{\sim} \mathbb{D}$. Combining the isomorphisms ι_1 and ι_2, we get the isomorphism

$$\iota = \iota_2 \circ \iota_1^{-1} : \mathbb{H}_2 \xrightarrow{\;\sim\;} \mathbb{D},$$

which is the inverse of the map $\mathbb{D} \to \mathbb{H}_2$ given in the beginning (§9.1) of this section.

Note that any reflection R_α $((\alpha, \overline{\alpha}) < 0)$ on $\mathbb{D}$ corresponds to gTg^{-1} for some $g \in U(2,2)$. In particular, the involution T on $\mathbb{H}_2$ corresponds to the reflection R_q on $\mathbb{D}$ with root $q = (0,0,0,0,1,1)$ of norm -2. Thus we have $\iota_2(\mathcal{F}) = q^\perp \cap \mathbb{D}$, and hence the isomorphism

$$\iota|_{\mathfrak{S}_2} : \mathfrak{S}_2 \xrightarrow{\;\sim\;} q^\perp \cap \mathbb{D}.$$

10. The Final Form of the Modular Interpretation

10.1. The Monodromy Group as a Congruence Subgroup on $\mathbb{H}_2$. The group $Aut(\mathbb{H}_2)$ of automorphisms of $\mathbb{H}_2$ is generated by

$$U(2,2) = \{g \in GL(4, \mathbb{C}) \mid g^* J g = J\}, \quad J = \begin{pmatrix} 0 & I_2 \\ -I_2 & 0 \end{pmatrix}$$

and T, which act on $\mathbb{H}_2$ according to

$$T \cdot W = {}^t W$$

and

$$g \cdot W = (AW + B)(CW + D)^{-1},$$

where

$$W \in \mathbb{H}_2 \quad \text{and} \quad g = \begin{pmatrix} A & B \\ C & D \end{pmatrix} \in U(2,2).$$

We have

$$Aut(\mathbb{H}_2) \simeq U(2,2) \rtimes \langle T \rangle, \quad \langle T \rangle = \{id, T\}.$$

This is the semidirect product of which $U(2,2)$ is the normal subgroup. Indeed, we have

LEMMA 10.1. $TgT = \overline{g}, \quad g \in U(2,2).$

PROOF. Following a straightforward computation,

$$(TgT) \cdot W = {}^t\{(A^tW + B)(C^tW + D)^{-1}\}$$
$$= (W^tC + {}^tD)^{-1}(W^tA + {}^tB)$$

and

$$\bar{g} \cdot W = (\overline{A}W + \overline{B})(\overline{C}W + \overline{D})^{-1},$$

we are led to

$$\{(TgT) \cdot W\}^{-1}\bar{g} \cdot W = (W^tA + {}^tB)^{-1}(W^tC + {}^tD)(\overline{A}W + \overline{B})(\overline{C}W + \overline{D})^{-1}.$$

On the other hand, the condition $g^*Jg = J$ dictates

$$-{}^tC\overline{A} + {}^tA\overline{C} = 0, \qquad -{}^tC\overline{B} + {}^tA\overline{D} = I_2,$$
$$-{}^tD\overline{A} + {}^tB\overline{C} = -I_2, \quad -{}^tD\overline{B} + {}^tB\overline{D} = 0.$$

These relations imply

$$(W^tC + {}^tD)(\overline{A}W + \overline{B}) = (W^tA + {}^tB)(\overline{C}W + \overline{D}),$$

which proves the lemma. $\square$

Let Γ be the *full modular group*

$$\Gamma = \{g \in PGL(4, \mathbb{Z}[i]) \mid g^*Jg = J\},$$

and let $\Gamma(1 + i)$ be the *congruence subgroup*

$$\Gamma(1 + i) := \{g \in \Gamma \mid g \equiv I_4 \mod (1 + i)\},$$

of level $(1 + i)$.

Since we know (§9.3) the isomorphism $Aut(\mathbb{H}_2) \xrightarrow{\sim} Aut(\mathbb{D})$ induced by the explicit isomorphism $\iota : \mathbb{H}_2 \xrightarrow{\sim} \mathbb{D}$, we can find, in principle, the subgroup of $Aut(\mathbb{H}_2)$ corresponding to a given subgroup of $Aut(\mathbb{D})$. I do not give an explicit formula for $Aut(\mathbb{D}) \xrightarrow{\sim} Aut(\mathbb{H}_2)$ because it is rather complicated. In any case you can prove

PROPOSITION 10.2. *Define*

$$\Gamma_T = \Gamma \rtimes \langle T \rangle,$$
$$\Gamma_T(1 + i) = \Gamma(1 + i) \rtimes \langle T \rangle,$$
$$\Gamma_M(1 + i) = \{g \in \Gamma(1 + i) \mid \det g = 1\}$$
$$\cup T\{g \in \Gamma(1 + i) \mid \det g = -1\} \subset \Gamma_T,$$

and

$$\Gamma_A = \{g \in PGL(6, \mathbb{Z}) \mid {}^t g A g = A, Con\mathbb{D}\} \quad \textit{(the full modular group)},$$
$$\Gamma_A(2) = \{g \in \Gamma_A \mid g \equiv I_6 \mod 2\} \quad \textit{(the congruence subgroup of level 2)}.$$

Then the isomorphism $\iota : \mathbb{H}_2 \xrightarrow{\sim} \mathbb{D}$ induces the following isomorphisms between discrete subgroups:

$$\Gamma_T \cong \Gamma_A, \quad \Gamma_T(1 + i) \cong \langle \Gamma_A(2), R_q \rangle, \quad \Gamma_M(1 + i) \cong \Gamma_A(2) = MG.$$

REMARK 10.1. We have the following inclusion relations:

$$\Gamma_T \overset{S_6}{\underset{\triangleright}{}} \Gamma_T(1 + i) \overset{\mathbb{Z}/2\mathbb{Z}}{\underset{\triangleright}{}} \Gamma_M(1 + i).$$

Recall how in §6 we defined the restriction of a group on a subspace of the space on which the group acts. Then we have

$$\Gamma|_{\mathfrak{S}_2} = \Gamma_T|_{\mathfrak{S}_2} \cong \Gamma_{\mathfrak{S}_2} := Sp(2, \mathbb{Z}),$$
$$\Gamma(1 + i)|_{\mathfrak{S}_2} = \Gamma_T(1 + i)|_{\mathfrak{S}_2} = \Gamma_M(1 + i)|_{\mathfrak{S}_2}$$
$$\cong \Gamma_{\mathfrak{S}_2}(2) := \{g \in \Gamma_{\mathfrak{S}_2} \mid g \equiv I_4 \mod 2\}.$$

10.2. A Paraphrase of Theorem 7.3. In this subsection we state a theorem giving a modular interpretation of the configuration space $X = X(3, 6)$ of six lines in $\mathbb{P}^2$ as well as those of $X/\langle * \rangle$, $X/\langle *, S_6 \rangle$ and $Q \subset X$.

In what follows I shall denote every map induced by the composition

$$\psi := \iota^{-1} \circ \phi : X \longrightarrow \mathcal{D} \xrightarrow{\sim} \mathbb{H}_2$$

by the same symbol ψ.

Recall that the variety Q is the set of fixed points of the involution $*$ on X, and that the space $\mathfrak{S}_2$ is the set of fixed points of the transposition T on $\mathbb{H}_2$. Since the map $\psi : X \to \mathbb{H}_2/\Gamma_M(1 + i)$ satisfies

$$\psi(*x) = T \cdot \psi(x)$$

(cf. Corollary 8.2 and §9.4), we can restrict the map ψ on Q and obtain the map

$$\psi|_Q : Q \to \mathfrak{S}_2.$$

Thus we have by Theorem 6.3 and Proposition 10.2

THEOREM 10.3. $\psi := \iota^{-1} \circ \phi : X \to \mathbb{H}_2$ *induces the isomorphisms*

$$
\begin{array}{ccc}
\overline{X} & \xrightarrow{\sim} & \bar{\mathbb{H}}_2/\Gamma_M(1+i) \\
\cup & & \cup \\
\overline{Q} & \xrightarrow{\sim} & \bar{\mathfrak{S}}_2/\Gamma_{\mathfrak{S}_2}(2),
\end{array}
\qquad
\begin{array}{ccc}
\overline{X}/\langle *\rangle & \xrightarrow{\sim} & \bar{\mathbb{H}}_2/\Gamma_T(1+i) \\
\cup & & \cup \\
\overline{Q} & \xrightarrow{\sim} & \bar{\mathfrak{S}}_2/\Gamma_{\mathfrak{S}_2}(2),
\end{array}
$$

and

$$
\begin{array}{ccc}
\overline{X}/\langle *, S_6\rangle & \xrightarrow{\sim} & \bar{\mathbb{H}}_2/\Gamma_T \\
\cup & & \cup \\
\overline{Q}/\langle S_6\rangle & \xrightarrow{\sim} & \bar{\mathfrak{S}}_2/\Gamma_{\mathfrak{S}_2}.
\end{array}
$$

Here $\bar{\mathbb{H}}_2$ and $\bar{\mathfrak{S}}_2$ are the rational closures of $\mathbb{H}_2$ and $\mathfrak{S}_2$ according to the following (cf. §6). First define

$$
f1 := \left\{ \begin{pmatrix} \tau & 0 \\ 0 & 1 \end{pmatrix} \mid \tau \in \mathbb{H} \right\} \subset \partial \mathbb{H}_2 \subset Gr(4,2),
$$

$$
f0 := I_2 \in \partial \mathbb{H}_2 \subset Gr(4,2),
$$

where $\partial \mathbb{H}_2$ is the topological boundary of $\mathbb{H}_2$ in $Gr(4,2)$. Then the rational closures are defined as

$$
\bar{\mathbb{H}}_2 := \mathbb{H}_2 \cup \Gamma_T \cdot f1 \cup \Gamma_T \cdot f0,
$$
$$
\bar{\mathfrak{S}}_2 := \mathfrak{S}_2 \cup \Gamma_{\mathfrak{S}_2} \cdot f1 \cup \Gamma_{\mathfrak{S}_2} \cdot f0.
$$

Of course the isomorphism $\iota : \mathbb{H}_2 \xrightarrow{\sim} \mathbb{D}$ extends naturally to the isomorphism $\bar{\mathbb{H}}_2 \xrightarrow{\sim} \bar{\mathbb{D}}$. The quotient spaces $\bar{\mathbb{H}}_2/\Gamma_T, \ldots$ and $\bar{\mathfrak{S}}_2/\Gamma_{\mathfrak{S}_2}, \ldots$ are called the Satake compactifications of $\mathbb{H}_2/\Gamma_T, \ldots$ and $\mathfrak{S}_2/\Gamma_{\mathfrak{S}_2}, \ldots$

REMARK 10.2. Let $x_1, \ldots, x_6$ be six distinct points on $\mathbb{P}^1$ and consider the family of curves

$$
S_x : \quad s^2 = \prod_{j=1}^{6}(t - x_j), \quad x \in X(2,6)
$$

of genus 2. This family is not considered in Chapter VI since the 6-tuple of $1/2$ is not an admissible sequence in the sense defined there. We have a more geometric reason for this, however. Since the family in question is the full family of curves of genus 2, the period space should not be the ball but the Siegel upper half space $\mathfrak{S}_2$ of degree 2. The period map of the family can be expressed as part of our argument as described below.

Let us fix a non-singular conic $F \simeq \mathbb{P}^1$ in $\mathbb{P}^2$, unique up to projective transformations; we regard the six points $x_1, \ldots, x_6$ as lying on F. The curve S_x is the double cover of F branched at these six points. Let l_x be the six tangent lines of F at the six points $x_1, \ldots, x_6$, and let $S(l_x)$ be the K3 surface defined in §1. (In this way we identify Q and $X(2,6)$.) Then it turns out that the surface $S(l_x)$ is a Kummer surface whose corresponding abelian surface is the jacobian of the curve S_x. Let $\omega_i(x)$ $(i = 1, 2)$ be two linearly independent holomorphic 1-forms on S_x and $c_j(x)$ $(j = 1, \ldots, 4)$ a standard basis of the first homology group of S_x. Define the periods by

$$u_{ij}(x) = \int_{c_j(x)} \omega_i, \quad (i = 1, 2, \ j = 1, \ldots, 4)$$

and the normalized period matrix z by

$$z(x) = \begin{pmatrix} u_{11} & u_{12} \\ u_{21} & u_{22} \end{pmatrix}^{-1} \begin{pmatrix} u_{13} & u_{14} \\ u_{23} & u_{24} \end{pmatrix} \in \mathfrak{S}_2.$$

It is not difficult to see that the normalized period map $X(2,6) \ni x \mapsto z(x) \in \mathfrak{S}_2$ of the curve coincides with our map $\psi|_Q : Q \ni x \mapsto \phi(l_x) \in \mathfrak{S}_2$.

11. The Structure of the Cusps

I think that you would probably like to have an intuitive idea of the action of the groups $\Gamma_T, \Gamma_T(1 + i)$ and $\Gamma_M(1 + i)$ on $\mathbb{H}_2$. Although this is a complex 4-dimensional space (real 8-dimensional), around a rational boundary component, we can actually "see" the action. This is what we are going to do in this section. You can think of this section as just amusement; please relax. The next (the last) section will be heavy.

11.1. Linear Parabolic Parts. In this subsection we study the action of the three discrete groups $\Gamma_T, \Gamma_T(1+i)$ and $\Gamma_M(1+i)$ at a rational boundary component, the point at infinity. Since any $W \in M(2,2)$ can be written as

$$W = \frac{W + W^*}{2} + i\frac{W - W^*}{2i}, \quad i = \sqrt{-1},$$

the domain $\mathbb{H}_2$ can be expressed as

$$\mathbb{H}_2 = H + iH^+,$$

where

$$H = \left\{ \begin{bmatrix} u & h \\ \bar{h} & v \end{bmatrix} \mid u, v \in \mathbb{R}, h \in \mathbb{C} \right\}$$

is the space of 2×2 hermitian matrices, and H^+ is the cone consisting of 2×2 positive definite hermitian matrices. When the imaginary part $(W - W^*)/2i$ of W becomes a very very positive hermitian matrix, W

tends to a boundary called $i\infty$. The maximal subgroup of $Aut(\mathbb{H}_2)$ which fixes $i\infty$ is given by

$$P = \left\{ \begin{pmatrix} A & B \\ 0 & D \end{pmatrix} \mid D^{-1} = A^*, {}^t B\bar{D} = {}^t D\bar{B} \right\} \rtimes \langle T \rangle.$$

Let us define the projection $pr : P \to GL(2, \mathbb{C}) \rtimes \langle T \rangle$ by

$$pr : \begin{pmatrix} A & B \\ 0 & D \end{pmatrix} \mapsto A, \quad T \mapsto T,$$

where the group $GL(2, \mathbb{C}) \rtimes \langle T \rangle$ is considered to act on H (preserving H^+) by

$$g \cdot X = gXg^*, \; T \cdot X = {}^t X, \quad g \in GL(2, \mathbb{C}), \; X \in H.$$

For a subgroup Λ of $Aut(\mathbb{H}_2)$, we define the subgroup $\pi(\Lambda)$ of $GL(2, \mathbb{C}) \rtimes \langle T \rangle$ as $pr(P \cap \Lambda)$. It is easy to see that

$$\pi(\Gamma_T) = G \rtimes \langle T \rangle,$$
$$\pi(\Gamma_T(1 + i)) = G(1 + i) \rtimes \langle T \rangle,$$
$$\pi(\Gamma_M(1 + i)) = \{g \in G(1 + i) \mid \det g^2 = 1\}$$
$$\cup T\{g \in G(1 + i) \mid \det g^2 = -1\},$$

where

$$G := GL(2, \mathbb{Z}[i]),$$
$$G(1 + i) := \{g \in G \mid g \equiv I_2 \mod (1 + i)\}.$$

11.2. Reflection Groups.

We define the inner product on H,

$$(X, X') := tr(\tilde{X}X') = uv' + vu' - (h\bar{h}' + \bar{h}h'),$$

where

$$X = \begin{bmatrix} u & h \\ \bar{h} & v \end{bmatrix}, \quad X' = \begin{bmatrix} u' & h' \\ \bar{h}' & v' \end{bmatrix},$$

and $\tilde{X}$ is the cofactor matrix of X. In particular, $(X, X) = 2 \det X$. Note that

$$T \circ g \circ T = \bar{g}, \quad g \circ T \circ g^{-1} = g\bar{g}^{-1} \circ T$$

and

$$(g \cdot X, g \cdot Y) = |\det g|^2 (X, Y).$$

These imply that the full group of isometries of H is given by

$$\{g \in GL(2, \mathbb{C}) \mid |\det g| = 1\} \rtimes \langle T \rangle.$$

For $\alpha \in H$ such that $(\alpha, \alpha) < 0$, we define the *reflection* R_α with *root* α by

$$X \longmapsto X - 2\frac{(\alpha, X)}{(\alpha, \alpha)}\alpha.$$

This is an isometry of H^+. Notice that

$$g \circ R_\alpha \circ g^{-1} = R_{g \cdot \alpha}.$$

The reflection R_α fixes pointwise the *mirror*

$$\alpha^\perp = \{X \in H \mid (\alpha, X) = 0\}.$$

Then notice that

$$(g \cdot \alpha)^\perp = g \cdot \alpha^\perp.$$

In particular, T is a reflection:

$$T = R_\beta, \quad \beta = \left\{\begin{matrix} 0 & i \\ -i & 0 \end{matrix}\right\};$$

its mirror is given by $h = \bar{h}$.

NOTATION. Elements of $GL(2, \mathbb{C})$, points of H, and also roots of reflections are 2×2-matrices. In order to avoid confusion, we use round brackets for elements of $GL(2, \mathbb{C})$, angle brackets for points in H, and curly brackets for roots.

We call

$$\alpha = \left\{\begin{matrix} \alpha_1 & \alpha_3 \\ \bar{\alpha}_3 & \alpha_2 \end{matrix}\right\} \in H, \quad (\alpha, \alpha) < 0$$

integral if

$$\alpha_1, \alpha_2 \in \mathbb{Z}, \quad (1 + i)\alpha_3 \in \mathbb{Z}[i].$$

When the *norm* of an integral root is -1 or -2, a computation leads to the following expression for R_α:

$$R_\alpha = (1 + i)\begin{pmatrix} \alpha_3 & -\alpha_1 \\ \alpha_2 & -\bar{\alpha}_3 \end{pmatrix} \circ T \quad \text{when} \quad (\alpha, \alpha) = -1$$

$$R_\alpha = \begin{pmatrix} \alpha_3 & -\alpha_1 \\ \alpha_2 & -\bar{\alpha}_3 \end{pmatrix} \circ T \quad \text{when} \quad (\alpha, \alpha) = -2.$$

Note that when an integral root α is of norm -2, then $\alpha_3 \in \mathbb{Z}[i]$ holds. Let us define two sets of roots,

$$R(-1) = \{\alpha \mid \text{integral root with } (\alpha, \alpha) = -1\},$$

$$R(-1, -2) = \{\alpha \mid \text{integral root with } (\alpha, \alpha) = -1, -2\},$$

and two reflection groups,

$$GR(-1) = \langle R_\alpha \mid \alpha \in R(-1)\rangle,$$
$$GR(-1,-2) = \langle R_\alpha \mid \alpha \in R(-1,-2)\rangle.$$

We can prove

PROPOSITION 11.1.

$$\pi(\Gamma_T) = GR(-1,-2),$$
$$\pi(\Gamma_M(1+i)) = GR(-1).$$

This proposition is the first step of a proof of Theorem 4.1, which I omitted. Note that it is obvious by definition that the right-hand sides are subsets of the left-hand sides.

For later use, we specify and name six integral roots of norm -1 and three integral roots of norm -2:

$$\alpha_1 = \begin{Bmatrix} 0 & (1-i)/2 \\ (1+i)/2 & 0 \end{Bmatrix}, \quad \alpha_2 = \begin{Bmatrix} 1 & (1-i)/2 \\ (1+i)/2 & 0 \end{Bmatrix},$$
$$\alpha_3 = \begin{Bmatrix} 0 & (1-i)/2 \\ (1+i)/2 & 1 \end{Bmatrix}, \quad \alpha_4 = \begin{Bmatrix} 0 & (1+i)/2 \\ (1-i)/2 & 1 \end{Bmatrix},$$
$$\alpha_5 = \begin{Bmatrix} 1 & (1+i)/2 \\ (1-i)/2 & 0 \end{Bmatrix}, \quad \alpha_6 = \begin{Bmatrix} 0 & (1+i)/2 \\ (1-i)/2 & 0 \end{Bmatrix},$$

and

$$\beta_1 = \begin{Bmatrix} 0 & i \\ -i & 0 \end{Bmatrix}, \quad \beta_2 = \begin{Bmatrix} 1 & 1 \\ 1 & 0 \end{Bmatrix}, \quad \beta_3 = \begin{Bmatrix} -1 & 0 \\ 0 & 1 \end{Bmatrix}.$$

11.3. Coxeter Graphs and Weyl Chambers. Let us study the action of the three groups $\pi(\Gamma_T), \pi(\Gamma_T(1+i))$ and $\pi(\Gamma_M(1+i))$ on the real 4-dimensional cone H^+. Since the action is linear on the cone, by projectivizing this action, we have a chance to actually "see" it in 3-dimensional space. We consider the quotient space $H^+/\mathbb{R}_{>0}$, which we identify with $\mathbb{C} \times \mathbb{R}_{>0}$ as follows:

$$H^+/\mathbb{R}_{>0} \ni \begin{bmatrix} u & h \\ \bar{h} & v \end{bmatrix} \longmapsto (z,t) = \left(\frac{h}{v}, \frac{\sqrt{uv - |h|^2}}{v}\right) \in \mathbb{C} \times \mathbb{R}_{>0}$$

$$H^+/\mathbb{R}_{>0} \ni \begin{bmatrix} u & h \\ \bar{h} & v \end{bmatrix} = \begin{bmatrix} |z|^2 + t^2 & z \\ \bar{z} & 1 \end{bmatrix} \longleftarrow (z,t) \in \mathbb{C} \times \mathbb{R}_{>0}.$$

The group $GL(2,\mathbb{C})$ and the transformation T act on $\mathbb{C} \times \mathbb{R}_{>0}$ as follows:

$$g = \begin{pmatrix} a & b \\ c & d \end{pmatrix} : (z,t) \mapsto \frac{(a\bar{c}(|z|^2 + t^2) + a\bar{d}z + b\bar{c}\bar{z} + b\bar{d}, \; |ad - bc|t)}{|c|^2(|z|^2 + t^2) + \Re(c\bar{d}z) + |d|^2},$$

$$T : (z,t) \mapsto (\bar{z},t).$$

Note that the mirror of the reflection $g \circ T \circ g^{-1}$ is given by

$$\Im\{a\bar{c}(|z|^2 + t^2) + a\bar{d}z + b\bar{c}\bar{z} + b\bar{d}\} = 0.$$

The following is the list of the nine roots defined above and the corresponding mirrors in $\mathbb{C} \times \mathbb{R}_{>0}$:

$$\alpha_1 \longleftrightarrow \Re z - \Im z = 0, \quad \alpha_2 \longleftrightarrow \Re z - \Im z - 1 = 0,$$

$$\alpha_3 \longleftrightarrow |z - (1 - i)/2|^2 + t^2 = 1/2,$$

$$\alpha_4 \longleftrightarrow |z - (1 + i)/2|^2 + t^2 = 1/2,$$

$$\alpha_5 \longleftrightarrow \Re z + \Im z - 1 = 0, \quad \alpha_6 \longleftrightarrow \Re z + \Im z = 0$$

and

$$\beta_1 \longleftrightarrow \Im z = 0, \quad \beta_2 \longleftrightarrow \Re z = 1/2, \quad \beta_3 \longleftrightarrow |z|^2 + t^2 = 1.$$

We regard $\mathbb{C} \times \mathbb{R}_{>0}$ as the half-space of the 3-space bounded by the wall defined by $t = 0$. Then these mirrors are planes and hemispheres meeting the wall perpendicularly (see Figure 11.1). Now, we need some intersection points of these mirrors (three on the boundary and four inside). We have the points

$$c_0 = i\infty, \quad c_1 = (0,0), \quad c_2 = (1,0),$$

on the boundary, and

$$e_1 = \frac{1}{2}(1 + i, \sqrt{2}), \quad e_2 = \frac{1}{2}(1 - i, \sqrt{2}), \quad e_3 = \frac{1}{2}(1, \sqrt{3}), \quad e_4 = (0,1)$$

inside.

PROPOSITION 11.2. *(1) The group $\pi(\Gamma_T)$ is a Coxeter group generated by four reflections with roots $\alpha_1, \beta_1, \beta_2$ and β_3, of which relations can be described by the Coxeter graph*

$$\beta_1 = \alpha_1 = \beta_2 - \beta_3.$$

The Weyl chamber with respect to these generating reflections is a tetrahedron, T_{small}, bounded by the four walls $\alpha_1^{\perp}, \beta_1^{\perp}, \beta_2^{\perp}$ and $\beta_3^{\perp}$ with the four vertices c_0, e_1, e_3 and e_4.

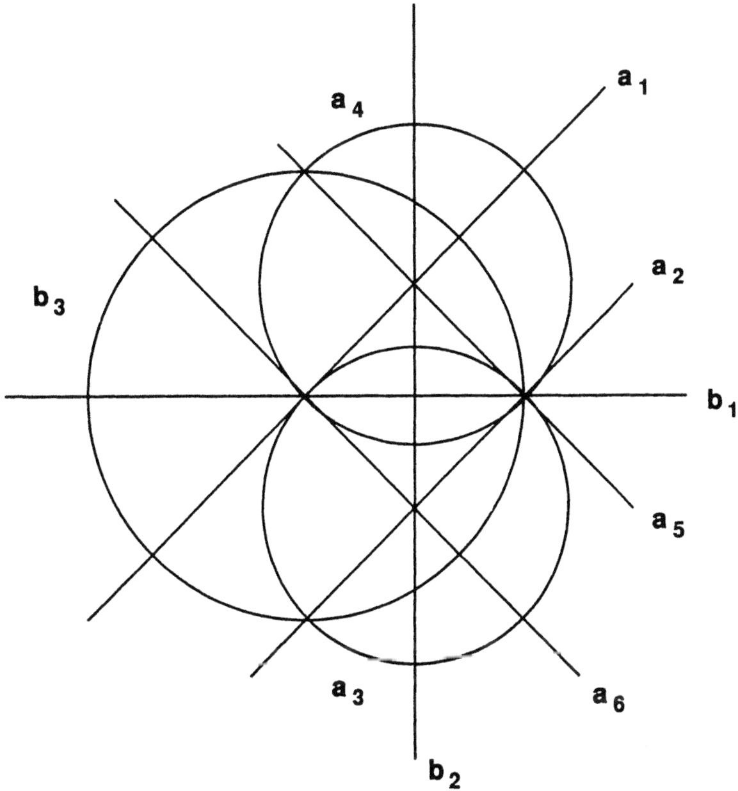

FIGURE 11.1. The wall defined by $t = 0$ intersecting the six planes
and the three hemispheres

*(2) The group $\pi(\Gamma_T(1+i))$ is a Coxeter group generated by four reflections
with roots $\alpha_1, \alpha_4, \alpha_5$ and β_1, of which relations can be described by the
Coxeter graph*

$$\begin{array}{c} \alpha_5 \\ \| \\ \alpha_1 \; = \; \beta_1 \; = \; \alpha_4. \end{array}$$

*The Weyl chamber with respect to these generating reflections is a tetra-
hedron, T_{big}, bounded by the four walls $\alpha_1^\perp, \alpha_4^\perp, \alpha_5^\perp$ and $\beta_1^\perp$ with the four
vertices c_0, c_1, c_2 and e_1.*
*(3) The group $\pi(\Gamma_M(1+i))$ is a Coxeter group generated by six reflections
with roots $\alpha_1, \ldots, \alpha_6$, of which relations can be described by the hexagonal
Coxeter graph*

$$\alpha_1 \overset{\infty}{-\!\!-} \alpha_2 \overset{\infty}{-\!\!-} \alpha_4 \overset{\infty}{-\!\!-} \alpha_6 \overset{\infty}{-\!\!-} \alpha_5 \overset{\infty}{-\!\!-} \alpha_3 \overset{\infty}{-\!\!-} \alpha_1.$$

The Weyl chamber with respect to these generating reflections is a double tetrahedron bounded by the six walls $\alpha_1^\perp, \ldots, \alpha_6^\perp$ with the five vertices c_0, c_1, c_2, e_1 and e_2.

Comment. Figure 11.2 hopefully gives a 3-dimensional view of the Weyl chamber of the group $\pi(\Gamma_M(1+i))$. You see four perpendicular walls $\alpha_1^\perp, \alpha_2^\perp, \alpha_5^\perp$ and $\alpha_6^\perp$, and two hemispheres $\alpha_3^\perp$ and $\alpha_4^\perp$. Put the two pictures very close to your eyes and gradually move them away. At a certain distance you will see the third picture, which should be 3-dimensional.

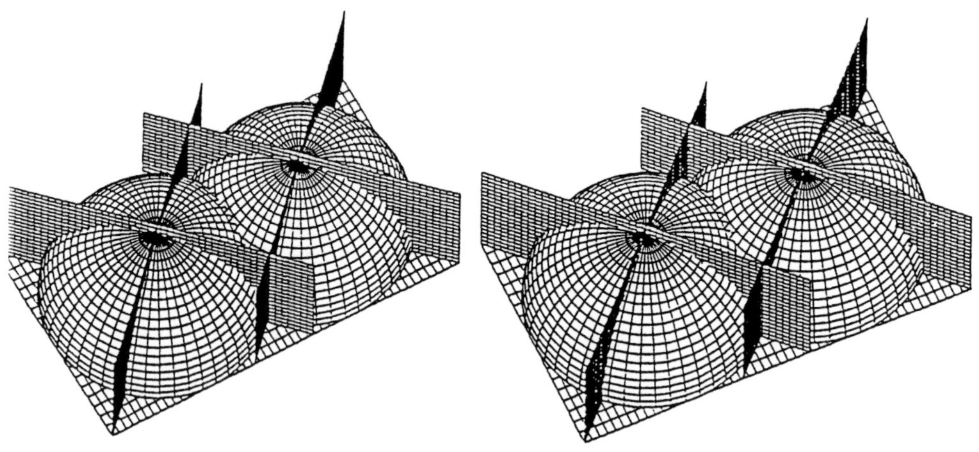

FIGURE 11.2. A 3D-picture of the Weyl chamber bounded by six walls

REMARK 11.1. A *Coxeter group* with *Coxeter graph C* is generated by the reflections R_α represented by the vertices α of C. If two vertices α and β are not connected, R_α and R_β are commutative, if they are connected by a single edge, $(R_\alpha R_\beta)^3 = 1$, and if they are connected by a double edge, $(R_\alpha R_\beta)^4 = 1$. These relations, together with the equality $R_\alpha^2 = 1 (\alpha \in C)$, generate all the relations.

A *Weyl chamber* is a fundamental domain bounded by the mirrors of the generating reflections. If two vertices α and β are not connected, the mirrors $\alpha^\perp$ and $\beta^\perp$ meet at an angle of $\pi/2$, if they are connected by a single edge, their mirrors meet at an angle of $\pi/3$, if they are connected by a double edge, their mirrors meet at an angle of $\pi/4$, and if they are connected by $\overset{\infty}{\text{—}}$, their mirrors meet (on the boundary) at an angle 0.

REMARK 11.2. The above three groups are examples of the so-called *hyperbolic Coxeter groups*, which are studied in [ImH].

The tetrahedron T_{big} can be subdivided into six tetrahedra, each isometric to the tetrahedron T_{small}, which are arranged as follows: Take the barycentric subdivision of the triangular face $\beta_1^\perp \cap T_{big} = (c_0, c_1, c_2)$ with barycenter e_3, and then make cones each whose apex is e_1 and whose base is one of the six triangles obtained by the above barycentric subdivision (see Figure 11.3).

The double tetrahedron is obtained by gluing the tetrahedron T_{big} to its mirror image with respect to the mirror $\beta_1^\perp$. This body can be subdivided into twelve tetrahedra, each isometric to the tetrahedron T_{small}. The above considerations show

$$\pi(\Gamma_T) \overset{S_3}{\underset{\triangleright}{}} \pi(\Gamma_T(1+i)) \overset{\mathbb{Z}/2\mathbb{Z}}{\underset{\triangleright}{}} \pi(\Gamma_M(1+i)).$$

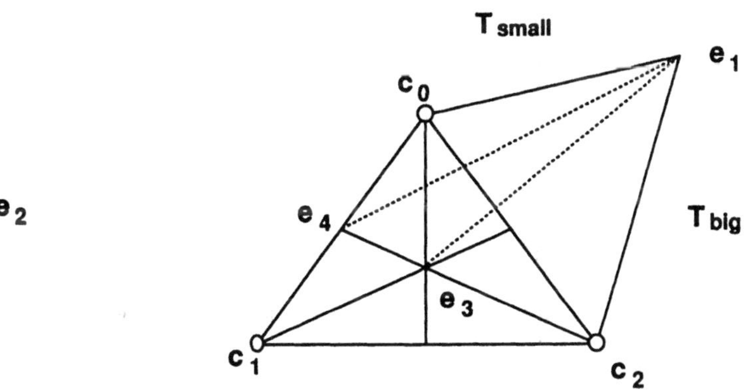

FIGURE 11.3. The Weyl chambers

12. Theta Functions on $\mathbb{H}_2$ Giving the Inverse of $\psi : X \to \mathbb{H}_2$

In this section we give the inverse of the map $X/\langle * \rangle \to \mathbb{H}_2/\Gamma_T(1+i)$ induced by $\psi : X \to \mathbb{H}_2$ in terms of "theta functions" on $\mathbb{H}_2$. This section will be expository; several facts are stated without proofs (but with references), which are too technical for this book.

12.1. Theta Functions Θ on $\mathbb{H}_2$. We studied in Chapter II the theta functions in two variables,

$$\theta_{ij}(z,\tau), \quad i,j \in \{0,1\}, \quad z \in \mathbb{C}, \quad \tau \in \mathbb{H},$$

and the theta functions (zero-values) in one variable,

$$\theta_{ij}(\tau) = \theta(0,\tau), \quad i,j \in \{0,1\}, \quad \tau \in \mathbb{H},$$

where $\theta_{11} = 0$. Since these thetas worked well, and since we are going to make an evolution of the theory presented in Chapter II, we want to define a kind of theta function on $\mathbb{H}_2$. We define theta functions on $\mathbb{H}_2$ directly as

$$\Theta\begin{pmatrix} a \\ b \end{pmatrix}(\tau) = \sum_{n \in \mathbb{Z}[i]^2} e[(n + \frac{a}{1+i})^* \tau (n + \frac{a}{1+i}) + 2\Re\{(\frac{b}{1+i})^* n\}],$$

where $e[x] = \exp(\pi i x)$ and $a,b \in \mathbb{Z}[i]^2$. These functions have the following properties:

LEMMA 12.1.

$(i) \quad \Theta\begin{pmatrix} a \\ b \end{pmatrix}(\tau) = \Theta\begin{pmatrix} a \\ b \end{pmatrix}({}^t\tau).$

$(ii) \quad \Theta\begin{pmatrix} \delta a \\ \varepsilon b \end{pmatrix}(\tau) = \Theta\begin{pmatrix} a \\ b \end{pmatrix}(\tau), \quad$ where δ and ε are units in $\mathbb{Z}[i]$.

$(iii) \quad \Theta\begin{pmatrix} a + r \\ b + s \end{pmatrix}(\tau) = e[\Re {}^t br] \Theta\begin{pmatrix} a \\ b \end{pmatrix}(\tau), \quad$ where $r,s \in (1+i)\mathbb{Z}[i]^2.$

$(iv) \quad$ If ${}^t ab \notin (1+i)\mathbb{Z}[i]$, then $\Theta\begin{pmatrix} a \\ b \end{pmatrix}(\tau)$ vanishes.

Recall the technique "pass your responsibility to the running n," which we used in §5.2 of Chapter II. Then (ii) can be shown by replacing n with $-n$ and $\pm in$, and (iii) by replacing n with $n + r/(1+i)$. Property (i) can be shown as follows:

$$\Theta\begin{pmatrix} a \\ b \end{pmatrix}(^t\tau) = \Theta\begin{pmatrix} \bar{a} + 2i\Im a \\ \bar{b} + 2i\Im b \end{pmatrix}(^t\tau) = \Theta\begin{pmatrix} \bar{a} \\ \bar{b} \end{pmatrix}(^t\tau)$$

$$= \sum_{n\in\mathbb{Z}[i]^2} e[\overline{(n + \frac{a}{1-i})^{*t}\tau(n + \frac{a}{1-i})} + 2\Re\{\overline{(\frac{b}{1-i})^*\bar{n}}\}]$$

$$= \sum_{n\in\mathbb{Z}[i]^2} e[(n + \frac{a}{1-i})^*\tau(n + \frac{a}{1-i}) + 2\Re\{(\frac{b}{1-i})^*n\}]$$

$$= \Theta\begin{pmatrix} a \\ b \end{pmatrix}(\tau).$$

Here we used the fact

$$\frac{a}{1-i} - \frac{a}{1+i}, \ \frac{b}{1-i} - \frac{b}{1+i} \in \mathbb{Z}[i]^2.$$

Finally, property (iv) can be proven as follows:

$$\Theta\begin{pmatrix} a \\ b \end{pmatrix}(\tau) = \Theta\begin{pmatrix} ia \\ ib \end{pmatrix}(\tau) = \Theta\begin{pmatrix} a + (i-1)a \\ b + (i-1)b \end{pmatrix}(\tau)$$

$$= e[\Re(b^*(-1i)a)]\Theta\begin{pmatrix} a \\ b \end{pmatrix}(\tau) = -\Theta\begin{pmatrix} a \\ b \end{pmatrix}(\tau)$$

$$= 0.$$

From this lemma we can easily see that there are only ten linearly independent theta functions.

12.2. Relations between Θ and Riemann's Theta Functions.
Riemann's theta functions $\theta_{c,d}$ on the Siegel upper half space

$$\mathfrak{S}_m = \{\tau \in M(m,m) \mid {}^t\tau = \tau, \ \Im\tau > 0\}$$

of degree m with $c, d \in \mathbb{Z}^m$ is defined as follows

$$\theta_{c,d}(\tau) = \sum_{n\in\mathbb{Z}^m} e[{}^t(n+c)\tau(n+c) + 2{}^t dn], \quad \tau \in \mathfrak{S}_m.$$

Note that when $m = 1$, these are the theta functions in Chapter II. There is a relation between our theta functions and Riemann's theta functions: Θ is the restriction of a theta function $\theta_{..}$ on a 4-dimensional subvariety of $\mathfrak{S}_4$, and the restriction of Θ on $\mathfrak{S}_2 \subset \mathbb{H}_2$ can be written as the product of two functions $\theta_{..}$. In fact, by their definition we see the following:

LEMMA 12.2.

(v) $\quad \Theta\begin{pmatrix} a \\ b \end{pmatrix}(\tau) = \theta_{\Re a, \Im a, \Re b, \Im b}(\tilde{\tau}), \quad \tilde{\tau} = \frac{1}{2}\begin{pmatrix} \tau + {}^t\tau & i(\tau - {}^t\tau) \\ -i(\tau - {}^t\tau) & \tau + {}^t\tau \end{pmatrix} \in \mathfrak{S}_4.$

(vi) $\quad \Theta\begin{pmatrix} a \\ b \end{pmatrix}(\tau) = \theta_{\Re a, \Re b}(\tau)\theta_{\Im a, \Im b}(\tau), \quad if \ \tau \in \mathfrak{S}_2 \subset \mathbb{H}_2.$

12.3. Transformation Formulae. The following formulae can be shown by standard arguments on Riemann's theta functions (cf. [Mum], [Mat2]) and a direct computation.

LEMMA 12.3. *(vii) If $g \in U(2,2)$ is of the form*

$$\begin{pmatrix} A & 0 \\ 0 & (A^*)^{-1} \end{pmatrix}, \quad A \in GL(2, \mathbb{Z}[i]),$$

then

$$\Theta\begin{pmatrix} a \\ b \end{pmatrix}(g \cdot \tau) = \Theta\begin{pmatrix} A^*a \\ A^{-1}b \end{pmatrix}(\tau).$$

(viii) If $g \in U(2,2)$ is of the form

$$\begin{pmatrix} I_2 & B \\ 0 & I_2 \end{pmatrix}, \quad B = B^* = (B_{jk}) \in M(2,2;\mathbb{Z}[i]),$$

then

$$\Theta\begin{pmatrix} a \\ b \end{pmatrix}(g \cdot \tau) = e[a^*Ba]\Theta\begin{pmatrix} a \\ \tilde{b} \end{pmatrix}(\tau),$$

where

$$\tilde{b} = b + Ba + \frac{1}{1+i}\begin{pmatrix} B_{11} \\ B_{22} \end{pmatrix}.$$

(ix)

$$\Theta\begin{pmatrix} a \\ b \end{pmatrix}(J \cdot \tau) = -e[\Re(a^*b)]\det\tau \ \Theta\begin{pmatrix} b \\ a \end{pmatrix}(\tau).$$

12.4. Quadratic Relations among the Theta Functions. By a standard method (cf. [Igu4], [Mat2]) of obtaining quadratic relations among the theta functions, we have the following identities among the ten linearly independent theta functions.

LEMMA 12.4. *For $c, d \in \{0,1\}^2$ satisfying ${}^tcd = 1$, we have*

$$\sum_{a,b\in\{0,1\}^2, \ {}^tab\in 2\mathbb{Z}} \Theta^2\begin{pmatrix} a \\ b \end{pmatrix}(\tau)\, e[{}^tca + {}^tdb] = 0.$$

12.5. Coding the Theta Functions. We are interested in the zero loci of the theta functions Θ. Let us make an observation. We apply the transformation formulae (i) and (vii) in the lemmas above to

$$\Theta\begin{pmatrix} a \\ b \end{pmatrix}, \quad a = (i,i), \ b = (i,i)$$

and

$$g = diag(1,i,1,i)\, T, \quad \text{and} \quad g = diag(1,-i,1,-i)\, T,$$

which correspond to the reflections $R_{\alpha(126)}$ and $R_{\alpha(345)}$. Then we have

$$\Theta\begin{pmatrix} a \\ b \end{pmatrix}(g \cdot \tau) = -\Theta\begin{pmatrix} a \\ b \end{pmatrix}(\tau).$$

This fact implies that this theta function vanishes on the mirrors $\alpha(126)^{\perp}$ and $\alpha(345)^{\perp}$. By the transformation formulae above, we can easily show that it also has zeros along the $\Gamma_M(1+i)$-orbits of these two mirrors (i.e. along the $\Gamma_T(1+i)$-orbit of one of the two mirrors). Moreover it is known (cf. [Fri]) that these zeros are simple and that they are the only zeros. This fact suggests that we code this theta function by

$$\begin{Bmatrix} 126 \\ 345 \end{Bmatrix} \quad \left(= \begin{Bmatrix} 345 \\ 126 \end{Bmatrix} \right).$$

The number of such labels is

$$\binom{6}{3}/2 = 10,$$

and the number of our theta functions is also ten – what a coincidence! It seems we should be able to label all ten thetas by this coding. In fact we can.

Let us denote the theta functions

$$\Theta\begin{pmatrix} a \\ b \end{pmatrix} \quad \text{by} \quad \Theta\begin{Bmatrix} ijk \\ lmn \end{Bmatrix}$$

when

$$\begin{pmatrix} a \\ b \end{pmatrix} = \begin{pmatrix} a_1 & a_2 \\ b_1 & b_2 \end{pmatrix} \quad \text{corresponds to} \quad \begin{Bmatrix} ijk \\ lmn \end{Bmatrix}$$

as follows:

$$\begin{pmatrix} i & 0 \\ 0 & i \end{pmatrix} \leftrightarrow \begin{Bmatrix} 123 \\ 456 \end{Bmatrix}, \quad \begin{pmatrix} i & 0 \\ 0 & 0 \end{pmatrix} \leftrightarrow \begin{Bmatrix} 124 \\ 356 \end{Bmatrix}, \quad \begin{pmatrix} i & i \\ 0 & 0 \end{pmatrix} \leftrightarrow \begin{Bmatrix} 125 \\ 346 \end{Bmatrix},$$

$$\begin{pmatrix} i & i \\ i & i \end{pmatrix} \leftrightarrow \begin{Bmatrix} 126 \\ 345 \end{Bmatrix}, \quad \begin{pmatrix} 0 & i \\ 0 & 0 \end{pmatrix} \leftrightarrow \begin{Bmatrix} 134 \\ 256 \end{Bmatrix}, \quad \begin{pmatrix} 0 & 0 \\ 0 & 0 \end{pmatrix} \leftrightarrow \begin{Bmatrix} 135 \\ 246 \end{Bmatrix},$$

$$\begin{pmatrix} 0 & 0 \\ i & i \end{pmatrix} \leftrightarrow \begin{Bmatrix} 136 \\ 245 \end{Bmatrix}, \quad \begin{pmatrix} 0 & 0 \\ 0 & i \end{pmatrix} \leftrightarrow \begin{Bmatrix} 145 \\ 236 \end{Bmatrix}, \quad \begin{pmatrix} 0 & 0 \\ i & 0 \end{pmatrix} \leftrightarrow \begin{Bmatrix} 146 \\ 235 \end{Bmatrix},$$

$$\begin{pmatrix} 0 & i \\ i & 0 \end{pmatrix} \leftrightarrow \begin{Bmatrix} 156 \\ 234 \end{Bmatrix}.$$

Then we have

PROPOSITION 12.5. *Each of the ten theta functions* $\Theta \begin{Bmatrix} ijk \\ lmn \end{Bmatrix}$ *has simple zeros exactly on the* $\Gamma_T(1+i)$*-orbits of the mirrors* $\alpha(ijk)^\perp$ *and* $\alpha(lmn)^\perp$.

12.6. Modular Forms on $\mathbb{H}_2$.

DEFINITION 12.1. A holomorphic function f on $\mathbb{H}_2$ is called a *modular form* of weight $2k$ relative to $\Gamma_T(1+i)$ (with the character det : $\Gamma(1+i) \to$ the group of units of $\mathbb{Z}[i]$) if $f(T\tau) = f(\tau)$ and the following condition is satisfied:

$$f(g\tau) = \det(g)^k \{\det(Cz + D)\}^{2k} f(\tau), \quad g = \begin{pmatrix} A & B \\ C & D \end{pmatrix} \in \Gamma(1+i).$$

PROPOSITION 12.6. *The square of every theta function is a modular form of weight 2 relative to* $\Gamma_T(1+i)$.

PROOF. It is sufficient to check the formula for the four types of g

$$T, \quad \begin{pmatrix} A & 0 \\ 0 & (A^*)^{-1} \end{pmatrix}, \quad \begin{pmatrix} I_2 & B \\ 0 & I_2 \end{pmatrix} \text{ and } J^{-1} \begin{pmatrix} I_2 & B \\ 0 & I_2 \end{pmatrix} J,$$

where

$$A \in GL(2, \mathbb{Z}[i]), \quad A \equiv I_2 \mod (1+i),$$
$$B = B^* \in M(2, 2, (1+i)\mathbb{Z}[i]).$$

This can be done using the transformation formulae (i) – (ix). $\square$

Let us adopt the following convention in order to define (the squares of) theta functions for any i, j, k, l, m and n:

$$\Theta^2 \left\{ \begin{matrix} ijk \\ lmn \end{matrix} \right\} := \Theta^2 \left\{ \begin{matrix} lmn \\ ijk \end{matrix} \right\},$$

$$\Theta^2 \left\{ \begin{matrix} ijk \\ lmn \end{matrix} \right\} := \operatorname{sign} \begin{pmatrix} ijk \\ pqr \end{pmatrix} \operatorname{sign} \begin{pmatrix} lmn \\ stu \end{pmatrix} \Theta^2 \left\{ \begin{matrix} pqr \\ stu \end{matrix} \right\},$$

for $\{i, j, k\} = \{p, q, r\}$, $\{l, m, n\} = \{s, t, u\}$.

REMARK 12.1. Under this convention, the quadratic relations of theta functions given above can be written as

$$\Theta^2 \left\{ \begin{matrix} ijk \\ lmn \end{matrix} \right\} - \Theta^2 \left\{ \begin{matrix} ijl \\ mnk \end{matrix} \right\} + \Theta^2 \left\{ \begin{matrix} ijm \\ nkl \end{matrix} \right\} - \Theta^2 \left\{ \begin{matrix} ijn \\ klm \end{matrix} \right\} = 0.$$

These give six linear relations among the squares of the ten theta functions; five relations among the six are linearly independent.

REMARK 12.2. Our coding of the theta functions gives the surjective homomorphism

$$\Gamma \ni g \longmapsto \sigma_g \in S_6$$

with kernel $\Gamma(1 + i)$ such that

$$\Theta^2 \left\{ \begin{matrix} ijk \\ lmn \end{matrix} \right\} (g \cdot \tau) = \{\det(C\tau + D)\}^2 \Theta^2 \left\{ \begin{matrix} \sigma_g(i)\sigma_g(j)\sigma_g(k) \\ \sigma_g(l)\sigma_g(m)\sigma_g(n) \end{matrix} \right\} (\tau),$$

where $g = \begin{pmatrix} A & B \\ C & D \end{pmatrix} \in \Gamma$.

12.7. Inverse of the Map $\psi : X/\langle * \rangle \to \mathbb{H}_2/\Gamma_T(1 + i)$. Now we are ready to state the result of this section.

THEOREM 12.7. *Let us define a map* $\Theta : \mathbb{H}_2/\Gamma_T(1 + i) \to \mathbb{P}^{10-1}$ *by*

$$\mathbb{H}_2 \ni \tau \longmapsto : \Theta^2 \left\{ \begin{matrix} ijk \\ lmn \end{matrix} \right\} (\tau) : \in \mathbb{P}^{10-1},$$

where the thetas are the ten thetas defined in §12.5. Then

(1) *The image under Θ is inside $Z \subset \mathbb{P}^{10-1}$ (defined in §1 of Chapter VII by the Plücker relations).*

(2) *Θ can be extended to the inverse map of the composition of the maps $\overline{p_r}^{-1} : Z \to \overline{X}/\langle * \rangle$ (given in §6 of Chapter VII) and $\psi : \overline{X}/\langle * \rangle \to \overline{\mathbb{H}}_2/\Gamma_T(1 + i)$ (given in Theorem 10.3).*

In fact, the five relations among the squares of the ten theta functions miraculously agree with the Plücker relations $\{Plk(ij)\}$ of the ten products $D(ijk)D(lmn)$ of the minors of $z \in M^*(3,6)$.

COROLLARY 12.8. *All the modular forms of even weight relative to* $\Gamma_T(1 + i)$ *form a graded ring. This ring is generated by squares of the theta functions, and thus it is freely generated by five of them.*

At long last, our journey has come to an end. It can be summarized by the following commutative diagram:

$$\begin{array}{ccc}
& \psi : \text{ solutions of } E(3,6;1/2) & \\
\overline{X}/\langle * \rangle & \longrightarrow & \bar{\mathbb{H}}/\Gamma_T(1+i) \\
& & \\
: D(ijk)D(lmn) : \searrow \ \overline{p_r} & & \Theta \ \nearrow : \Theta^2 \left\{ \begin{matrix} ijk \\ lmn \end{matrix} \right\} : \\
& Z \subset \mathbb{P}^{10-1} &
\end{array}$$

REMARK 12.3. The restriction of Θ on $\mathfrak{S}_2$ gives an embedding of $\mathfrak{S}_2/\Gamma_{\mathfrak{S}_2}(2)$ into $Z \cong \mathbb{P}^4$ whose image is the quartic Q (Proposition 4.9 in Chapter VII). This reproduces the modular variety found in [Igu1–3]:

$$\begin{array}{ccc}
& \psi|_{\overline{Q}} & \\
\overline{Q} \subset \overline{X}/\langle * \rangle & \longrightarrow & \bar{\mathfrak{S}}_2/\Gamma_{\mathfrak{S}_2}(2) \\
& & \\
\overline{p_r}|_{\overline{Q}} \searrow & & \Theta|_{\bar{\mathfrak{S}}_2} \ \nearrow : \theta^4_{..} : \\
& \overline{Q} \subset Z &
\end{array}$$

REMARK 12.4. We can give the inverse of $X \to \mathbb{H}_2/\Gamma_M(1+i)$ in terms of slightly more general "theta functions". For this purpose we need a function on $\mathbb{H}_2$ which changes sign under the operation T (see [Mat2]).

Bibliography

[AK]   P. Appell et Kampé de Feriet, *Fonctions hypergeométriques et hypersphériques – polynômes d'Hermite*, Gauthier-Villars, 1926.

[Aom]   K. Aomoto, On the structure of integrals of power products of linear functions, Sci. Papers, Coll. Gen. Education, Univ. Tokyo **27**(1977), 49–61.

[AoK]   K. Aomoto and M. Kita, *Hypergeometric functions* (in Japanese), Springer-Verlag, Tokyo, 1994.

[Apé]   F. Apéry, *Models of the Real Projective Plane.* Vieweg Verlag, 1987.

[AVY1]   V. Almazar, Ma. T. Valena and M. Yoshida, Democratic projective embedding of the configuration space of 5 points on the projective line, Matimyas Math. **15**(1992), 19–34.

[AVY2]   –, Projective embedding of the configuration space of 6 points on the projective line, Matimyas Math. **16**(1993), 1–9.

[AY]   F. Apéry and M. Yoshida, Pentagonal structure of the configuration space of five points in the real projective line, preprint, Kyushu-MPS-1996-12.

[BHH]   G. Barthel, F. Hirzebruch and T. Höfer, *Geradenkonfigurationen und Algebraische Flächen*, Vieweg, Wiesbaden, 1987.

[CH]   P. Cohen and F. Hirzebruch, Book review on [DM2], Bulletin of the AMS **32**(1995), 88–105.

[Cho]   K. Cho, A note on vanishing theorems of cohomology groups of local systems, preprint 1994.

[CM]   K. Cho and K. Matsumoto, Intersection theory for twisted cohomologies and twisted Riemann's period relation I, Nagoya

Math. J. **139**(1995), 67–86.

[CMY]   K. Cho, K. Matsumoto and M. Yoshida, Combinatorial structure of the configuration space of 6 points on the projective plane. Mem. Fac. Sci., Kyushu Univ. 47(1993), 119–146.

[Cox]   H.S.M. Coxeter, *Regular complex polytopes,* Cambridge Univ. Press, 1974.

[CW]   P. Cohen and J. Wolfart, Fonctions hypergéométriques en plusieurs variables et espace des modules de variétés abéliennes, Ann. scient. Ec. Norm. sup., **26**(1993), 665–690.

[CY]   K. Cho and M. Yoshida, Comparison of (co)homologies of branched covering spaces and twisted ones of base spaces I, Kyushu J. Math. 48(1994), 111–122.

[Del]   P. Deligne, Équations différentielles à points singuliers réguliers, Lecture Notes in Math. 163, Springer-Verlag, 1970.

[DM1]   P. Deligne and G. D. Mostow, Monodromy of hypergeometric functions and non-lattice integral monodromy, I.H.E.S. Publ. Math. **63** (1986), 5–89.

[DM2]   –, *Commensurabilities among lattices in $PU(1,n)$,* Annals of Math. Studies 132, Princeton UP, 1993.

[DO]   I. Dolgachev and D. Ortland, *Point sets in projective spaces and theta functions,* Astérisque **165**(1988).

[Erd]   A. Erdélyi (Editor), *Higher transcendental functions, vol. 1,* McGraw-Hill, 1953.

[Ext]   H. Exton, *Multiple hypergeometric functions and applications,* Ellis Horwood Ltd, addr Chichester, 1976.

[Fri]   E. Freitag, Modulformen zweiten Grades zum rationalen und Gaussschen Zahlkörper, Sitzungsber. Heidelb. Akad. Wiss. 1(1967), 1–49.

[Fuj]   M. Fujiwara, *Ordinary differential equations,* Iwanami, 1930.

[Gel]   I.M. Gelfand, General theory of hypergeometric functions, Soviet Math. Dokl. **33**(1986), 573–577.

[GGR]   I.M. Gelfand, M.I. Graev and V.S. Retakh, General hypergeometric systems of equations and series of hypergeometric type, Russian Math. Surveys **47** (1992), 1–88.

[GGr2] I.M. Gelfand and M.I. Graev, Hypergeometric functions associated with the Grassmannian $G_{3,6}$, Math. Sbornik USSR **66**(1990), 1–40.

[Gon]   G. Gonzalez-Dies, Loci of curves which are prime Galois coverings of $\mathbb{P}^1$, Proc. London Math. Soc. **62**(1991), 496–489.

[Gun]   R. C. Gunning, *On uniformization of complex manifolds: The role of connections*, Princeton Univ. Press, 1978.

[Hat]   A. Hattori, Topology of $\mathbb{C}^n$ minus a finite number of affine hyperplanes in general position, J. Fac. Univ. Tokyo, Sci. IA **22** (1975), 205–219.

[Hol1]   R-P. Holzapfel, *Geometry and Arithmetic around Euler partial differential equations*, D. Reidel Publishing Company, Boston, 1986.

[Hol2]   –, *The Ball and Some Hilbert Problems*. Lect. in Math. ETH Zürich, Birkhäuser Verlag, 1995.

[Hol3]   –, *Ball and Surface Arithmetics*. Vieweg Verlag, Wiesbaden, 1997.

[HSY]   M. Hara, T. Sasaki and M. Yoshida, Tensor products of linear differential equations – a study of exterior products of hypergeometric equations, Funkcial. Ekvac. **32**(1989), 455–477.

[Hun1]   R. Hunt, A Siegel modular 3-fold that is a Picard Modular 3-fold, Compositio Math. **76**(1990), 203–242.

[Hun2]   –, The geometry of some special quotients of bounded symmetric domains, Habilitationsschrift Bonn Univ., 1995.

[HY]   M. Hanamura and M. Yoshida, Twisted Riemann inequalities, in preparation.

[ImH]   H. C. Im Hof, A class of hyperbolic Coxeter groups, Expo. Math. **3**(1985), 179–186.

[Igu1]   J. Igusa, On Siegel modular forms of genus two I, Amer. J. Math. **84**(1962), 175–200.

[Igu2]   –, On Siegel modular forms of genus two II, Amer. J. Math. **86**(1964), 392–412.

[Igu3]   –, On the graded ring of theta-constants, Amer. J. Math. **86**(1964), 219–246.

[Igu4]   –, *Theta functions*, Springer, 1972.

[IK1]   K. Iwasaki and M. Kita, Exterior power structure of the twisted de Rham cohomology associated with hypergeometric function of type $(n+1, m+1)$, to appear in J. Math. Pures Appl.

[IK2]   –, Twisted homology of the configuration space of n points with application to the hypergeometric functions, U.T.M.S. 94-11(1994), 1–68.

[IKSY] K. Iwasaki, H. Kimura, S. Shimomura and M. Yoshida, *From Gauss to Painlevé – A modern theory of special functions*, Vieweg Verlag, Wiesbaden, 1991.

[KHT] H. Kimura, Y. Haraoka and K. Takano, The generalized confluent hypergeometric functions, Proc. Japan Acad. **68A** (1992), 290–295.

[Kit1] M. Kita, On hypergeometric functions in several variables I, Japan J. Math. **18**(1992), 25–74.

[Kit2] –, ibid. II – On the Wronskian of the hypergeometric functions of type $(n+1, m+1)$, J. Math. Soc. Japan **45**(1993), 645–669.

[Kit3] –, On vanishing of the twisted rational de Rham cohomology associated with hypergeometric functions, Nagoya Math. J. **135** (1994), 55–85.

[KM] M. Kita and K. Matsumoto, Duality of hypergeometric functions and invariant Gauss-Manin systems, preprint 1995.

[KN] M. Kita and M. Noumi, On the structure of cohomology groups attached to the integral of certain many-valued analytic functions, Japan J. Math. **9** (1983), 113–157.

[KY1] M. Kita and M. Yoshida, Intersection theory for twisted cycles, Math. Nachr. 166(1994), 287-304.

[KY2] –, Intersection theory for twisted cycles II – Degenerate arrangements, Math. Nachr. 168(1994), 171-190.

[Kod] K. Kodaira, On the structure of compact complex analytic surfaces I, Amer. J. Math. **86**(1964), 751–798.

[Mat1] K. Matsumoto, On modular functions in 2 variables attached to a family of hyperelliptic curves of genus 3. Ann. Sc. Norm. Sup. Pisa XVI(1989), 557–578.

[Mat2] –, Theta functions on the bounded symmetric domain of type $I_{2,2}$ and the period map of a 4-parameter family of K3 surfaces, Math. Ann. **295** (1993), 383–409.

[MFK] D. Mumford, J. Fogarty and F. Kirwan, *Geometric invariant theory*, Springer-Verlag, Berlin, 1994.

[MSTY1] K. Matsumoto, T. Sasaki, N. Takayama and M. Yoshida, Monodromy of the hypergeometric differential equation of type (3,6) I, Duke Math. J. 71(1993), 403-426.

[MSTY2] –, Monodromy of the hypergeometric differential equation of type (3,6) II – The unitary reflection group of order $2^9 \cdot 3^7 \cdot 5 \cdot 7$, Ann. Sc. Norm. Sup. Pisa XX(1993), 617-631.

[MSY1]   K. Matsumoto, T. Sasaki and M. Yoshida, The monodromy of the period map of a 4-parameter family of K3 surfaces and the hypergeometric function of type $(3,6)$, Internat. J. Math. **3** (1992), 1–164.

[MSY2]   –, Recent progress of Gauss-Schwarz theory and related Geometric structures, Mem. Fac. Sci., Kyushu Univ. 47(1993), 283-381.

[Mum]   D. Mumford, *Tata lectures on theta I, II*, Birkhäuser, Boston 1983, 1984.

[MY]   K. Matsumoto and M. Yoshida, Configuration space of 8 points on the projective line and a 5-dimensional Picard modular group, Compositio Math. **86** (1993), 265–280.

[Nam]   M. Namba, *A 3-act-play on complex functions*, (in Japanese) Sugaku Books 10 Asakura, 1990.

[Oda]   T. Oda, A review of the transcendental theory of moduli of algebraic surfaces with trivial canonical divisor (in Japanese), RIMS Ko-kyu-roku **909**(1995), 190–200.

[OT]   P. Orlik and H. Terao, *Arrangements of Hyperplanes*, Springer-Verlag, 1991.

[Sas1]   T. Sasaki, On the projective geometry of hypersurfaces, Equations differentielles dans le champ complexe (Colloque Franco-Japonais 1985), vol 3, ed. R.Gérard and K.Okamoto, IRMA, Univ. L. Pasteur, 1988, 115–161.

[Sas2]   –, *Projective Differential Geometry and Linear Homogeneous differential equations*, Lecture notes at Brown Univ. 1989.

[Sas3]   –, *On the Veronese embedding and related system of differential equations*, Lecture Notes in Math. 1481, Springer-Verlag, (1991), 210–247.

[Sek1]   J. Sekiguchi, The versal deformation of the E_6-singularity and a family of cubic surfaces, J. Math. Soc. Japan **46**(1994), 355–383.

[Sek2]   –, Cross ratio varieties for root systems, Kyushu J. Math. **48**(1994), 123–168.

[SekY]   J. Sekiguchi and M. Yoshida, $W(E_6)$-action on the configuration space of six lines on the real projective plane, preprint, Kyushu-MPS-1996-2.

[SG]   G. Sansone and J. Gerretsen, *Lectures on the theory of functions of a complex variable*, Wolters-Noordhoff Publishing,

Groningen, 1969.

[Shi] H. Shiga, One attempt to the $K3$ modular functions I, Ann. Sc. Norm. Sup. Pisa, Serie IV-6 (1979), 609–635.

[SST] M. Saito, B. Sturmfels and N. Takayama, Hypergeometric polynomials and integer programming, preprint, 1997.

[ST] J. Sekiguchi and N. Takayama, Compactifications of the configuration space of 6 points of the projective plane and fundamental solutions of the hypergeometric system of type $(3,6)$, to appear in Tohoku Math. J.

[StiY] J. Stienstra and M. Yoshida, The arrangements of six colored lines on the real projective plane, in preparation.

[SY1] T. Sasaki and M. Yoshida, Linear differential equations in two variables of rank 4, I, Math. Ann. **282**(1988), 69-93.

[SY2] –, Linear differential equations in two variables of rank 4, II – The uniformizing equation of a Hilbert modular orbifold, Math. Ann. **282**(1988), 95–111.

[SY3] –, Linear differential equations modeled after hyperquadrics Tohoku Math. J. **41**(1989), 321-348.

[SY4] –, Tensor products of linear differential equations II – New formulae for hypergeometric functions, Funkcial. Ekvac. **33**(1990), 527-549.

[SYY] T. Sasaki, K. Yamaguchi and M. Yoshida, On the rigidity of differential systems modeled on hermitian symmetric spaces and disproofs of a conjecture concerning modular interpretations of configuration spaces, preprint, Kyushu-MPS-1996-1.

[Take] K. Takeuchi, Arithmetic triangle groups, J. Math. Soc. Japan **29**(1977), 91–106.

[Trd] T. Terada, Fonctions hypergéometriques F_1 et fonctions automorphes I, II, Math. Soc. Japan **35**(1983), 451–475; **37**(1985), 173–185.

[Tsm] T. Terasoma, Exponential Kummer coverings and determinants of hypergeometric functions, Tokyo J. Math. **16**(1993), 497–508.

[TY] N. Takayama and M. Yoshida, $\mathbb{CR}$-geometry on the configuration space of 5 points on the projective line, Funkcial. Ekvac. **39**(1996), 165–181.

[W1] E. J. Wilczynski, *Projective Differential Geometry of Curves and Ruled Surfaces*, Teubner, 1906 (reprinted by Chelsea Publ.

Co.).

[W2]   –, Projective differential geometry of curved surfaces (first memoir), Trans. Amer. Math. Soc. **8**(1907), 233–260.

[Yos1] M. Yoshida, *Fuchsian Differential Equations – with special emphasis on the Gauss-Schwarz theory,* Vieweg Verlag, Wiesbaden, 1987.

[Yos2] –, *Configuration spaces, elliptic curves and the hypergeometric differential equation.* Lecture Note, University of the Philippines, 1992.

[Yos3] –, A presentation of the fundamental group of the configuration space of 5 points on the projective line, Kyushu J. Math. 48(1994), 283-289.

[YY]   T. Yamazaki and M. Yoshida, On Hirzebruch's examples of surfaces with $c_1^2 = 3c_2$, Math. Ann. **266**(1984), 421–431.

Sources of Figures

- Figure 8.1 in Chapter III.   [SG]: Figures 14.2 – 4,8,11 and 15.
- Figure 8.2 in Chapter III.   [Cox]: Figure 2.4 A.
- Figure 8.3 in Chapter III.   [Fuj]: Figures on page 200.
- Figures 8.4 – 8.6 in Chapter III.   These are due to Professor H. Doi.
- Figures 11.2 in Chapter IX.   This is due to Professor K. Matsumoto.

Index

Index of Symbols

General

$\mathbb{Z}$: ring of integers

$\mathbb{Q}$: field of rational numbers

$\mathbb{R}$: field of real numbers

$\mathbb{R}_{>0}$: set of positive real numbers

$\mathbb{C}$: field of complex numbers

$K^{\times}$: multiplicative group $K - \{0\}$

K^n : n-dimensional affine space over a field K

$\mathbb{P}^n_K$: n-dimensional projective space over a field K

$\Re x$: real part of x

$\Im x$: imaginary part of x

δ_{ij} : Kronecker's symbol

$diag\,(a_1, \ldots, a_n)$: diagonal matrix $(\delta_{ij} a_i)$

$GL(n, R)$: general linear group of size n over a ring R

$PGL(n, R)$: projectivization of $GL(n, R)$

$SL(n, R) := \{g \in GL(n, R) \mid \det g = 1\}$

$PSL(n, R)$: projectivization of $SL(n, R)$

I_r : identity matrix of size r

S_r : symmetric group on r letters

$Aut(X)$: group of analytic automorphisms of a complex manifold X

$\langle a, b, \ldots \rangle$: group generated by $a, b, \ldots$

$\pi_1(X, b)$: fundamental group of a manifold X with base point b

Chapter I

$\exp$: exponential function

$\log C_z$: logarithmic function

$\arg C_z$: argument function

$M(k, n)$: set of $k \times n$ complex matrices

$M^*(k, n) := \{x \in M(k, n) \mid \text{ no } k\text{-minor vanishes}\}$

$X(n) = X(2, n)$: configuration space of n points on $\mathbb{P}^1$

$D_x(ij)$: (i, j)-minor of $x \in M(2, 4)$

$Y := \{y_1 : y_2 : y_3 \in \mathbb{P}^2 \mid y_1 - y_2 + y_3 = 0\}$

$Y_0 := \{y_1 : y_2 : y_3 \in Y \mid y_1 y_2 y_3 \neq 0\}$

$f : X(4) \rightarrow Y_0$: a realization of $X(4)$

$\overline{X}(4)$: compactification of $X(4)$ (obtained by adding three points)

$X\{n\} := X(n)/S_n$: configuration space of n unlabeled points on $\mathbb{P}^1$
$j : X\{4\} \to \mathbb{C}$: a realization of $X\{4\}$

Chapter II

$L = L(\omega_1, \omega_2) := \mathbb{Z}\omega_1 + \mathbb{Z}\omega_2$: lattice in $\mathbb{C}$
$E = E(\omega_1, \omega_2) := \mathbb{C}/L(\omega_1, \omega_2)$: elliptic curve
$E(\tau) := E(\tau, 1)$
$\mathbb{H} := \{\tau \in \mathbb{C} \mid \Im\tau > 0\}$: upper half plane
$\Gamma(2) := \{g \in SL(2, \mathbb{Z}) \mid g \equiv I_2 \mod 2\}$: congruence subgroup of level 2
$\wp$: Weierstrass $\wp$ function
$\lambda : \mathbb{H}/\Gamma(2) \to X(4)$: an isomorphism
$J : \mathbb{H}/SL(2, \mathbb{Z}) \to X\{4\}$: an isomorphism
$\theta(z, \tau)$: theta function in two variables
$\theta_{ij}(z, \tau)$: theta functions
$\theta_{ij}(\tau) := \theta_{ij}(0, \tau)$: theta functions (zero values)

Chapter III

$F(a, b, c; x)$: hypergeometric series
$(a, n) := a(a + 1) \cdots (a + n - 1)$
$D := x d/dx$: Euler operator
$E(a, b, c)$: hypergeometric differential equation
$\gamma_* u$: analytic continuation of u along a path γ

Chapter IV

$P = P(x, D)$: hypergeometric differential operator
reg : regularization operator for loaded cycles
$T(x) := \mathbb{C} - \{x_1, \ldots, x_n\}$
α_j : exponents
I_j : loaded cycle with support (x_j, x_{j+1})
$\check{I}_j$: dual of I_j
$c_j = \exp 2\pi i \alpha_j$
$c_{ij\ldots} = c_i c_j \cdots$
$d_j = c_j - 1$
$d_{ij\ldots} = c_{ij\ldots} - 1$
H : invariant form
$Int(n + 1, \alpha)$: intersection matrix for I_j

$h(ijk; lmn)$: coordinates on $\mathbb{P}^{10-1}$
Y : 4-dimensional subvariety of $\mathbb{P}^{30-1}$
$f : X \to Y$: embedding of X
$X_3^{ijk}, X_{2\alpha}^{ijk;lmn}, \dots$: strata of $\overline{X}$
$X_3 := \cup X_3^{ijk}, X_{2\alpha} = \cup X_{2\alpha}^{ijk;lmn}, \dots$
$Z^{abc} := p(\overline{X}_3^{ijk}), Z^{ab} := p(\overline{X}_{1\alpha}^{ab})$: strata of Z

Chapter VIII

L_j : linear form in t with coefficients x_{ij}
$E(k, n; \alpha)$: hypergeometric system of type (k, n)
$\iota : X(2, 6) \to Q \subset X$: embedding induced by the Veronese embedding
$\overset{\bullet}{x} \in X$: base arrangement
T : affine r-space coordinatized by $t_1, \dots, t_r$
$T_{\mathbb{R}} := T \cap \mathbb{R}^r$
D_P : chambers in $T_{\mathbb{R}}$
$P(j) := $ cardinality of $\{i \mid p_i < j\}$
S : affine line coordinatized by s
H_j : hyperplane in T defined by $L_j = 0$
$reg D_P$: regularization of D_P
$D_P \cdot \check{D}_Q$: intersection number
$Int(3, 6; \alpha)$: intersection matrix
$H(\alpha) := d_6 d_{12345} Int(3, 6; \alpha)^{-1}$: invariant form
$\rho(1, \dots, r + 1; \alpha)$: loop around X_3^{ijk} in X
$M(1, \dots, r + 1; \alpha)$: circuit matrix along $\rho(1, \dots, r + 1; \alpha)$
$a(ijk)$: roots of generating quasi-reflections

Chapter IX

$l = (l_j)$: system of six lines on $\mathbb{P}^2$
$S(l)$: K3 surface, the double cover of $\mathbb{P}^2$ branching along l
$\chi(S)$: Euler characteristic of a manifold S
$L = H_2(S, \mathbb{Z})$: lattice
$c_j(l), c_j'(l)$: 2-cycles of $S(l)$
$A := U \oplus U \oplus (-I_2)$
$U := \begin{pmatrix} 0 & 1 \\ 1 & 0 \end{pmatrix}$
$M(ijk) := M(ijk; 1/2)$
$\alpha(ijk)$: normalized roots

R_α : reflection with root α

MG : normalized monodromy group generated by $R_{\alpha(ijk)}$

$\alpha^\perp$: mirror of the reflection R_α

$(z, w) := zA^t w$

$Quad$: quadratic hypersurface of $\mathbb{P}^5$ defined by A

$\mathbb{H}^2 := \mathbb{H} \times \mathbb{H}$

$\mathbb{D}$: a component of $\{x \in \mathbb{C}^6 \mid (x, x)_A = 0, (x, \bar{x}) > 0\}/\mathbb{C}^\times \subset Quad$

$\Gamma_A := \{Y \in GL(6, \mathbb{Z}) \mid {}^t Y A Y = A, \ \mathrm{Con}\mathbb{D}\}$

$\Gamma_A(2) := \{Y \in \Gamma_A \mid Y \equiv I_6 \mod 2\}$

$R_A(-1) = \{\alpha \in \mathbb{Z}^6 \mid (\alpha, \alpha) = -1\}$

$R_A(-1, -2) = \{\alpha \in \mathbb{Z}^6 \mid (\alpha, \alpha) = -1, -2\}$

$GR_A(-1) = $ group generated by $\{R_\alpha \mid \alpha \in R_A(-1)\}$

$GR_A(-1, -2) = $ group generated by $\{R_\alpha \mid \alpha \in R_A(-1, -2)\}$

$R(ijk)$: $\Gamma_A(2)$-orbit of $\alpha(ijk)$

$\phi : X \ni l \mapsto u_1(l) : \cdots : u_6(l) \in \mathbb{D}$

$\varphi : X \to \mathbb{D}/\Gamma_A(2)$

$\overline{X}' := \overline{X} - \{X_{1a} \cup X_{0a}\}$

$\bar{\mathbb{H}}, \bar{\mathbb{D}}, \bar{\mathbb{H}}_2, \bar{\mathfrak{S}}_2$: rational closures

$F1, F0$: rational boundary components

$F_1^{ij}, F_0^{ij;kl;mn}$: $\Gamma_A(2)$-orbits of rational boundaries

$\mathbb{H}_2 := \{\tau \in M(2, 2) \mid (\tau - \tau^*)/2i > 0\}$

$J = \begin{pmatrix} 0 & I_2 \\ -I_2 & 0 \end{pmatrix}$

$U(2, 2) := \{g \in GL(4, \mathbb{C}) \mid g^* J g = J\}$

$Sp(2, \mathbb{R}) := \{g \in GL(4, \mathbb{R}) \mid {}^t g J g = J\}$

$Gr(4, 2)$: Grassmannian

$\iota : \mathbb{H}_2 \to \mathbb{D}$: isomorphism

$\Gamma := \{g \in PGL(4, \mathbb{Z}[i]) \mid g^* J g = J\}$

$\Gamma(1 + i) := \{g \in \Gamma) \mid g \equiv I_4 \mod (1 + i)\}$

$\Gamma_T := \Gamma \rtimes \langle T \rangle$

$\Gamma_T(1 + i) := \Gamma(1 + i) \rtimes \langle T \rangle$

$\Gamma_M(1 + i)$: a subgroup of Γ_T of index 2

$\mathfrak{S}_g$: Siegel upper half space of degree g

$\Gamma_{\mathfrak{S}_g}$: Siegel modular group $Sp(g, \mathbb{Z})$

$\psi := \iota^{-1} \circ \psi : X \to \mathbb{H}$ or any map induced by this map

θ_{ij} : theta functions on $\mathfrak{S}_g$

Θ : theta function on $\mathbb{H}_2$